MW00845562

Achieving sustainable cultivation of potatoes

Volume 1: Breeding improved varieties

It is widely recognised that agriculture is a significant contributor to global warming and climate change. Agriculture needs to reduce its environmental impact and adapt to current climate change whilst still feeding a growing population, i.e. become more 'climate-smart'. Burleigh Dodds Science Publishing is playing its part in achieving this by bringing together key research on making the production of the world's most important crops and livestock products more sustainable.

Based on extensive research, our publications specifically target the challenge of climate-smart agriculture. In this way we are using 'smart publishing' to help achieve climate-smart agriculture.

Burleigh Dodds Science Publishing is an independent and innovative publisher delivering high quality customer-focused agricultural science content in both print and online formats for the academic and research communities. Our aim is to build a foundation of knowledge on which researchers can build to meet the challenge of climate-smart agriculture.

For more information about Burleigh Dodds Science Publishing simply call us on +44 (0) 1223 839365, email info@bdspublishing.com or alternatively please visit our website at www.bdspublishing.com.

Related titles:

Achieving sustainable cultivation of potatoes Volume 2: Production, storage and crop protection
Print (ISBN 978-1-78676-128-6); Online (ISBN 978-1-78676-131-6, 978-1-78676-130-9)

Achieving sustainable cultivation of cassava Volume 1: Cultivation techniques
Print (ISBN 978-1-78676-000-5); Online (ISBN 978-1-78676-002-9, 978-1-78676-003-6)

Achieving sustainable cultivation of cassava Volume 2: Genetics, breeding, pests and diseases
Print (ISBN 978-1-78676-004-3); Online (ISBN 978-1-78676-006-7, 978-1-78676-007-4)

Chapters are available individually from our online bookshop: https://shop.bdspublishing.com

BURLEIGH DODDS SERIES IN AGRICULTURAL SCIENCE

NUMBER 26

Achieving sustainable cultivation of potatoes

Volume 1: Breeding improved varieties

Edited by Prof. Gefu Wang-Pruski
Dalhousie University, Canada

Published by Burleigh Dodds Science Publishing Limited
82 High Street, Sawston, Cambridge CB22 3HJ, UK
www.bdspublishing.com

Burleigh Dodds Science Publishing, 1518 Walnut Street, Suite 900, Philadelphia, PA 19102-3406, USA

First published 2018 by Burleigh Dodds Science Publishing Limited
© Burleigh Dodds Science Publishing, 2018 except the following: Chapters 1, 3 and 9 were prepared by U.S. Department of Agriculture employees as part of their official duties and are therefore in the public domain. All rights reserved.

Notice
No responsibility is assumed by the publisher for any injury and/or damage to persons or property as a matter of product liability, negligence or otherwise, or from any use or operation of any methods, products, instructions or ideas contained in the material herein.

Library of Congress Control Number: 2017960358

British Library Cataloguing in Publication Data
A catalogue record for this book is available from the British Library

ISBN 978-1-78676-100-2 (Print)
ISBN 978-1-78676-103-3 (PDF)
ISBN 978-1-78676-102-6 (ePub)
ISSN 2059-6936 (print)
ISSN 2059-6944 (online)

DOI 10.19103/AS.2017.0016

Typeset by Deanta Global Publishing Services, Chennai, India
Printed by Lightning Source

Contents

**Part 3 Translating research into practice: improving cultivation
in the developing world**

Actionaid International Malawi, Malawi; Dianah Ngonyama, Association of African Agricultural Professionals in the Diaspora, USA; and Paul Demo, International Potato Center (CIP), Malawi

Series list

Title	Series number
Achieving sustainable cultivation of maize - Vol 1 From improved varieties to local applications Edited by: Dr Dave Watson, CGIAR Maize Research Program Manager, CIMMYT, Mexico	001
Achieving sustainable cultivation of maize - Vol 2 Cultivation techniques, pest and disease control Edited by: Dr Dave Watson, CGIAR Maize Research Program Manager, CIMMYT, Mexico	002
Achieving sustainable cultivation of rice - Vol 1 Breeding for higher yield and quality Edited by: Prof. Takuji Sasaki, Tokyo University of Agriculture, Japan	003
Achieving sustainable cultivation of rice - Vol 2 Cultivation, pest and disease management Edited by: Prof. Takuji Sasaki, Tokyo University of Agriculture, Japan	004
Achieving sustainable cultivation of wheat - Vol 1 Breeding, quality traits, pests and diseases Edited by: Prof. Peter Langridge, The University of Adelaide, Australia	005
Achieving sustainable cultivation of wheat - Vol 2 Cultivation techniques Edited by: Prof. Peter Langridge, The University of Adelaide, Australia	006
Achieving sustainable cultivation of tomatoes Edited by: Dr Autar Mattoo, USDA-ARS, USA & Prof. Avtar Handa, Purdue University, USA	007
Achieving sustainable production of milk - Vol 1 Milk composition, genetics and breeding Edited by: Dr Nico van Belzen, International Dairy Federation (IDF), Belgium	008
Achieving sustainable production of milk - Vol 2 Safety, quality and sustainability Edited by: Dr Nico van Belzen, International Dairy Federation (IDF), Belgium	009
Achieving sustainable production of milk - Vol 3 Dairy herd management and welfare Edited by: Prof. John Webster, University of Bristol, UK	010
Ensuring safety and quality in the production of beef - Vol 1 Safety Edited by: Prof. Gary Acuff, Texas A&M University, USA & Prof. James Dickson, Iowa State University, USA	011
Ensuring safety and quality in the production of beef - Vol 2 Quality Edited by: Prof. Michael Dikeman, Kansas State University, USA	012
Achieving sustainable production of poultry meat - Vol 1 Safety, quality and sustainability Edited by: Prof. Steven C. Ricke, University of Arkansas, USA	013
Achieving sustainable production of poultry meat - Vol 2 Breeding and nutrition Edited by: Prof. Todd Applegate, University of Georgia, USA	014
Achieving sustainable production of poultry meat - Vol 3 Health and welfare Edited by: Prof. Todd Applegate, University of Georgia, USA	015

Acknowledgement

We wish to acknowledge the following for their help in reviewing a particular chapter:

- Chapter 13: Dr Netsayi Mudege, International Potato Center (CIP), Peru

Introduction

Potatoes are one of the world's key food crops. Their nutritional value, and the fact that they can be grown with relatively few inputs in a wide range of environments, makes them an important food security crop. However, yields in developing countries are held back by factors such as poor cultivation practices and the impact of pests and diseases, whilst more intensive systems need to become more 'climate smart' to minimise environmental impact and adapt to climate change. These challenges are addressed in the two volumes of *Achieving sustainable cultivation of potatoes*:

- Volume 1 Breeding improved varieties
- Volume 2 Production, storage and crop protection

Volume 1 reviews general developments in breeding, research on improving particular traits, from stress resistance to nutritional quality, as well the challenges facing potato cultivation in particular regions. The volume reviews the latest research on understanding potato plant physiology and genetic variety. It discusses major advances in conventional, hybrid and marker-assisted breeding as well as their application in improved varieties, before focusing on ways of supporting smallholders in regions such as Africa and Latin America. Although a separate species, the book also includes selective coverage of research on sweet potato. The book is accompanied by a second volume which looks at ways of improving potato cultivation as well as advances in pest and disease management.

Part 1   Plant physiology and breeding

The first part of the volume assesses recent research on plant physiology and genetic diversity and their implications for conventional, hybrid and marker-assisted breeding. The subject of Chapter 1 is advances in understanding potato plant physiology and growth. The chapter looks at what we know about initial crop development and the factors affecting the subsequent development of the potato plant. The chapter examines in particular potato responses to water and heat stresses as well as nutrient availability and other factors.

Given its important implications for shelf-life, Chapter 2 reviews research on understanding ageing processes in seed potatoes. The physiological quality of seed tubers is very important for the performance of the crop grown from them, and interacts strongly with seed tuber size. Physiological quality consists of two components: dormancy and physiological age. The chapter reviews the conditions which influence both dormancy and physiological age, as well as the effects of seed quality on various aspects of crop performance. After considering seed tubers as the main propagules in potato production, the chapter deals with the quality characteristics of seed tubers, dormancy and physiological age, and the importance of seed quality as a yield-limiting and quality-determining factor. The chapter focuses on the importance of understanding dormancy, bud activation, initial sprout growth and apical dominance, as well as understanding aging in sprouts and mother tubers. The chapter provides an analysis of the dynamic development of physiological age and resulting crop performance, as well as assessing the causes of variation in physiological age and options for manipulation.

The subject of Chapter 3 is the importance of ensuring the genetic diversity of potatoes. The opportunities for advances in the potato crop through genetics are significant, since potato has many needs for improvement, and related species with the traits required are available. The chapter discusses the special challenges, opportunities, and recent developments and accomplishments for potato genebanks in the areas of acquisition, classification, preservation, evaluation, and distribution of genetic stocks and information, as well as discussing key issues in access to genetic material.

Moving on the subject of breeding techniques, the subject of Chapter 4 is advances in conventional potato-breeding techniques. Potato is highly heterozygous and, in order to maintain productivity, improved potato varieties must be developed by inter-mating desired parental lines and selecting superior clones from the progeny. Since potato is vegetatively propagated, any selected genotype can be fixed with all its intra- and inter-locus interactions responsible for phenotypic expression, and multiplied for commercial cultivation if desired. Recent advances in molecular breeding provide opportunities for rapid genetic gain. Nevertheless, phenotypic selection remains the common practice in conventional potato breeding programmes. Nearly all new varieties of potato still emerge from a process free from use of molecular technologies. The chapter reviews the progress and advances made in phenotypic selection techniques of conventional potato breeding. The chapter describes the role of molecular approaches in improving phenotypic selection.

Complementing the preceding chapter's theme of potato breeding, Chapter 5 looks at hybrid potato breeding for improved varieties. Hybrid potato breeding promises to create new cultivars within a few years. This would facilitate the introgression of genes by marker assisted selection, and hybrid cultivars could then be made available as true seeds, free of soil-borne pathogens, quick to multiply and easy to transport and store. What were previously thought to be prohibiting factors for hybrid potato breeding have recently been overcome: nearly homozygous inbred lines have been created and the first experimental hybrids have been evaluated in the field. The chapter reviews the scientific basis for hybrid potato breeding and highlights the key features of a strategy for creating an inbred, line-based, hybrid potato crop that can be propagated through seed. The chapter discusses the recent progress made towards the development of useful hybrid varieties, and considers how the hybrid potato breeding technology platform will need to be adapted and optimized for different production systems.

Part 2   Improving particular traits

Building on Part 1, the second part of the book looks at ways developments in breeding have been used to improve particular traits. The focus of Chapter 6 is on advances in the development of potato varieties resistant to abiotic stress. Abiotic stresses such as drought, high or low temperature, salinity, submergence and nutrient deficiency can significantly impact potato yields. These suboptimal conditions restrict potato plant performance so that the plants do not reach their full genetic potential. The chapter examines different abiotic stress improvement targets in the potato as well as the variety of tools and techniques being developed and used for crop improvement for abiotic stresses. The chapter reviews technological advances to develop abiotic stress resistance in potatoes and tolerant varieties, especially through genetic engineering.

Chapter 7 examines the challenge of developing early-maturing, stress-resistant potato varieties. The chapter describes the selection of germplasm and traits for breeding early maturing varieties of potato, exploring genetic aspects of earliness as a trait. The chapter looks at early tuber initiation, high dry matter partitioning efficiency and basic factors that need to be taken into account when breeding for earliness in potato. The chapter includes a detailed case study of developing an early-maturing, late blight-resistant Kufri Khyati potato variety for cultivation in India.

As a point of comparison, Chapter 8 deals with developing new sweet potato varieties with improved performance. Novel sweet potato varieties with improved traits are needed, especially for marginal lands and disease-prone regions. However, the high degree of heterozygosity, high male sterility, and self- and interspecific incompatibility of the sweet potato plant results in strong segregation of hybrid progenies. Molecular breeding provides a promising approach for the development of new varieties with value-added traits. The chapter reviews the development and application of genetic transformation and trait improvement to sweet potato, including the development of sweet potato plants which are resistant to disease and abiotic stress, and sweet potatoes with improved starch quality and higher anthocyanin content.

Chapter 9 begins by considering the nutritional properties and enhancement and bio-fortification of potatoes. There are a number of factors that make potatoes a logical focus for nutritional breeding efforts. As one of the world's staple foods, they have a key role to play in improving global food security, largely due to their nutritional value, storability, affordability and high yield. Recent years have also witnessed greatly increased consumer concern for healthy food choices, leading to high demand for an increase in the nutritional value of foods that have previously been subject to negative health publicity. The chapter reviews the nutritional composition of potatoes from diverse germplasm including vitamin C, B vitamins, potassium, carotenoids, phenylpropanoids and glycoalkaloids. In each case, the chapter discusses the feasibility and health benefits of increasing these nutrients through traditional and precision breeding programmes.

Chapter 10 deals with improving the breeding, cultivation and use of sweet potato in Africa. Sweet potato is a low input crop with significant potential for improving public health and nutrition and developing food security in Sub-Saharan Africa. The chapter examines the nutritional contribution made by OFSP (orange-fleshed sweet potato) in poor rural communities in Malawi, Ghana, Nigeria and Burkina Faso, sustainable breeding and seed systems and effective commercialisation and marketing to benefit the communities concerned. The chapter includes detailed case studies from Ghana and Malawi.

Part 3   Translating research into practice: improving cultivation in the developing world

The book's third section looks at ways of supporting smallholders in regions such as Asia, Africa and Latin America to improve potato cultivation. Chapter 11 offers an overview of potato production and breeding in China. The chapter outlines current potato production and consumption, identifying key trends and challenges. The chapter explores challenges associated with germplasm material, breeding objectives and development of new varieties, and the types of new variety available. The chapter considers the possibility of virus-free seed potato production.

The subject of Chapter 12 is the challenge of improving potato cultivation to promote food self-sufficiency in Africa. Demand for potato in sub-Saharan Africa is growing, but the projected growth in demand is not matched by the projected growth rate in local potato production. An interplay of factors ensures the production gains achieved are small and slow. The chapter reviews the current state of potato production in Africa, and the challenges it faces. The chapter describes the development and promotion of suitable potato varieties, considers crop improvement initiatives and programs, and emphasises the potential of potato to contribute to food security and poverty reduction.

The volume's final chapter, Chapter 13, addresses the importance of supporting smallholder women farmers in potato cultivation. It is clear that women farmers have a vital role to play in shaping and maximising this growth, safeguarding potatoes as a primary food security crop. The chapter offers a summary of the current state of potato cultivation and the role of women, with a focus on sub-Saharan Africa. The chapter highlights the enormous potential of women farmers in promoting the broader goals of development and food security in these areas, before examining some of the challenges women face in making their voices count. The chapter presents different strategies for supporting women smallholders, with a particular emphasis on ensuring that women benefit from agricultural training and have the opportunity to apply their knowledge and resources.

Part 1

Plant physiology and breeding

Chapter 1

Advances in understanding potato plant physiology and growth

Curtis M. Frederick, University of Wisconsin, USA; Masahiko Mori, Obihiro University of Agriculture and Veterinary Medicine, Japan; and Paul C. Bethke, USDA-ARS and University of Wisconsin, USA

1 Introduction

Potato is grown on over 19 million hectares worldwide under a wide range of production environments and at farm scales from less than one hectare to several thousand hectares (FAO 2009; Walker et al. 1999). This global adoption of potato on small and large farms reflects the wide range of productive environments where potato can be grown successfully, the high nutritive value of the harvested tubers and the benefits of potato as a cash crop (FAO 2009; Guenthner 2010). As described below and throughout this volume, the growth and productivity of potato is highly variable, depending strongly on variety, location, environment and production methods. The need to manage this variability and produce a crop with acceptable risk interacts at multiple levels with goals for economic, environmental and social sustainability (Lutaladio and Castaidi 2009; Allen and Sachs 1991; Van Evert et al. 2013; Haverkort et al. 2013).

http://dx.doi.org/10.19103/AS.2017.0016.01

Published by Burleigh Dodds Science Publishing Limited, 2018.

2 Crop rotation, planting and initial crop development

2.1 Potato production systems

Potato is typically grown in rotation with other crops as part of an integrated production system. Rotating with other crops is used to minimize accumulation of pests and diseases that can be transmitted from one generation of potatoes to the next (Larkin et al. 2010, 2011; Nelson et al. 2009; Stark and Porter 2005). Rotation crops that carry out nitrogen fixation and green manures planted between crops are used to maintain soil productivity and may contribute to overall farm income (Hoshide et al. 2006). Soil loss from erosion is influenced by the order of crop rotation and may be greater when potato is followed by a spring-sown crop that develops slowly (Fiener and Auerswald 2007).

Rotation crops are used to control soilborne pathogens, weeds and insect pests, as well as to improve soil quality (Myers et al. 2008; Bohl and Johnson 2010; Stark and Love 2003). Most rotation crops are harvested and sold at maturity, although a few are used as green manures (Grandya et al. 2002) or as alternatives to chemical fumigants (McGuire 2003) and are incorporated into the soil. Crop rotations in which potato is grown one year out of every three or more years are recommended in many parts of North America and elsewhere as part of an integrated pest management system (Myers et al. 2008; Svubure et al. 2015). Three-year rotations reduce pressure from soilborne diseases including *Rhizoctonia*, common scab and *Verticillium albo-atrum* (Johnson and Dung 2010). Longer rotations are required to control *Verticillium dahliae* (Johnson and Dung 2010; Stark and Love 2003) and other diseases that have resting propagules that persist in soil for many years. Crop rotation can also be used to decrease pressure from insect pests including Colorado potato beetle (Huseth et al. 2015). Potato production in the developed world typically requires multiple passes through each field with heavy equipment. This degrades soil structure, compacts soils and encourages erosion (Angers et al. 1999; Grandya et al. 2002; Bohl and Johnson 2010; Nelson et al. 2009). Rotation crops contribute to maintenance of soil structure through the incorporation of organic matter (Grandya et al. 2002; Angers et al. 1999; Nelson et al. 2009).

A wide variety of cropping systems are used for potato production. Examples include barley (*Hordeum vulgare* L.) underseeded with red clover (*Trifolium pratense* L.) followed by potato the next year in Presque Isle, Maine (Larkin et al. 2011); soybeans, maize, small grains and vegetables in various sequences in the Upper Midwest of the United States (Huseth et al. 2015); wheat, sugar beet, alfalfa, barley, bean, peas, maize and onion in rotations of two to over six years in Idaho (Stark and Love 2003); and maize, small grains, sugar beet and oilseed rape in various sequences in Northern Germany (Steinmann and Dobers 2013). In Hokkaido, Japan, a four-year cropping system with potato or sweet corn, winter wheat, sugar beet and soybean, adzuki bean (*Vigna angularis*) or kidney bean (*Phaseolus vulgaris*) is typical (Koga 2008, 2013). In regions such as West Bengal, India, potato may be grown as part of a cropping system that also includes rice, wheat, sesame, rapeseed and jute (Biswas et al. 2006). The choice of rotation crops depends on grower experience, market factors, local environmental conditions and management goals of the cropping system.

The frequency with which potato is planted in a crop rotation depends on many factors. Long-term improvements in soil health and disease suppression must be weighed against the need for profitability (Stark and Love 2003; Biswas et al. 2006). Potato is often the most profitable crop produced within a cropping system, and this encourages growers to

Published by Burleigh Dodds Science Publishing Limited, 2018.

adopt shorter rotations. In some cases, short rotations may increase the risk of production and decrease profitability over the long term (Myers et al. 2008). Larger farms often make substantial investments in specialized production equipment. Reducing the length of the crop rotation is one way to reduce the cost of production per hectare produced, especially when land available for potato production is restricted (Vos 1992; Myers et al. 2008). Two-year crop rotations are used by some growers in parts of North America and Europe (Huseth et al. 2015; Larkin et al. 2011; Stark and Love 2003), and this suggests that there are substantial economic forces that discourage growers from realizing the potential benefits offered by longer rotation systems.

2.2 Planting and early growth of the crop

Potatoes are most commonly propagated from seed tubers. Potato tubers from the prior year's harvest are size-selected or cut into seed pieces of 20–80 g, with each piece having one or more 'eyes'. Each eye contains several buds and one or more of these may sprout to produce a stem that is genetically identical to the mother tuber. Potato varieties that produce tubers containing numerous, evenly spaced eyes are highly desirable where seed potato costs are relatively high, or when there is restricted availability of high-quality seed tubers, since each potato can be cut into a large number of equal-sized seed pieces. As described in Section 3, seed potato quality has important ramifications for the cost of production and yield potential of the crop.

Seed potatoes are planted 5–20 cm deep in holes or, most commonly, in furrows. Planting into furrows and mounding of soil over the seed pieces to produce linear ridges, referred to as hills, has been widely adopted in much of the world (Bohl and Johnson 2010; Stark and Love 2003). Furrows between hills can be used to distribute irrigation water, and potatoes that develop in hills are less likely to be submerged in standing water after heavy rains. Hilling has the added benefit of contributing to early season weed control. Earth heaped up to establish the final shape and height of the hill also covers weed seedlings. Timing and frequency of hilling operations impact water run-off and soil erosion, especially when potatoes are grown on sloping land (Xing et al. 2011).

An alternative to traditional hill and furrow planting is production in raised beds or flat planting. Beds may promote retention of soil water in the root zone (Essah and Honeycutt 2004) and even distribution of water (Robinson 1999). This applies to rain-fed production under dry conditions as well as irrigated production. In areas with sprinkler irrigation, the traditional planting configuration of hills and furrows is not needed for irrigation water distribution, and may be counterproductive to efficient water management (Tarkalson et al. 2011). Under these conditions, bed planting may increase water use efficiency (King et al. 2011) while maintaining (Tarkalson et al. 2011) or increasing yields (King et al. 2011).

The timing and extent of tuber sprouting has important consequences for potato production and value. Seed potatoes should sprout uniformly and quickly after planting to generate a stand of vines that develops and matures uniformly. The dormancy of potato tuber buds is under physiological, environmental and hormonal control (Hartmann et al. 2011; Sonnewald and Sonnewald 2013). In some landraces of potato, including those of *Solanum tuberosum* Group Phureja, tuber dormancy is weak or not established. Such lines are useful because they allow for nearly continuous cropping in equatorial regions where distinct seasonal variation in climate is lacking. Most potatoes are dormant at harvest, however, and dormancy is lost during storage (Suttle 2004). Seed tubers that are physiologically older produce a greater number of sprouts than younger tubers do

Published by Burleigh Dodds Science Publishing Limited, 2018.

(Coleman 2000). As tuber age increases further, sprout vigour begins to decline, even as sprout number continues to increase (Coleman 2000).

Optimum planting density depends on the capacity of mother tubers to establish multiple main stems. The number of sprouts determines stem number and this in turn determines the number and size distribution of harvested tubers (Knowles and Knowles 2006; Haverkort et al. 1990). As the number of stems per planted area increases, tuber number increases and average tuber size decreases (Bussan et al. 2007). Growing conditions during the seed crop's period of tuber initiation have been identified as a large contributor to variation in stem number (Wurr et al. 2001). Seed storage conditions have an influence on the length of dormancy (Van Ittersum and Scholte 1993) and it has been proposed that storage conditions can be used to manage stem number in a crop (Oliveira et al. 2012). Implementing aging treatments on seed lots from northern and southern growing regions successfully altered stem numbers and days to emergence (Knowles and Knowles 2006). Warmer storage temperatures increased stem numbers and decreased emergence time in a predictable fashion. For some varieties, treatment of seed tubers with gibberellin can increase stem and tuber number per hectare (Blauer et al. 2013). Since tubers in different size categories have different value in the marketplace, seed of an appropriate age can contribute to profitability and economic sustainability. Producers can adjust planting rates to achieve optimum stem densities for the target market (Wurr et al. 1993; Love and Thompson-Johns 1999; Bussan et al. 2007).

Potatoes may be grown from true potato seed harvested from potato fruit, although this is less common than the use of seed tubers (Almekinders et al. 2009; Struik and Wiersema 1999). In cases where true seed results from self-pollination or outcrossing with neighbouring plants, progeny are likely to be inferior to either parent. The *Chacasone* F2 population developed by the International Potato Center (CIP) in the 1990s is an exception to this rule (Alexopoulos et al. 2006). These seeds produce relatively uniform plants that have a single stem and mature in approximately 120 days. Tuber yields tend to be depressed relative to yields from conventional seed potatoes, and the best use of the harvested tubers may be as seed tubers for the following year (Almekinders et al. 2009). Such seed tubers are likely to have a lower incidence of viral infection than seed tubers propagated in the field through many generations (Struik and Wiersema 1999).

2.3 Tuber initiation

Initial growth of root and shoot systems is followed by production of stolons, stems that initiate underground and elongate horizontally. Tubers initiate near the tip of stolons (Reeve et al. 1969; Xu et al. 1998) in response to environmental and developmental cues (Abelenda et al. 2014). High temperatures inhibit tuber formation and promote continued stolon elongation (Struik et al. 1989; Jackson 1999). Stolons that grow through the side of hills often produce vegetative shoots, sometimes referred to as heat runners, rather than tubers. Stolon elongation and inhibition of tuber formation by high temperatures are promoted by gibberellins (Hannapel et al. 2004; Martinez-Garcia et al. 2001; Xu et al. 1998).

The length of the photoperiod is an important signal for tuber initiation. Wild species relatives of potato and cultivated landraces from equatorial regions of South America require short days for tuber initiation (Jackson 1999). These genotypes produce tubers late in the growing season or not at all when grown at higher latitudes. As potato production moved from the centre of origin in the Andes Mountains into southern Chile, Europe and

North America (Ames and Spooner 2008), cultivated potato developed the capacity to initiate tubers during the long days of early summer (Kloosterman et al. 2013). Genetic evidence from crosses between short- and long-day potato suggests that tuber formation under long-day photoperiods is controlled by two dominant genes present in cultivated potato (Kittipadukal et al. 2012).

Photoperiod in potato is sensed primarily by the phytochrome B photoreceptor (Martinez-Garcia et al. 2001) and the day length signal is relayed from leaves to stolon tips in the form of phloem-mobile signalling molecules. The best characterized is StBel5 (Hannapel 2010) which acts in concert with miRNA172 (Martin et al. 2009) and miRNA156 (Bhogale et al. 2014) to promote tuber formation. The signal transduction pathway used to initiate tuber formation shares many features with the pathway that is used to promote flowering (Lagercrantz 2009; Rodriguez-Falcon et al. 2006; Abelenda et al. 2014).

3 Development of the potato plant

3.1 Carbon assimilation and distribution

A sustainable potato production system should maximize light interception and the conversion of light energy to photosynthate. Photosynthetic carbon should be partitioned between shoot, root and tubers to maximize carbon accumulation in tubers at plant maturity. Relative growth rates of source and sink tissues determine how likely it is that individual genotypes will achieve this goal. Constraints of the production environment, especially the length of the growing season, strongly restrict the developmental strategies that are likely to favour high yields.

Carbon acquisition and utilization in potatoes can be divided into in five steps: photosynthesis, partitioning of carbon in leaf between sugar and starch, phloem loading of sucrose, long-distance transport and phloem unloading and utilization by the sink (Dwelle 1990). Fundamentally, the amount of photosynthate transported to sink tissues is dependent on the light energy that is absorbed by the source tissues. It has been argued that gains in tuber yield could be made by increasing levels of energy capture itself (Allen and Scott 1980). An overview of carbon metabolism and metabolites in leaves and developing potatoes can be found in Winter and Huber (2000), Tetlow (2004), Kolbe and Beckmann (1997) and Vos and Oyarzún (1987). Here we highlight some of the more notable research findings with potato that suggest opportunities for increasing tuber number or yield by augmenting or altering carbon flow pathways.

Research on source tissue has included understanding limitations to photosynthesis, enzymes involved with starch synthesis in leaves and the loading of sucrose into the phloem. The classical view is that the sucrose gradient between source and sink is the primary driver of the carbon distribution (Lalonde et al. 2003). CO_2 concentration within the chloroplast often limits photosynthetic rate. Inserting glycolate dehydrogenase genes from *Escherichia coli* into potato increased chloroplast CO_2 concentration, lowered photorespiration and increased tuber yield 2.3-fold compared with controls (Nölke et al. 2014). Heterologous expression of a spinach sucrose transporter, *SoSUT1*, in potato decreased sugar content in leaves and increased sugar content in the tubers (Leggewie et al. 2003). Starch accumulates in leaf chloroplasts during daytime to store excess carbohydrate from photosynthesis. This starch is broken down to sugars as needed to satisfy sink demands. Modifying the starch breakdown pathway in leaves by overexpressing a maize sucrose-phosphate synthase (Ishimaru et al. 2008) or redirecting photoassimilates from leaf starch formation to sink organs resulted in higher sucrose transport to tubers, greater tuber starch concentration and higher tuber yield than the controls (Jonik et al. 2012).

The ability of tubers to import and utilize sucrose from the phloem for starch synthesis has also been modified with success. Increasing ATP/ADP levels in the amyloplast through modulation of plastidial adenylate kinase activity increased tuber yield by 80% and starch concentration by 60% (Regierer et al. 2002). Targeted increases in glucose-6-phosphate availability through the overexpression of amyloplast membrane translocators also resulted in increased starch and tuber yields (Jonik et al. 2012). Increasing the expression of sucrose synthetase in tubers increased yield and starch content relative to controls (Baroja-Fernández et al. 2009).

Molecular manipulations of potato may lead to increases in yield. However, sustainable production takes into account the values of the people in the growing areas and the cost of developing new technology (P. Allen et al. 1991; The Royal Society 2009). Even when a transgene is highly favourable for the production environment, the cost of introducing a transgenic variety often prohibits its entry into the marketplace (Qaim et al. 2009). More generally, research that utilizes biotechnology can identify genes that limit productivity and demonstrate what happens when their activity in specific cell or tissue types is altered. This information could permit the identification of favourable alleles that are present naturally and can be used in traditional plant breeding.

3.2 Light interception and tuber yield

The amount of photosynthetically active radiation that a plant receives during the growing season is the primary contributor to the growth rate and yield of the below ground tuber crop (Kooman et al. 1996a). Light interception is necessary for crop growth but is not an easy quality to measure in field research settings. Radiation capture is commonly estimated using either leaf area index (LAI) or per cent ground cover (Burstall and Harris 1983; Haverkort et al. 1991). Haulm index, a combination of stem number, canopy height and ground cover measured in mid-season, has been shown to reasonably predict dry matter yield in breeding populations (Moll and Klemke 1990). Tuber dry matter accumulation per MJ of photosynthetically active radiation received varies with genotype and production environment (Kooman et al. 1996a; Kooman et al. 1996b). In northern Europe, harvested yields were between 2.29 and 2.95 g dry weight per MJ of light energy received (Khurana and McLaren 1982; Spitters 1988; Kooman and Rabbinge 1996). In tropical environments, however, yields ranged from 0.64 to 3.5 g dry weight per MJ (Kooman et al. 1996a; Haverkort and Harris 1986). This wide range of production efficiencies depended on factors such as planting date, moisture stress (Jefferies et al. 1989), temperature and maturity class as well as LAI.

3.3 Architecture of the shoot system

The ability of the canopy to absorb solar energy is dependent on total leaf area. Thus, the appearance and longevity of leaves is the most important component of canopy development. Differences in total leaf area and in how frequently and uniformly leaves develop on stems other than the main stem is genetically determined (Firman et al. 1995). Moderate temperatures, high levels of nitrogen fertilizer and optimum planting

Published by Burleigh Dodds Science Publishing Limited, 2018.

dates favour development of the canopy. As temperature conditions increase beyond a favourable range, the proportion of leaves on basal stems increases. Although this alters canopy architecture, total leaf area remains the primary factor that is associated with photosynthetic rate (Fleisher et al. 2006). It is unclear how leaf development on different stem types might impact tuber quality traits. Leaf longevity contributes to the maintenance of total leaf area (Kooman and Rabbinge 1996). Continued production of new leaves, however, plays a greater role in canopy maintenance than leaf longevity in late maturity lines (Firman et al. 1995).

Canopy traits are altered by drought stress and may influence drought stress tolerance. Cultivars with a high stem to leaf proportion by weight had greater dry matter yield under moderate and high drought stress than cultivars with a lower proportion (Schittenhelm et al. 2006). Leaf area in relation to stem mass has been found to vary with cultivar, growing location and growing conditions (Fleisher et al. 2006; Lactin and Holliday 1995). Cultivars that have performed well under drought stress have fewer stems (Moll and Klemke 1990). For later maturing varieties, a reduction in stem height under drought stress was associated with maintenance of dry matter accumulation (Deblonde and Ledent 2001). Clones that are better able to maintain a high LAI maintained yields better under drought than other cultivars did (Abdullah-Al-Mahmud et al. 2014). Leaf characteristics of the canopy appear to be important for stress avoidance, but exactly how this affects photosynthetic efficiency on a wide range of cultivars is yet to be identified.

3.4 Canopy growth period

The rate and extent of canopy development is an important aspect of the relationship between above ground biomass and tuber growth. Historically, researchers have classified potato canopy development by maturity class, a measure of vine vigour at the end of a growing season (Haga et al. 2012). This method is subjective, however, and is usually based on a single time point of canopy development. Maturity class can be assessed more accurately by utilizing a canopy growth curve index (Khan et al. 2013). The best predictor of dry matter accumulation across all cultivars is the length of the canopy growth period, defined as the time from emergence to the time that leaf senescence reduces the canopy to 50% of the planted area (Kooman et al. 1996a). Relationships between the canopy growth period in degree-days and dry matter partitioning and light interception are illustrated in Fig. 1. When selecting cultivars that are adapted to a growing region, matching the canopy growth curve to the growing season is important for maximizing production (Silva and Pinto 2005; Rodrigues et al. 2009). Although several different methods of measuring canopy growth period have been developed, a key aspect of each is that they attempt to model light interception. Understanding relationships between tuber development and canopy growth dynamics in potato holds promise for improvements in breeding and crop modelling applications.

Canopy development can be divided into three phases: (i) from emergence until tuber initiation, (ii) completion of canopy growth and tuber bulking and (iii) tuber maturation and canopy decline (Khan et al. 2013; Kooman and Rabbinge 1996). These are indicated in Fig. 1. In the first phase, the potato seed piece must sprout after planting and rapidly form leaves. It was observed that cultivars with a rapid development of leaf area led to a higher dry matter accumulation in the crop (Khurana and McLaren 1982). There is an optimum temperature and day length that allows the canopy to develop most extensively before tubers begin competing effectively with vines as carbohydrate sinks (Kooman et al.

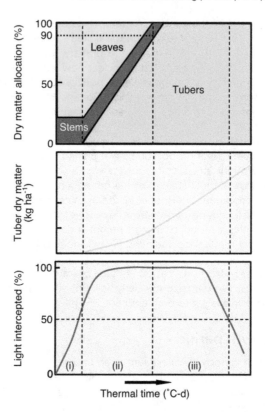

Figure 1 Relationships between potato development, in units of thermal time, and dry matter partitioning between leaves, stems and tubers (upper panel), dry matter accumulation in tubers (middle panel) and percentage of incident light intercepted by the canopy. Dashed vertical lines partition each figure panel into growth phases, where (i) is the period from emergence to tuber initiation, (ii) is the period from tuber initiation until tubers are acquiring 90% of available dry matter and (iii) is the period that ends when light interception by the canopy decreases to 50%. Adapted from Kooman et al., 1996b.

1996a; Geremew et al. 2007). Early maturing cultivars were found to initiate tubers earlier and this resulted in competition for photoassimilates between leaves and tubers (Kooman and Rabbinge 1996). At a given location, genotypes that were most successful in terms of maximizing dry matter yield were those with a larger canopy before tuber initiation (Kooman et al. 1996a; Jefferies et al. 1989).

In the second phase of canopy development, from tuber initiation to the end of net leaf growth, canopy growth rate begins to slow as tubers accumulate an increasingly larger share of photoassimilates. The end of this period is also defined as the time when tubers receive 90% of daily assimilate production (Kooman and Rabbinge 1996) or when the ground cover growth curve has a negative slope (Khan et al. 2013). The length of this phase is determined by an interaction between genotype, temperature and day length (Kooman et al. 1996b). This phase is longer for medium and late maturity types, but they have a slower canopy growth rate than early cultivars during this developmental period

Published by Burleigh Dodds Science Publishing Limited, 2018.

(Khan 2012; Silva and Pinto 2005). This characteristic of early cultivars, a rapid gain in canopy size followed by a rapid reduction in size, results in a high harvest index (Spitters 1988).

The final phase of canopy development begins when net leaf growth rate becomes zero and ends when canopy ground cover is reduced to 50% (Kooman and Rabbinge 1996). Tubers are very strong sinks during this phase and all available photoassimilates are allocated to them. Differences in the length of this phase are attributed to temperature, water and nitrogen availability (Kooman et al. 1996b). Soil moisture deficit in this phase has a sizable impact on dry matter yield (Jefferies et al. 1989). Generally, longer maturation times result in a longer end of life phase. Such longer-season genotypes accumulate more dry matter than their shorter-season counterparts, exhibiting a 'stay green' trait similar to that of cereal crops (Silva and Pinto 2005; Rolando et al. 2015). However, it is noted that high temperatures and light intensities hasten leaf senescence and might make some cultivars unsuitable for tropical spring environments (Kooman and Rabbinge 1996).

3.5 Root architecture

The root system of potato comprises adventitious roots that initiate from stem nodes in the soil and lateral roots that branch from each adventitious root. The growth direction and length of adventitious roots determine the main framework of the root system. The branching pattern of lateral roots establishes the fine root system. As illustrated in Fig. 2, adventitious roots are classified into four types of roots that differ in their site of origin: (i) basal roots presenting at the junction of the stem and mother tuber, (ii) stolon roots arising from the stolon-stem junction, (iii) stolon node roots that emerge from nodes on stolons and (iv) tuber roots produced on new tubers (Kratzke and Palta 1985; Wishart et al. 2013; Villordon et al. 2014). Functionally there are some differences among each type of adventitious root. Basal roots, which mainly elongate downwards, play an important role in water uptake from deep in the soil, while stolon roots, which spread horizontally near the soil surface, become the primary source for nutrient uptake to the leaves (Kratzke and Palta 1985; Wishart et al. 2013). Stolon and tuber roots supply water and nutrients to the tuber, although the total root mass of these roots is very small (Kratzke and Palta 1985; Busse and Palta 2006).

The root system of potato is often distributed primarily in the upper 30 cm of the soil profile (Lesczynski and Tanner 1976; Asfary et al. 1983; Parker et al. 1989; Opena and Porter 1999; Stalham and Allen 2001), although a few roots reach depths of up to 100 cm (Stone 1982; Opena and Porter 1999; Stalham and Allen 2001; Iwama 2008). Extensive variation in total root length per unit area has been reported. Vos and Groenwold (1986) measured 4 to 7 km m^{-2}, whereas 8 to 24 km m^{-2} was observed in the studies by Iwama et al. (1993). This large variation in values is, in most cases, caused by differences in environmental conditions including soil type, precipitation amount and cultivation system. Yamaguchi and Tanaka (1990) made a comparison of root systems between potato and other field crops. The results showed clearly that potato has a shallow root system and low root length density compared with other field crops. Therefore, potato is considered a drought-sensitive crop (Yuan et al. 2003).

To enhance drought tolerance in potato, some researchers have focused on expanding the root system. Under water-deficit conditions, genotypes with larger root masses had higher tuber yields when compared with genotypes having smaller root masses (Iwama

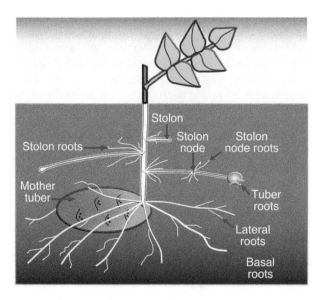

Figure 2 Sites of origin for roots in a young potato plant. Basal roots, stolon roots, stolon node roots and tuber roots are adventitious roots originating from underground stems. Lateral roots branch off of adventitious roots.

2008; Deguchi et al. 2010). Deeper root systems may also help potato to avoid drought stress (Lahlou and Ledent 2005; Deguchi et al. 2010). Significant variation was detected among genotypes for root characteristics (Iwama 2008; Wishart et al. 2013). Further developments in potato root research would benefit substantially from new systems for measuring root traits. Wishart et al. (2013) developed a systematic measuring system to evaluate root traits such as total root length, total root number and total root weight. This phenotyping method may accelerate identification of QTLs associated with individual root traits as part of an effort to breed potato for more efficient capture of resources from the soil.

4 Potato responses to water and heat stresses

Potato plants exhibit a high degree of plasticity in their responses to the growth environment. Vine development, harvested yields and tuber quality are strongly influenced by availability of water and nutrients and by the severity of abiotic stress. Water stress and heat stress, acting independently or in combination, are two of the most damaging, recurring environmental stresses experienced by the potato crop.

4.1 Potato yield responses to water deficits

Water availability is often the single largest limitation on potato productivity. Limited water available can occur daily, episodically or seasonally. In each case, the effects of

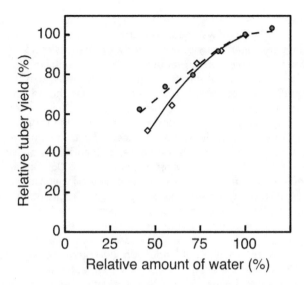

Figure 3 Representative yield responses of potato to increased availability of water. Data are for cultivars 'Russet Norkotah' (filled circles, King et al, 2011) and 'Andasureddo' (diamond symbols, Yuan et al, 2003).

insufficient water decrease the harvested yield of tubers. Water stress reduces stomatal conductance and decreases photosynthetic rates during the stress (Liu et al. 2005; Kopka et al. 1997; Rolando et al. 2015), and this decreases the availability of photosynthate for vine and tuber growth (Munns and Pearson 1974). Water stress also decreases vine, root and tuber water pressure potentials (Liu et al. 2005; Bethke et al. 2009; Heuer and Nadler 1998), which can slow or stop growth (Weisz et al. 1994). Water stress periods are cumulative. Reduced photosynthesis and growth at one time results in smaller canopy size and reduced photosynthetic potential relative to unstressed plants. With each additional stress period, the ability to fix and store carbon falls farther below the genetic potential. A summary of how drought impacts the morphological and physiological traits of potato can be found in Obidiegwu et al. (2015).

Potato is more susceptible to water stress than most crops (Weisz et al. 1994) and total available soil water should not be depleted by more than 30–50% (Steduto et al. 2012). Potato cultivars differ in their ability to tolerate insufficient water, but none is known to be resistant to drought. Some wild species relatives of potato live in drought-prone regions. This observation suggests that genetic adaptations for survival under drought may exist in potato, an inference supported by experimental studies (Vasquez-Robinet et al. 2008). For the cultivated *S. tuberosum*, harvested yields are often proportional to the fraction of water applied relative to the amount of water needed for maximum yield (Yuan et al. 2003; Fabeiro et al. 2001; AndasureddoKing et al. 2011). This relationship can be seen clearly in Fig. 3, where yields for cultivar increased from 550 g plant^{-1} to 1100 g plant^{-1} as water applied increased from 150 mm to 350 mm. In experiments using a range of water deficits and nitrogen application rates, water lost by evaporation from the crop was proportional to soil water availability (Ferreira and Carr 2002). In that study, reduced nitrogen treatments decreased canopy size and transpiration but increased evaporation from soil (Ferreira and

Carr 2002). For water applied through overhead irrigation or furrow irrigation, tuber yields range from 4 kg/m^3 to 7 kg/m^3 of water applied (Erdem et al. 2006; Ati et al. 2012) whether plants receive adequate water or are grown under water deficits of up to 50%.

One strategy for sustainable potato production in regions with limited water supplies is to irrigate some fields with water sufficient for high yields rather than to apply less water over a larger area (Steduto et al. 2012). This approach has the advantage of reducing cost of land per tonne harvested, but this benefit needs to be balanced against the risk associated with highly localized disease or pest outbreaks. Drip irrigation, which has the potential of applying water exactly where it is needed with minimal movement out of the root zone, has a higher water utilization efficiency than furrow or sprinkler irrigation, with harvested yield of 5–11.5 kg/m^3 water (Erdem et al. 2006; Onder et al. 2005; Ati et al. 2012). In general, drip irrigation does not increase total yields compared with furrow irrigation (Ati et al. 2012). Implementation of drip irrigation for potato on a commercial scale is a challenging task and a topic of current research.

Experiments in which controlled moisture deficits have been imposed at discrete times of production or throughout the production season have invariably shown that even modest reductions in water result in yield reductions (Obidiegwu et al. 2015; Fabeiro et al. 2001; Pavlista 2015). Where the economics of deficit irrigation have been evaluated, the usual finding is that the cost savings from reduced irrigation do not offset the income lost from reductions in yield (Shock et al. 1998; King et al. 2011). The most favourable time to impose water deficits has been examined on numerous occasions, and the results have been somewhat mixed. Reduced water during early season vegetative growth is most favourable for yield maintenance in some cases (Fabeiro et al. 2001), but reducing water late in the season has been found to be most beneficial in other cases (Pavlista 2015). Computational modelling of plant growth and yield under drought stress highlighted the complex interaction between timing of water deficit, transpiration rate and genotype maturity (Spitters and Schapendonk 1990). Key findings were that short periods of drought reduced the yield of late genotypes less than that of early genotypes, and that late drought affected early genotypes less because of escape.

Early maturing genotypes are well suited for production environments where water is available early in the season and becomes reduced or lacking at later times. Mediterranean climates and tropical regions planted following a rainy season are some of the examples. In this case, the unpredictable adverse effects of reduced water availability are avoided by targeting production to the most favourable times (Levy et al. 2013).

4.2 Tuber quality defects resulting from water stress

In addition to decreasing yield, water deficits, especially transient water deficits, can decrease tuber quality. Dry periods followed by sufficient water can cause tuber defects including hollow heart, growth cracks, knobby tubers and misshapen tubers (Mikitzel 2014). Hollow heart and growth cracks are believed to occur when rapid, turgor-pressure-driven tuber expansion follows a period when water deficits have reduced turgor pressure and slowed tuber growth. In the case of hollow heart, adjacent cells within the tuber separate as rapid growth resumes. The initial crack that forms is subsequently pulled open during subsequent tuber expansion to produce a hollow cavity. Growth cracks are thought to occur for a similar reason, with the initial tissue separation occurring at the periphery of the tuber. Tubers with growth crack are difficult to market and tubers that have hollow heart develop unsightly brown necrotic tissue on the surface of the enlarging cavity.

Published by Burleigh Dodds Science Publishing Limited, 2018.

Susceptibility to hollow heart and growth cracks is influenced by genetic factors, since potato varieties develop these defects to different degrees. The physiological details underpinning those genetic differences, however, have not been characterized.

Knobby tubers, which result when tissues surrounding tuber eyes begin to expand, and misshapen tubers with hourglass shapes or pointed ends are also more difficult to market and in the latter case may be difficult to store. Some genotypes, such as the North American variety Russet Burbank, develop 'glassy ends' when exposed to strong transient water deficits. The ends of these tubers appear translucent because they have very low starch contents. Such tubers are damaged easily when handled and this makes them more likely to be infected with pathogens that promote tuber rot (Thompson et al. 2008).

4.3 Heat stress

Potato grows best at cool temperatures, although some cultivars are more tolerant of heat stress than others (Rykaczewska 2015; Reynolds et al. 1990). Optimum temperatures for vine growth are in the range of 20–25°C, with lower temperatures of 15–20°C optimal for tuber growth (Rykaczewska 2015). Potato is grown successfully in regions where daytime temperatures regularly exceed 30°C, but productivity is improved when night-time temperatures drop to below 20°C. Increased temperatures resulting from climate change are likely to impact the sustainability of agriculture in many ways (Bita and Gerats 2013). Many potato-producing regions may experience losses in productivity if regional temperatures increase (Hijmans 2003).

Moderate heat stress of approximately 30°C increases partitioning of carbon to shoots instead of tubers (Prange et al. 1990; Gawronska et al. 1992). Photosynthetic rates at high temperature have been shown to increase in some cases (Hancock et al. 2014) and decrease in others (Prange et al. 1990; Fleisher et al. 2006). Under extreme high-temperature

Table 1 Effects of soil temperature during bulking and maturation on yield and average tuber weight of 'Premier Russet' and 'Ranger Russet' tubers (Zommick et al. 2014)

Variety	Growth stage	Days after planting	Soil temp. (°C)	Total yield (MT ha^{-1})	g tuber^{-1}
Premier Russet	Bulking and maturation	111–180	16	64.3	233
	Bulking	111–164	23	75.3	230
	Bulking	111–164	29	7.6	107
	Maturation	151–180	23	67.7	194
	Maturation	151–180	29	34.3	192
Ranger Russet	Bulking and maturation	111–180	16	79.8	286
	Bulking	111–180	23	94.2	253
	Bulking	111–164	29	3.7	76
	Maturation	151–150	23	74.2	215
	Maturation	151–180	29	56.7	226

conditions, photosynthesis is impaired because of decreased stomatal conductance and, in some clones, decreased photosynthetic ability (Reynolds et al. 1990).

High temperatures impair the conversion of sugars to starch in tubers (Hancock et al. 2014; Ewing 1981; Krauss and Marschner 1984). Tuber growth is strongly reduced or inhibited at tuber temperatures of approximately 30°C or more (Zommick et al. 2014; Krauss and Marschner 1984; Struik et al. 1989). Heat stress during early tuber bulking has been shown to cause greater yield reductions than heat stress in midseason or late season (Rykaczewska 2015). For some cultivars, transient heat stress early (Thompson et al. 2008) or late in the growing season (Zommick et al. 2014; Wang et al. 2012) contributes to tuber processing quality defects. Table 1 highlights the profound effects of increased soil temperature on tuber yield and size. In this study, heat cables buried in the soil were used to elevate temperature around developing tubers. Yields were strongly depressed at 23°C and nearly abolished at 29°C.

5 Potato responses to nutrient availability

5.1 Potato yield responses to nitrogen and nitrogen use efficiency

Nitrogen is a required macronutrient and potato has a moderately high need for nitrogen. The amount of nitrogen required for a crop in the tropics and subtropics may fall in the range of 80–150 kg N ha^{-1} (Roy et al. 2006), but rates twice as large or more are recommended in highly productive locations (Alva et al. 2009; Mikkelsen and Hopkins 2009). As indicated in Fig. 4, harvested yields increase with available nitrogen over a wide range (Badr et al. 2012; Joern and Vitosh 1995; Jatav et al. 2013; Feibert et al. 1998; Zebarth et al. 2004b), although there have been instances where a strong yield response to nitrogen was not observed (Joern and Vitosh 1995).

For most efficient use, nitrogen availability should match crop growth needs (Mikkelsen and Hopkins 2009). Potato varieties differ in their requirements for total nitrogen and optimum time of nitrogen application (Love et al. 2005). Efforts to increase the sustainability of potato production would benefit from use of varieties with comparatively low nitrogen requirements (Tiemens-Hulscher et al. 2014; Ospina et al. 2014). Nitrogen for the potato crop can be supplied through residual soil nitrogen, especially when potato follows clover or alfalfa in the crop rotation (Feibert et al. 1998; Porter and Sisson 1993; Neeteson 1989). In Fig. 4, for example, residual nitrogen in the soil was sufficient for yields of 30–40 t ha^{-1}. Implementation of this approach may be economically favourable where potato production can be integrated with feed production for dairy cattle (Hoshide et al. 2006).

Nitrogen can be readily leached from sandy soils following periods of heavy rainfall. Nitrogen flushed out of the root zone is unavailable for crop growth and is a potential source of groundwater contamination. Timed applications of nitrogen minimize potential nitrogen leaching and increase nitrogen use efficiency by decreasing the amount of surplus nitrogen present at any one time. A simple approach towards this end is the use of split nitrogen applications, with some applied at planting and additions made at one or more times throughout the growing season (Love et al. 2005). Synthetic slow-release fertilizers (Wilson et al. 2010), as well as incorporation of green or composted manures, are other ways in which distributed nitrogen applications can be made. The challenge

Published by Burleigh Dodds Science Publishing Limited, 2018.

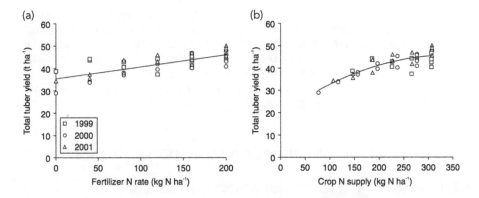

Figure 4 Harvested yields of 'Russet Burbank' in response to (a) added nitrogen and (b) nitrogen availability in the soil. Figure reproduced from Zebarth, B. J. et al. (2004a).

with these approaches is to match crop needs to nutrient availability under a range of soil, moisture and temperature conditions since these influence nitrogen release rates (Zebarth and Rosen 2007).

Nitrogen availability strongly affects plant growth and maturity. Higher availability of nitrogen late in the growing season delays vine senescence and allows for greater accumulation of carbohydrate and higher yields (Hope et al. 1960). Tuber chemical maturity and skin set, however, are delayed under these conditions (Hope et al. 1960). Late-season nitrogen availability should be managed so that tubers have well-set skins and optimum dry matter contents at harvest, although this can be problematic to implement (Sabba et al. 2007).

5.2 Potato growth as influenced by potassium, phosphorus and calcium

Potassium and phosphorus are macronutrients required for vigorous plant growth, and potato plants have a relatively high requirement for both (Mikkelsen and Hopkins 2009; Rosen et al. 2014; Panique et al. 1997). Soil tests are the most common method to determine the need for additional phosphorus, but soil type, region, measurement method and variety interact to determine if supplemental phosphorus is needed (Rosen et al. 2014). Phosphorus availability to potato plants can also be increased by changing soil properties to increase pH, incorporating organic acids and management practices that increase the effective size of the root zone (Hopkins et al. 2014). Breeding efforts that produce cultivars with larger root systems are likely to have a secondary benefit of improving potassium and phosphorus acquisition (Thornton et al. 2014).

The need for adequate phosphorus nutrition to crops must be balanced against potential environmental impacts that result when phosphorus runs off agricultural fields into lakes and streams (Ruark et al. 2014). Mitigating these concerns requires careful evaluation of phosphorus need and source, application method and timing, and proximity to surface water (Ruark et al. 2014).

Potato tubers are susceptible to mild calcium deficiencies. Calcium translocated in the xylem from the main root system supplies much of the plant with calcium. Tubers, however, receive most of their water from the phloem, and calcium concentrations in the phloem sap are very low. Small roots on the stolons and tubers are a major source of calcium for potato tubers and calcium concentrations in tubers can be increased by making soluble calcium available throughout the tuber bulking period (Palta 2010). Such applications have the potential to alter tuber number and minimize tuber quality defects including hollow heart and bruising (Palta 2010). Where soils are deficient in calcium, improvements in tuber yield can be achieved by application of gypsum at planting.

6 Additional factors affecting sustainable production

6.1 Physiological tolerance of pests and pathogens

Diseases and pests that affect the growing vines or harvested tubers may have large effects on the yield of marketable tubers. The most significant of these all over the world is the late blight pathogen, *Phytophthora infestans*. Also important are early blight (*Alternaria solani*), bacterial diseases that rot vines or tubers (*Pectobacterium* spp.), herbivory from Colorado Potato Beetles and other insects, tuber blemish diseases such as scab (*Streptomyces scabies*), nematodes and viral infections. Mechanisms of resistance to these pests and pathogens are discussed in detail later in this volume, but it is important to note that the physiological state of the affected organs can have a strong influence on disease incidence and progression. An example is the late blight pathogen, which more readily infects mature rather than young leaves (Panter and Jones 2002). Likewise, symptom development from the early blight pathogen, which causes necrosis and accelerated senescence of leaves, is rarely observed in young plants. Once leaves mature or begin to senesce, early blight lesions increase in frequency and size. This observation is striking when genotypes of potato with different maturity classes are planted in adjacent rows. Although the amount of inoculum reaching each plant is likely to be similar, late maturing plants can be free of early blight symptoms at a time when appreciable disease is present on the early maturing plants. Likewise, tuber susceptibility to common scab disease is greatest shortly after tuber initiation (Khatri et al. 2011).

6.2 Post-harvest storage

The potato crop can be harvested successfully in many ways. Traditional potato forks, with tines spaced to lift tubers larger than a minimum size, have been used for centuries with good effect on small farms. Mechanized harvest allows for larger farms to complete the harvest quickly, before cold temperatures or heavy rains damage the crop by freezing or decaying. When possible, fields should be planted with a cover crop shorty after harvest to minimize soil loss caused by wind or water erosion.

Harvest should be delayed until physiological maturity is achieved and the skin is firmly attached to tubers. Tubers that are harvested prior to adequate skin set, as for immature 'new potatoes', must be handled very gently to prevent periderm removal and should be marketed quickly. Skinning degrades appearance (Neubauer et al. 2013; Sabba and Lulai 2002), promotes water loss (Burton 1978; Lulai and Orr 1995) and increases likelihood of disease (Lulai and Corsini 1998) since the skin is the primary barrier to pathogen entry. Skin set is a developmental process that begins as vines begin to senesce and requires several weeks to complete (Wiltshire et al. 2005; Bethke and Busse 2010). Seed potatoes should not be harvested before the skin is well set because they need to avoid spoilage for an extended time.

In most potato-producing regions, supply is large and prices are low during and shortly after harvest of the main crop (Moazzem and Fujita 2004; Chandran and Pandey 2007; Fuglie et al. 1997). Storage of the harvested crop is beneficial in that it makes fresh tubers available for consumption, sale or use over an extended period of time. In many markets, prices for stored tubers exceed those for tubers sold during the harvest period and this has the potential for increased profitability (Guenthner 1995; Fuglie 1995). Potato storage in developed countries is done in dedicated storage buildings that often have temperature and ventilation control systems designed for that purpose (Gottschalk 2011; Cunnington 2008; Brook et al. 1995). Such facilities are not available to small farmers, where the amount of material to be stored is much less and there is little economic justification for the capital outlay needed to build sophisticated potato storages. Simpler potato storages, however, may be adequate for grower needs as long as they provide sufficient fresh air to the tubers and avoid wet conditions that are conducive to storage rot (Pringle et al. 2009; Bethke 2014). The simplest and least costly of these methods is ground storage, where potatoes are not harvested until needed. Diffuse light storage systems are another low-cost alternative and are suitable for regions that do not experience temperatures below freezing (Babarinsa and Williams 2015; Muthoni et al. 2013). In these structures, tubers are stored in flat racks that prevent accumulation of moisture and are exposed to dim light that promotes the development of short, tough sprouts. Diffuse light systems may be particularly attractive for seed storage in tropical regions.

Allowances must be made to prevent sprouting when storing fresh market or processing potatoes. Developing sprouts dramatically increase the rate of water loss from tubers (Burton 1978) and are undesirable to consumers. Thus sprouted tubers decrease the value of the crop. Furthermore, sprouts are easily damaged and the wounds produced allow for entry of pathogens. Selection of genotypes that retain dormancy after harvest is one effective method to prevent sprouting, as is storage at low temperatures, preferably between 3 and 6°C. Sprout suppressant can be applied to the stored crop, although doing this effectively and safely requires specialized equipment and training. Chemistries for sprout inhibition include both synthetic and natural products (Gomez-Castillo et al. 2013; Mehta et al. 2010; Wills et al. 2004; Kleinkopf et al. 2003).

6.3 Yield stability as a component of sustainable production

Minimizing variation in harvested crop yields is one way to reduce the financial risks associated with potato production and may contribute to economic sustainability. Trait stability is a measure of how a particular genotype performs across multiple environments. There are three types of stability. Type I stability is when the trait value for a genotype does not change across environments. Type II is when the trait value for a genotype is in the same position relative to the overall trait mean of the population. Type III stability is how accurately the observed trait values for a genotype across the environments can be modelled (Bernardo 2010; Eberhart and Russell 1966). Potato is well suited for trait stability analysis because it is clonally propagated and therefore identical genotypes can be planted in multiple locations. Across studies in northern growing regions, genotypes of elite breeding material were evaluated under a set of environments. Some individuals embodied one of the three stability types and others displayed characteristics of two or more stability types (Tai and Young 1972;

Haynes et al. 2010). In stability estimates in more southern and tropical environments, genotypes also displayed a range of stability characteristics for the traits of tuber yield, dry matter content and specific gravity (Tsegaw 2011; Cotes et al. 2002; Mulema et al. 2008; Hassanpanah and Azimi 2010). Interestingly, chipping potatoes also have a genotype by storage environment interaction for chip colour, and stability types I, II and III were found across varying storage temperatures (Rak et al. 2013). It is evident that stability is a factor that potato breeders must be aware of when developing lines for different growing areas where genotype by environment interactions may be significant. Ignoring this can have an effect on the variation of yield from year to year and can reduce selection efficiency in breeding (Paget et al. 2015).

7 Summary

The outcome of each potato production year depends on interactions between a potato variety and the production environment. These interactions depend on the physiology of the plant. Outcomes essential for sustainable potato production include marketable yield, maintenance of environmental resources and economic return. Potato production becomes more sustainable when variety selection and crop management decisions are carefully matched with the abilities of the grower, the characteristics of the land and the need to adjust agronomic practice in response to year-to-year variation in environmental stresses and market conditions.

8 Future trends

Sustainable potato production requires a balanced approach that incorporates grower profitability, environmental management and societal concerns. To achieve these goals, potato varieties tailored to the needs and challenges of specific growing regions are required. These varieties should reduce the economic risks of potato production by having tolerance to endemic diseases and regional abiotic stresses. Identifying varieties suitable for farms that differ widely in scale will be a challenging task. Breeding programmes and seed companies focused on large-scale agriculture in the developed world may not invest resources in evaluating varietal performance on smallholder farms. Engaging smallholders in participatory variety evaluation trials is an attractive approach for overcoming this limitation (Almekinders et al. 2014; Morris and Bellon 2004), although there are difficulties in implementing this approach successfully (Almekinders et al. 2007; Thiele et al. 2001). A challenge for sustainable production applicable to all scales of production is the need to balance the inputs applied to the crop to maximize the benefit obtained from each input. Water, nutrients and pest and disease control measures should be applied as needed by the crop in a way that maximizes overall profitability or nutritive value. An excess of one input relative to others results in waste of that nutrient since limitations from other inputs prevent its utilization. As an example, optimal nitrogen application rate is likely to increase as water availability increases. Developing the on-farm knowledge required to tailor crop input amount and timing to crop needs will be a substantial challenge for the future, given the wide range of environments under which potato is grown.

Published by Burleigh Dodds Science Publishing Limited, 2018.

9 Where to look for further information

Lutaladio, N. and Castaidi, L., 2009. Potato: The hidden treasure. *Journal of Food Composition and Analysis*, 22(6), pp. 491–3.

Bohl, W. and Johnson, S. B. eds., 2010. *Commercial Potato Production in North America*, The Potato Association of America.

Obidiegwu, J. E. et al., 2015. Coping with drought: Stress and adaptive responses in potato and perspectives for improvement. *Frontiers in Plant Science*, 6, p.542.

Mikkelsen, R. and Hopkins, B., 2009. *Fertilizer MBPS*, Norcross, GA.

Recommended conferences: Potato Association of America annual meeting; European Association for Potato Research Triennial conference.

10 References

Abdullah-Al-Mahmud, Hossain, M. A., Abdullah-Al-Mamun, M., Shamimuzzaman, M., Rahaman, E. H. M. S., Khan, M. S. A., and Bazzaz, M. M. 2014. Plant canopy, tuber yield and growth analysis of potato under moderate and severe drought condition. *Journal of Plant Sciences*, 2(5), 201–8.

Abelenda, J. A., Navarro, C., and Prat, S. 2014. Flowering and tuberization: a tale of two nightshades. *Trends in Plant Science*, 19(2), 115–22.

Alexopoulos, A. A., Akoumianakis, K. A., and Passam, H. C. 2006. Effect of plant growth regulators on the tuberisation and physiological age of potato (*Solanum tuberosum* L.) tubers grown from true potato seed. *Canadian Journal of Plant Science*, 86(4), 1217–25.

Allen, E. J. and Scott, R. K. 1980. An analysis of growth of the potato crop. *Journal of Agricultural Science*, 94(03), 583–606.

Allen, P. and Sachs, C. 1991. What do we want to sustain? Developing a comprehensive vision of sustainable agriculture. Brown, M. (ed), Available at: https://escholarship.org/uc/item/4qf8p2qk.pdf [Accessed 12 March 2016].

Allen, P., Van Dusen, D., Lundy, J., and Gliessman, S. 1991. Expanding the definition of sustainable agriculture. Available at: http://eprints.cdlib.org/uc/item/6cd573mh [Accessed 12 March 2016].

Almekinders, C. J. M., Mertens, L., van Loon, J. P., and Lammerts van Bueren, E. T. 2014. Potato breeding in the Netherlands: a successful participatory model with collaboration between farmers and commercial breeders. *Food Security*, 6(4), 515–24.

Almekinders, C. J. M., Chujoy, E., and Thiele, G. 2009. The use of true potato seed as pro-poor technology: The efforts of an international agricultural research institute to innovating potato production. *Potato Research*, 52(4), 275–93.

Almekinders, C. J. M., Thiele, G., and Danial, D. L. 2007. Can cultivars from participatory plant breeding improve seed provision to small-scale farmers? *Euphytica*, 153(3), 363–72.

Alva, A. K., Collins, H. P. and Boydston, R. A. 2009. Nitrogen management for irrigated potato production under conventional and reduced tillage. *Soil Science Society of America Journal*, 73(5), 1496–503.

Ames, M. and Spooner, D. M. 2008. DNA from herbarium specimens settles a controversy about origins of the European potato. *American Journal of Botany*, 95(2), 252–7.

Angers, D. A., Edwards, L. M., Sanderson, J. B., and Bissonnette, N. 1999. Soil organic matter quality and aggregate stability under eight potato cropping sequences in a fine sandy loam of Prince Edward Island. *Canadian Journal of Soil Science*, 79(3), 411–7.

Asfary, A. F., Wild, A., and Harris, P. M. 1983. Growth, mineral nutrition and water use by potato crops. *Journal of Agricultural Science*, 100(01), 87–101.

Ati, A. S., Iyada, A. D., and Najim, S. M. 2012. Water use efficiency of potato (*Solanum tuberosum* L.) under different irrigation methods and potassium fertilizer rates. *Annals of Agricultural Sciences*, 57(2), 99–103.

Published by Burleigh Dodds Science Publishing Limited, 2018.

Babarinsa, F. A. and Williams, J. O. 2015. Development of a diffuse light store for 'seed' potato storage. *International Journal of Agriculture and Earth Science*, 1(4).

Badr, M. A., El-Tohamy, W. A., and Zaghloul, A. M. 2012. Yield and water use efficiency of potato grown under different irrigation and nitrogen levels in an arid region. *Agricultural Water Management*, 110, 9–15.

Baroja-Fernández, E, Muñoz, F. J., Montero, M., Etxeberria, E., Sesma, M. T., Ovecka, M., et al. 2009. Enhancing sucrose synthase activity in transgenic potato (*Solanum tuberosum* L.) tubers results in increased levels of starch, ADPglucose and UDPglucose and total yield. *Plant and Cell Physiology*, 50(9), 1651–62.

Bernardo, R. 2010. *Breeding for quantitative traits in plants*. 2nd ed., Woodbury, MN: Stemma Press.

Bethke, P. C. 2014. Postharvest storage and physiology. In Pavek, M. J. and Navarre, D. A. (eds), *The Potato: Botany, Production and Uses*. CABI, pp. 255–71.

Bethke, P. C. and Busse, J. S. 2010. Vine-kill treatment and harvest date have persistent effects on tuber physiology. *American Journal of Potato Research*, 87(3), 299–309.

Bethke, P. C., Sabba, R., and Bussan, A. J. 2009. Tuber water and pressure potentials decrease and sucrose contents increase in response to moderate drought and heat stress. *American Journal of Potato Research*, 86(6), 519–32.

Bhogale, S., Mahajan, A. S., Natarajan, B., Rajabhoj, M., Thulasiram, H. V., and Banerjee, A. K. 2014. MicroRNA156: A Potential graft-transmissible microrna that modulates plant architecture and tuberization in *Solanum tuberosum* ssp. *andigena*. *Plant Physiology*, 164(2), 1011–27.

Biswas, B., Ghosh, D. C., Dasgupta, M. K., Trivedi, N., Timsina, J., and Dobermann, A. 2006. Integrated assessment of cropping systems in the Eastern Indo-Gangetic plain. *Field Crops Research*, 99(1), 35–47.

Bita, C. E. and Gerats, T. 2013. Plant tolerance to high temperature in a changing environment: scientific fundamentals and production of heat stress-tolerant crops. *Frontiers in Plant Science*, 4, 273.

Blauer, J. M., Knowles, L. O., and Knowles, N. R. 2013. Manipulating stem number, tuber set and size distribution in specialty potato cultivars. *American Journal of Potato Research*, 90(5), 470–96.

Bohl, W. and Johnson, S. B. eds. 2010. *Commercial Potato Production in North America*, The Potato Association of America.

Brook, R. C., Fick, R. J., and Forbush, T. D. 1995. Potato storage design and management. *American Journal of Potato Research*, 72(8), 463–80.

Burstall, L. and Harris, P. M. 1983. The estimation of percentage light interception from leaf area index and percentage ground cover in potatoes. *Journal of Agricultural Science*, 100(01), 241–4.

Burton, W. 1978. Post-harvest behaviour and storage of potatoes. In Coaker, T. H. (ed.), *Applied Biology*. Academic Press, pp. 86–228.

Bussan, A. J., Mitchell, P. D., Copas, M. E., and Drilias, M. J. 2007. Evaluation of the effect of density on potato yield and tuber size distribution. *Crop Science*, 47(6), 2462–72.

Busse, J. S. and Palta, J. P. 2006. Investigating the in vivo calcium transport path to developing potato tuber using ^{45}Ca: a new concept in potato tuber calcium nutrition. *Physiologia Plantarum*, 128(2), 313–23.

Chandran, K. P. and Pandey, N. K. 2007. Potato price forecasting using seasonal ARIMA approach. *Potato Journal*, 34(1–2), 137–8.

Coleman, W. 2000. Physiological ageing of potato tubers: A review. *Annals of Applied Biology*, 137(2), 189–99.

Cotes, J. M., Ñustez, C. E., Martinez, R. and Estrada, N. 2002. Analyzing genotype by environment interaction in potato using yield-stability index. *American Journal of Potato Research*, 79(3), 211–18.

Cunnington, A. C. 2008. Developments in Potato Storage in Great Britain. *European Potato Journal*, 51(3–4), 403–10.

Deblonde, P. M. K. and Ledent, J. F. 2001. Effects of moderate drought conditions on green leaf number, stem height, leaf length and tuber yield of potato cultivars. *European Journal of Agronomy*, 14(1), 31–41.

Published by Burleigh Dodds Science Publishing Limited, 2018.

Deguchi, T., Naya, T., Wangchuk, P., Itoh, E., Matsumoto, M., Zheng, X., et al. 2010. Aboveground characteristics, yield potential and drought tolerance in 'Konyu' potato cultivars with large root mass. Potato Research, 53(4), 331–40.

Dwelle, R. 1990. Source sink relationships during tuber growth. American Potato Journal, 67(12), 829–33.

Eberhart, S. A. and Russell, W. A. 1966. Stability Parameters for Comparing Varieties. Crop Science, 6(1), 36–40.

Erdem, T., Erdem, Y., Orta, H., and Okursoy, H. 2006. Water-yield relationships of potato under different irrigation methods and regimens. Scientia Agricola, 63(3), 226–31.

Essah, S. Y. C. and Honeycutt, C. W. 2004. Tillage and seed-sprouting strategies to improve potato yield and quality in short season climates. American Journal of Potato Research, 81(3), 177–86.

Ewing, E. 1981. Heat stress and the tuberization stimulus. American Potato Journal, 58(1), 31–49.

Fabeiro, C., de Santa Olalla, F. M., and De Juan, J. A., 2001. Yield and size of deficit irrigated potatoes. Agricultural and Water Management, 48(3), 255–66.

FAO, 2009. International Year of the Potato 2008: New light on a hidden treasure. Experimental Agriculture, 45(03), 376. https://doi.org/10.1017/s0014479709007686.

Feibert, E. B. G., Shock, C. C., and Saunders, L. D., 1998. Nitrogen Fertilizer Requirements of Potatoes Using Carefully Scheduled Sprinkler Irrigation. HortScience, 33(2), 262–5.

Ferreira, T. C. and Carr, M., 2002. Responses of potatoes (Solanum tuberosum L.) to irrigation and nitrogen in a hot, dry climate: I. Water use. Field Crops Research, 78(1), 51–64.

Fiener, P. and Auerswald, K. 2007. Rotation effects of potato, maize, and winter wheat on soil erosion by water. Soil Science Society of America Journal, 71(6), 1919–25.

Firman, D. M., O'Brien, P. J., and Allen, E. J., 1995. Appearance and growth of individual leaves in the canopies of several potato cultivars. Journal of Agricultural Science, 125(03), 379–94.

Fleisher, D. H., Timlin, D. J., and Reddy, V. R., 2006. Temperature influence on potato leaf and branch distribution and on canopy photosynthetic rate. Agronomy Journal, 98(6), 1442–52.

Fuglie, K., Khatana, V. S., Ilangantileke, S. G., Singh, J., Kumar, D. and Scott, G. J., 1997. Economics of potato storage in India. Social Sciences Department Working Paper No. 1997-5. International Potato Center (CIP). Lima Peru.

Fuglie, K. O., 1995. Measuring welfare benefits from improvements in storage technology with an application to Tunisian potatoes. American Journal of Agricultural Economics, 77(1), 162–73.

Gawronska, H., Thornton, M. K. and Dwelle, R. B., 1992. Influence of heat stress on dry matter production and photo-assimilate partitioning by four potato clones. American Potato Journal, 69(10), 653–65.

Geremew, E. B., Steyn, J. M. and Annandale, J. G., 2007. Evaluation of growth performance and dry matter partitioning of four processing potato (Solanum tuberosum) cultivars. New Zealand Journal of Crop and Horticultural Science, 35, 385–93.

Gomez-Castillo, D., Cruz, E., Iguaz, A., Arroqui, C., and Virseda, P. 2013. Effects of essential oils on sprout suppression and quality of potato cultivars. Postharvest Biology and Technology, 82, 15–21.

Gottschalk, K., 2011. Recent developments in potato storage in Europe. Potato Journal, 38(2), 85–99.

Grandya, A. S., Porter, G. A. and Erichb, M. S., 2002. Organic amendment and rotation crop effects on the recovery of soil organic matter and aggregation in potato cropping systems. Soil Science Society of America Journal, 66(4), 1311–19.

Guenthner, J. F., 1995. Economics of potato storage. American Potato Journal, 72(8), 493–502.

Guenthner, J. F., 2010. Past, present and future of world potato markets: an overview. Potato Journal, 37(1–2).

Haga, E., Weber, B. and Jansky, S., 2012. Examination of potential measures of vine maturity in potato. American Journal of Plant Sciences, 3, 495–505.

Hancock, R. D., Morris, W. L., Ducreux, L. J. M., Morris, J. A., Usman, M., Verrall, S. R., et al. 2014. Physiological, biochemical and molecular responses of the potato (Solanum tuberosum L.) plant to moderately elevated temperature. Plant, Cell and Environment, 37(2), 439–50.

Hannapel, D. J., 2010. A model system of development regulated by the long-distance transport of mRNA. *Journal of Integrative Plant Biology*, 52(1), 40–52.

Hannapel, D. J., Chen, H., Rosin, F., Banerjee, A. and Davies, P. 2004. Molecular controls of tuberization. *American Journal of Potato Research*, 81(4), 263–74.

Hartmann, A., Senning, M., Hedden, P., Sonnewald, U., and Sonnewald, S. 2011. Reactivation of meristem activity and sprout growth in potato tubers require both cytokinin and gibberellin. *Plant Physiology*, 155(2), 776–96.

Hassanpanah, D. and Azimi, J., 2010. Yield stability analysis of potato cultivars in spring cultivation and after barley harvest cultivation. *American-Eurasian Journal of Agricultural and Environmental Science*, 9(2), 140–4.

Haverkort, A. J. and Harris, P. M., 1986. Conversion coefficients between intercepted solar radiation and tuber yeilds of potato crops under tropical highland conditions. *Potato Research*, 29(4), 529–33.

Haverkort, A. J., Uenk, D., Veroude, H., and Van De Waart, M. 1991. Relationships between ground cover, intercepted solar radiation, leaf area index and infrared reflectance of potato crops. *Potato Research*, 34(1), 113–21.

Haverkort, A. J., De Ruijter, F. J., Van Evert, F. K., Conijn, J. G., and Rutgers, B. 2013. Worldwide sustainability hotspots in potato cultivation. 1. Identification and mapping. *Potato Research*, 56(4), 343–53.

Haverkort, A. J., Waart, M. and Bodlaender, K. B. A., 1990. Interrelationships of the number of initial sprouts, stems, stolons and tubers per potato plant. *Potato Research*, 33(2), 269–74.

Haynes, K. G., Clevidence, B. A., Rao, D., Vinyard, B. T. and White, J. M. 2010. Genotype x environment interactions for potato tuber carotenoid content. *Journal of The American Society for Horticultural Science*, 135(3), 250–8.

Heuer, B. and Nadler, A., 1998. Physiological response of potato plants to soil salinity and water deficit. *Plant Science*, 137(1), 43–51.

Hijmans, R. J., 2003. The effect of climate change on global potato production. *American Potato Journal*, 80(4), 271–9.

Hope, G. W., MacKay, D. C. and Townsend, L. R., 1960. The effect of harvest date and rate of nitrogen fertilization on the maturity, yield and chipping quality of potatoes. *American Potato Journal*, 37(1), 28–33.

Hopkins, B. G., Horneck, D. A. and MacGuidwin, A. E., 2014. Improving phosphorus use efficiency through potato rhizosphere modification and extension. *American Journal of Potato Research*, 91(2), 161–74.

Hoshide, A. K., Dalton, T. J. and Smith, S. N., 2006. Profitability of coupled potato and dairy farms in Maine. *Renewable Agriculture and Food Systems*, 21(4), 261–72.

Huseth, A. S., Petersen, J. D., Poveda, K., Szendrei, Z., Nault, B. A., Kennedy, G. G. and Groves, R. L. 2015. Spatial and temporal potato intensification drives insecticide resistance in the specialist herbivore, *Leptinotarsa decemlineata*. *PLoS ONE*, 10(6), e0127576.

Ishimaru, K., Hirotsu, N. and Kashiwagi, T., 2008. Overexpression of a maize SPS gene improves yield characters of potato under field conditions. *Plant Production Science*, 11(1), 104–7.

Iwama, K., 2008. Physiology of the potato: New Insights into root system and repercussions for crop management. *Potato Research*, 51(3–4), 333–53.

Iwama, K., Hukushima, T., Yoshimura, T. and Nakaseko, K. 1993. Influence of planting density on root growth and yield in potato. *Japanese Journal of Crop Science*, 62(4), 628–35.

Jackson, S., 1999. Multiple signaling pathways control tuber induction in potato. *Plant Physiology*, 119(1), 1–8.

Jatav, M. K., Kumar, M., Trehan, S. P., Dua, V. K. and Kumar, S. 2013. Effect of nitrogen and varieties of potato on yield and agronomic N use efficiency in North-Western plains of India. *Potato Journal*, 40(1), 55–9.

Jefferies, R. A., Heilbronn, T. D. and MacKerron, D. K. L., 1989. Estimating tuber dry matter concentration from accumulated thermal time and soil moisture. *Potato Research*, 32(4), 411–17.

Published by Burleigh Dodds Science Publishing Limited, 2018.

Joern, B. C. and Vitosh, M. L., 1995. Influence of applied nitrogen on potato part I: Yield, quality, and nitrogen uptake. *American Potato Journal*, 72(1), 51–63.

Johnson, D. A. and Dung, J. K. S., 2010. Verticillium wilt of potato – The pathogen, disease and management. *Canadian Journal of Plant Pathology*, 32(1), 58–67.

Jonik, C., Sonnewald, U., Hajirezaei, M.-R., Flügge, U.-I. and Ludewig, F. 2012. Simultaneous boosting of source and sink capacities doubles tuber starch yield of potato plants. *Plant Biotechnology Journal*, 10(9), 1088–98.

Khan, M. S., 2012. *Assessing genetic variation in growth and development of potato*. PhD Thesis, Wageningen University, Wageningen, the Netherlands ISBN 978–94–6173–359–7

Khan, M. S., van Eck, H. J. and Struik, P. C., 2013. Model-based evaluation of maturity type of potato using a diverse set of standard cultivars and a segregating diploid population. *Potato Research*, 56(2), 127–46.

Khatri, B. B., Tegg, R. S., Brown, P. H., and Wilson, C. R. 2011. Temporal association of potato tuber development with susceptibility to common scab and Streptomyces scabiei-induced responses in the potato periderm. *Plant Pathology*, 60(4), 776–86.

Khurana, S. C. and McLaren, J. S., 1982. The influence of leaf area, light interception and season on potato growth and yield. *Potato Research*, 25(4), 329–42.

King, B. A., Tarkalson, D. D., Bjorneberg, D. L., and Taberna, J. P., Jr. 2011. Planting system effect on yield response of Russet Norkotah to irrigation and nitrogen under high intensity sprinkle irrigation. *American Journal of Potato Research*, 88(2), 121–34.

Kittipadukal, P., Bethke, P. C. and Jansky, S. H., 2012. The effect of photoperiod on tuberisation in cultivated x wild potato species hybrids. *Potato Research*, 55(1), 27–40.

Kleinkopf, G., Oberg, N. and Olsen, N., 2003. Sprout inhibition in storage: Current status, new chemistries and natural compounds. *American Journal of Potato Research*, 80(5), 317–27.

Kloosterman, B., Abelenda, J. A., del Mar Carretero Gomez, M., Oortwijn, M., de Boer, J. M., Kowitwanich, K., et al. 2013. Naturally occurring allele diversity allows potato cultivation in northern latitudes. *Nature*, 1–7.

Knowles, N. and Knowles, L., 2006. Manipulating stem number, tuber set, and yield relationships for northern- and southern-grown potato seed lots. *Crop Science*, 46(1), 284–96.

Koga, N., 2008. An energy balance under a conventional crop rotation system in northern Japan: Perspectives on fuel ethanol production from sugar beet. *Agriculture, Ecosystems and Environment*, 125(1–4), 101–10.

Koga, N., 2013. Nitrous oxide emissions under a four-year crop rotation system in northern Japan: impacts of reduced tillage, composted cattle manure application and increased plant residue input. *Soil Science and Plant Nutrition*, 59(1), 56–68.

Kolbe, H. and Beckmann, S., 1997. Development, growth and chemical composition of the potato crop (*Solanum tuberosum* L.). 2. Tuber and whole plant. *Potato Research*, 40(2), 135–53.

Kooman, P. and Rabbinge, R., 1996. An analysis of the relation between dry matter allocation to the tuber and earliness of a potato crop. *Annals of Botany*, 77(3), 235–42.

Kooman, P., Fahem, M., Tegera, P. and Haverkort, A., 1996a. Effects of climate on different potato genotypes. 1. Radiation interception, total and tuber dry matter production. *European Journal of Agronomy*, 5, 193–205.

Kooman, P., Fahem, M., Tegera, P. and Haverkort, A., 1996b. Effects of climate on different potato genotypes .2. Dry matter allocation and duration of the growth cycle. *European Journal of Agronomy*, 5, 207–17.

Kopka, J., Provart, N. and MullerRober, B., 1997. Potato guard cells respond to drying soil by a complex change in the expression of genes related to carbon metabolism and turgor regulation. *Plant Journal*, 11(4), 871–82.

Kratzke, M. G. and Palta, J. P., 1985. Evidence for the existence of functional roots on potato tubers and stolons: Significance in water transport to the tuber. *American Potato Journal*, 62(5), 227–36.

Krauss, A. and Marschner, H., 1984. Growth rate and carbohydrate metabolism of potato tubers exposed to high temperatures. *Potato Research*, 27(3), 297–303.

Lactin, D. J. and Holliday, N. J., 1995. Spatiotemporally variable area: Mass relationship of Russet Burbank potato leaflets: Implications to potato production models. *American Potato Journal*, 72(5), 287–97.

Lagercrantz, U., 2009. At the end of the day: a common molecular mechanism for photoperiod responses in plants? *Journal of Experimental Botany*, 60(9), 2501–15.

Lahlou, O. and Ledent, J.-F., 2005. Root mass and depth, stolons and roots formed on stolons in four cultivars of potato under water stress. *European Journal of Agronomy*, 22(2), 159–73.

Lalonde, S., Tegeder, M., Throne-Holst, M., Frommer, W. B., and Patrick, J. W. 2003. Phloem loading and unloading of sugars and amino acids. *Plant, Cell and Environment*, 26(1), 37–56.

Larkin, R. P., Honeycutt, C. W., Griffin, T. S., Olanya, O. M., Halloran, J. M., and He, Z. 2011. Effects of different potato cropping system approaches and water management on soilborne diseases and soil microbial communities. *Phytopathology*, 101(1), 58–67.

Larkin, R. P., Griffin, T. S. and Honeycutt, C. W., 2010. Rotation and cover crop effects on soilborne potato diseases, tuber yield, and soil microbial communities. *Plant Disease*, 94(12), 1491–502.

Leggewie, G., Kolbe, A., Lemoine, R., Roessner, U., Lytovchenko, A., Zuther, E., et al. 2003. Overexpression of the sucrose transporter *SoSUT1* in potato results in alterations in leaf carbon partitioning and in tuber metabolism but has little impact on tuber morphology. *Planta*, 217(1), 158–67.

Lesczynski, D. B. and Tanner, C. B., 1976. Seasonal variation of root distribution of irrigated, field-grown Russet Burbank potato. *American Potato Journal*, 53(2), 69–78.

Levy, D., Coleman, W. K. and Veilleux, R. E., 2013. Adaptation of potato to water shortage: Irrigation management and enhancement of tolerance to drought and salinity. *American Journal of Potato Research*, 90(2), 186–206.

Liu, F., Jensen, C., Shahanzari, A., Andersen, M., and Jacobsen, S. 2005. ABA regulated stomatal control and photosynthetic water use efficiency of potato (*Solanum tuberosum* L.) during progressive soil drying. *Plant Science*, 168(3), 831–6.

Love, S. L. and Thompson-Johns, A., 1999. Seed piece spacing influences yield, tuber size distribution, stem and tuber density, and net returns of three processing potato cultivars. *HortScience*, 34(4), 629–33.

Love, S. L., Stark, J. C. and Salaiz, T., 2005. Response of four potato cultivars to rate and timing of nitrogen fertilizer. *American Journal of Potato Research*, 82(1), 21–30.

Lulai, E. and Corsini, D., 1998. Differential deposition of suberin phenolic and aliphatic domains and their roles in resistance to infection during potato tuber (*Solanum tuberosum* L.) wound-healing. *Physiological and Molecular Plant Pathology*, 53(4), 209–22.

Lulai, E. and Orr, P., 1995. Porometric measurements indicate wound severity and tuber maturity affect the early stages of wound-healing. *American Potato Journal*, 72(4), 225–41.

Lutaladio, N. and Castaidi, L., 2009. Potato: The hidden treasure. *Journal of Food Composition and Analysis*, 22(6), 491–3.

Martin, A., Adam, H., Diaz-Mendoza, M., Zurczak, M., Gonzalez-Schain, N. D., and Suarez-Lopez, P. 2009. Graft-transmissible induction of potato tuberization by the microRNA miR172. *Development (Cambridge, England)*, 136(17), 2873–81.

Martinez-Garcia, J., Garcia-Martinez, J., Bou, J., and Prat, S. 2001. The interaction of gibberellins and photoperiod in the control of potato tuberization. *Journal of Plant Growth Regulation*, 20(4), 377–86.

McGuire, A. M., 2003. Mustard green manures replace fumigant and improve infiltration in potato cropping system. *Crop Management*, 22 Aug 2003, pp.1–6. Available at: https://dl.sciencesocieties.org/publications/cm/abstracts/2/1/2003-0822-01-RS [Accessed 10 May 2016]. https://doi.org/10.1094/cm-2003-0822-01-rs

Mehta, A., Singh, B., Ezekiel, R., and Kumar, D. 2010. Effect of CIPC on sprout inhibition and processing quality of potatoes stored under traditional storage systems in India. *Potato Research*, 53(1), 1–15.

Mikitzel, L., 2014. Tuber physiological disorders. In Pavek, M. J. and Navarre, D. A. (eds), *The Potato: Botany, Production and Uses*. Wallingford: CABI, pp. 237–54.

Published by Burleigh Dodds Science Publishing Limited, 2018.

Mikkelsen, R. and Hopkins, B., 2009. *Fertilizer MBPS*, Norcross, GA.

Moazzem, K. G. and Fujita, K., 2004. The potato marketing system and its changes in Bangladesh: from the perspective of a village study in Comilla district. *The Developing Economies*, 42(1), 63–94.

Moll, A. and Klemke, T., 1990. A simple model for the evaluation of haulm characters in potato breeding. *Archiv für Züchtungsforschung*, 20(2), 151–8.

Morris, M. L. and Bellon, M. R., 2004. Participatory plant breeding research: Opportunities and challenges for the international crop improvement system. *Euphytica*, 136(1), 21–35.

Mulema, J. M. K., Adipala, E., Olanya, O. M., and Wagoire, W. 2008. Yield stability analysis of late blight resistant potato selections. *Experimental Agriculture*, 44(02), 145–55.

Munns, R. and Pearson, C. J., 1974. Effect of water deficit on translocation of carbohydrate in *Solanum tuberosum*. *Austrailian Journal of Plant Physiology*, 1(4), 529–37.

Muthoni, J., Kabira, J. N., Kipkoech, D., Abong, G. O., and Nderitu, J. H. 2013. Yield Performance of Potato Seed Tubers After Storage in a Diffuse Light Store (DLS). *Journal of Agricultural Science*, 6(1), 1–8.

Myers, P., McIntosh, C. S., Patterson, P. E., Taylor, R. G. and Hopkins, B. G., 2008. Optimal crop rotation of Idaho potatoes. *American Journal of Potato Research*, 85(3), 183–97.

Neeteson, J. J., 1989. Effects of legumes on soil mineral nitrogen and response of potatoes to nitrogen fertilization. In Vos, J., van Loon, C., and Bollen, G. J., (eds),. *Effects of Crop Rotation on Potato Production in the Temperate Zones*. 89–93.

Nelson, K. L., Lynch, D. H. and Boiteau, G., 2009. Assessment of changes in soil health throughout organic potato rotation sequences. *Agriculture, Ecosystems and Environment*, 131(3–4), 220–8.

Neubauer, J. D., Lulai, E. C., Thompson, A. L., Suttle, J. C., Bolton, M. D., and Campbell, L. G. 2013. Molecular and cytological aspects of native periderm maturation in potato tubers. *Journal of Plant Physiology*, 170(4), 413–23.

Nölke, G., Houdelet, M., Kreuzaler, F., Peterhänsel, C., and Schillberg, S. 2014. The expression of a recombinant glycolate dehydrogenase polyprotein in potato (*Solanum tuberosum*) plastids strongly enhances photosynthesis and tuber yield. *Plant Biotechnology Journal*, 12(6), 734–42.

Obidiegwu, J. E., Bryan, G. J., Jones, H. G., and Prashar, A. 2015. Coping with drought: Stress and adaptive responses in potato and perspectives for improvement. *Frontiers in Plant Science*, 6, 542.

Oliveira, J. S., Moot, D., Brown, H. E., Gash, A., and Sinton, S. 2012. Sprout development of seed potato tuber after different storage conditions. *Agronomy New Zealand*, 42, 53–8.

Onder, S., Caliskan, M. E., Onder, D., and Caliskan, S. 2005. Different irrigation methods and water stress effects on potato yield and yield components. *Agricultural Water Management*, 73(1), 73–86.

Opena, G. B. and Porter, G. A., 1999. Soil management and supplemental irrigation effects on potato: ii. Root growth. *Agronomy Journal*, 91(3), 426–31.

Ospina, C. A., van Bueren, E. T. L., Allefs, J. J. H. M., Engel, B., van der Putten, P. E. L., van der Linden, C. G., and Struik, P. C. 2014. Diversity of crop development traits and nitrogen use efficiency among potato cultivars grown under contrasting nitrogen regimes. *Euphytica*, 199(1–2), 13–29.

Paget, M. F., Apiolaza, L. A., Anderson, J. A. D., Genet, R. A., and Alspach, P. A. 2015. Appraisal of test location and variety performance for the selection of tuber yield in a potato breeding program. *Crop Science*, 55(5), 1957–1968.

Palta, J. P., 2010. Improving potato tuber quality and production by targeted calcium nutrition: The discovery of tuber roots leading to a new concept in potato nutrition. *Potato Research*, 53(4), 267–75.

Panique, E., Kelling, K. A., Schulte, E. E., Hero, D. E., Stevenson, W. R., and James, R. V. 1997. Potassium rate and source effects on potato yield, quality, and disease interaction. *American Potato Journal*, 74(6), 379–98.

Panter, S. N. and Jones, D. A., 2002. Age-related resistance to plant pathogens. *Advances in Botanical Research*, 38, 251–80.

Parker, C. J., Carr, M. K. V., Jarvis, N. J., Evans, M. T. B., and Lee, V. H. 1989. Effects of subsoil loosening and irrigation on soil physical properties, root distribution and water uptake of potatoes (Solanum tuberosum). Soil and Tillage Research, 13(3), 267–85.

Pavlista, A. D., 2015. Scheduling reduced irrigation on 'Atlantic' potato for minimal effect. American Journal of Potato Research, 92(6), 673–83.

Porter, G. A. and Sisson, J. A., 1993. Yield, market quality and petiole nitrate concentration of non-irrigated Russet Burbank and Shepody potatoes in response to sidedressed nitrogen. American Potato Journal, 70(2), 101–16.

Prange, R. K., McRae, K. B., Midmore, D. J., and Deng, R. 1990. Reduction in potato growth at high temperature: role of photosynthesis and dark respiration. American Potato Journal, 67(6), 357–69.

Pringle, R. T., Bishop, C. F. H. and Clayton, R. C., 2009. Potatoes Postharvest First, CABI.

Qaim, M., Subramanian, A. and Sadashivappa, P., 2009. Commercialized GM crops and yield. Nature Biotechnology, 27(9), 803–4.

Rak, K., Navarro, F. M. and Palta, J. P., 2013. Genotype × storage environment interaction and stability of potato chip color: implications in breeding for cold storage chip quality. Crop Science, 53, 1944–52.

Reeve, R. M., Hautala, E. and Weaver, M. L., 1969. Anatomy and compositional variation within potatoes. American Potato Journal, 46(10), 361–73.

Regierer, B., Fernie, A. R., Springer, F., Perez-Melis, A., Leisse, A., Koehl, K., et al. 2002. Starch content and yield increase as a result of altering adenylate pools in transgenic plants. Nature Biotechnology, 20(12), 1256–60.

Reynolds, M. P., Ewing, E. E. and Owens, T. G., 1990. Photosynthesis at high temperature in tuber-bearing Solanum species a comparison between accessions of contrasting heat tolerance. Plant Physiology, 93(2), 791–7.

Robinson, D., 1999. A comparison of soil-water distribution under ridge and bed cultivated potatoes. Agricultural Water Management, 42(2), 189–204.

Rodrigues, G. B., Pinto, C. A. B., Benites, F. R., and Melo, D. S. 2009. Seleção para duração do ciclo vegetativo em batata e relação com a produtividade de tubérculos. Horticultura Brasileira, 27, 280–5.

Rodriguez-Falcon, M., Bou, J. and Prat, S., 2006. Seasonal control of tuberization in potato: Conserved elements with the flowering response. Annual Review of Plant Biology, 57, 151–80.

Rolando, J. L., Ramírez, D. A., Yactayo, W., Monneveux, P., and Quiroz, R. 2015. Leaf greenness as a drought tolerance related trait in potato (Solanum tuberosum L.). Environmental and Experimental Botany, 110, 27–35.

Rosen, C. J., Kelling, K. A., Stark, J. C., and Porter, G. A. 2014. Optimizing phosphorus fertilizer management in potato production. American Journal of Potato Research, 91(2), 145–60.

Roy, R. N., Finck, A., Blair, G. J., and Tandon, H. 2006. Plant Nutrition for Food Security. 2006. Fertilizer and Plant Nutrition Bulletin #16. Food and Agriculture Organization of the United Nations, Rome. ISBN 92-5-105490-8.

Ruark, M. D., Kelling, K. A. and Good, L. W., 2014. Environmental concerns of phosphorus management in potato production. American Journal of Potato Research, 91(2), 132–44.

Rykaczewska, K., 2015. The effect of high temperature occurring in subsequent stages of plant development on potato yield and tuber physiological defects. American Journal of Potato Research, 92(3), 339–49.

Sabba, R. P. and Lulai, E., 2002. Histological analysis of the maturation of native and wound periderm in potato (Solanum tuberosum L.) tuber. Annals of Botany, 90(1), 1–10.

Sabba, R. P., Bussan, A., Michaelis, B. A., Hughes, R., Drillias, M., and Glynn, M. T. 2007. Effect of planting and vine-kill timing on sugars, specific gravity and skin set in processing potato cultivars. American Journal of Potato Research, 84(3), 205–15.

Schittenhelm, S., Sourell, H. and Löpmeier, F.-J., 2006. Drought resistance of potato cultivars with contrasting canopy architecture. European Journal of Agronomy, 24(3), 193–202.

Shock, C., Feibert, E. and Saunders, L., 1998. Potato yield and quality response to deficit irrigation. *HortScience*, 33(4), 655–9.

Silva, L. A. S. and Pinto, C. A. B. P., 2005. Duration of the growth cycle and the yield potential of potato genotypes. *Crop Breeding and Applied Biotechnology*, 5(1), 20–8.

Sonnewald, S. and Sonnewald, U., 2013. Regulation of potato tuber sprouting. *Planta*, 239, 27–38.

Spitters, C. J. T., 1988. An analysis of variation in yield among potato cultivars in terms of light absorption, light utilization and dry matter partitioning. *Acta Horticulturae*, (214), 71–84.

Spitters, C. J. T. and Schapendonk, A. H. C. M., 1990. Evaluation of breeding strategies for drought tolerance in potato by means of crop growth simulation. *Plant and Soil*, 123(2), 193–203.

Stalham, M. A. and Allen, E. J., 2001. Effect of variety, irrigation regime and planting date on depth, rate, duration and density of root growth in the potato (*Solanum tuberosum*) crop. *Journal of Agricultural Science*, 137(03), 251–70.

Stark, J. and Love, S., 2003. *Potato production systems: A Comprehensive Guide for Potato Production*, University of Idaho Extension.

Stark, J. C. and Porter, G. A., 2005. Potato nutrient management in sustainable cropping systems. *American Journal of Potato Research*, 82(4), 329–38.

Steduto, P., Hsiao, T. C., Fereres, E., and Raes, D. 2012. *Crop yield response to water*. FAO Irrigation and Drainage Paper 66. Food and Agriculture Organization of the United Nations. Rome.

Steinmann, H. H. and Dobers, E. S., 2013. Spatio-temporal analysis of crop rotations and crop sequence patterns in Northern Germany: Potential implications on plant health and crop protection. *Journal of Plant Diseases and Protection*, 120(2), 85–94.

Stone, D. A., 1982. The effects of subsoil loosening and deep incorporation of nutrients on yield of broad beans, cabbage, leck, potatoes and red beet. *Journal of Agricultural Science*, 98(02), 297–306.

Struik, P. C. and Wiersema, S. G., 1999. *Seed Potato Technology*, Wageningen Academic Press.

Struik, P. C., Geertsema, J. and Custers, C. H. M. G., 1989. Effects of shoot, root and stolon temperature on the development of the potato (*Solanum tuberosum* L.) plant. III. Development of tubers. *Potato Research*, 32(2), 151–8.

Suttle, J. C., 2004. Physiological regulation of potato tuber dormancy. *American Potato Journal*, 81(4), 253–62.

Svubure, O., Struik, P. C., Haverkort, A. J., and Steyn, J. M. 2015. Yield gap analysis and resource footprints of Irish potato production systems in Zimbabwe. *Field Crops Research*, 178, 77–90.

Tai, G. C. C. and Young, D. A., 1972. Genotypic stability analysis of eight potato varieties tested in a series of ten trials. *American Potato Journal*, 49(4), 138–50.

Tarkalson, D. D., King, B. A., Bjorneberg, D. L., and Taberna, J. P., Jr. 2011. Evaluation of in-row plant spacing and planting configuration for three irrigated potato cultivars. *American Journal of Potato Research*, 88(3), 207–17.

Tetlow, I. J., 2004. Recent developments in understanding the regulation of starch metabolism in higher plants. *Journal of Experimental Botany*, 55(406), 2131–45.

The Royal Society, 2009. *Reaping the benefits: Science and the sustainable intensification of global agriculture.*

Thiele, G., van de Fliert, E. and Campilan, D., 2001. What happened to participatory research at the International Potato Center? *Agriculture and Human Values*, 18(4), 429–46.

Thompson, A. L., Love, S. L., Sowokinos, J. R., Thornton, M. K., and Shock, C. C. 2008. Review of the sugar end disorder in potato (*Solanum tuberosum*, L.). *American Journal of Potato Research*, 85(5), 375–86.

Thornton, M. K., Novy, R. G. and Stark, J. C., 2014. Improving phosphorus use efficiency in the future. *American Journal of Potato Research*, 91(2), 175–9.

Tiemens-Hulscher, M., van Bueren, E. T. L. and Struik, P. C., 2014. Identifying nitrogen-efficient potato cultivars for organic farming. *Euphytica*, 199(1–2), 137–54.

Tsegaw, T., 2011. Genotype x environment interaction for tuber yield, dry matter content and specific gravity in elite tetraploid potato (*Solanum tuberosum* L.) genotypes. *East African Journal of Science*, 5(1), 1–5.

Published by Burleigh Dodds Science Publishing Limited, 2018.

Van Evert, F. K., De Ruijter, F. J., Conijn, J. G., Rutgers, B., and Haverkort, A. J. 2013. Worldwide sustainability hotspots in potato cultivation. 2. Areas with improvement opportunities. *Potato Research*, 56(4), 355–68.

Van Ittersum, M. K. and Scholte, K., 1993. Shortening dormancy of seed potatoes by a haulm application of gibberellic acid and storage temperature regimes. *American Journal of Potato Research*, 70(1), 7–19.

Vasquez-Robinet, C., Mane, S. P., Ulanov, A. V., Watkinson, J. I., Stromberg, V. K., De Koeyer, D., et al. 2008. Physiological and molecular adaptations to drought in Andean potato genotypes. *Journal of Experimental Botany*, 59(8), 2109–23.

Villordon, A. Q., Ginzberg, I. and Firon, N., 2014. Root architecture and root and tuber crop productivity. *Trends in Plant Science*, 19(7), 419–25.

Vos, J., 1992. A case history: Hundred years of potato production in Europe with special reference to The Netherlands. *American Journal of Potato Research*, 69(11), 731–51.

Vos, J. and Groenwold, J., 1986. Root growth of potato crops on a marine-clay soil. *Plant and Soil*, 94(1), 17–33.

Vos, J. and Oyarzún, P. J., 1987. Photosynthesis and stomatal conductance of potato leaves-effects of leaf age, irradiance, and leaf water potential. *Photosynthesis research*, 11(3), 253–64.

Walker, T. S., Schmiediche, P. E. and Hijmans, R. J., 1999. World trends and patterns in the potato crop: An economic and geographic survey. *European Potato Journal*, 42(2), 241–64.

Wang, Y., Bussan, A. J. and Bethke, P. C., 2012. Stem-end defect in chipping potatoes (*Solanum tuberosum* L.) as influenced by mild environmental stresses. *American Potato Journal*, 89(5), 392–9.

Weisz, R., Kaminski, J. and Smilowitz, Z., 1994. Water deficit effects on potato leaf growth and transpiration: Utilizing fraction extractable soil water for comparison with other crops. *American Potato Journal*, 71(12), 829–40.

Wills, R. B. H., Warton, M. A. and Kim, J. K., 2004. Effect of low levels of ethylene on sprouting of potatoes in storage. *HortScience*, 39(1), 136–7. Available at: http://hortsci.ashspublications.org/cgi/content/abstract/39/1/136.

Wilson, M. L., Rosen, C. J. and Moncrief, J. F., 2010. Effects of polymer-coated urea on nitrate leaching and nitrogen uptake by potato. *Journal of Environment Quality*, 39(2), 492–8.

Wiltshire, J., Milne, F. and Peters, J., 2006. *Improving the understanding and management of skin set and bloom in potatoes*, British Potato Council Report 2006/1. British Potato Council, Oxford, UK

Winter, H. and Huber, S., 2000. Regulation of sucrose metabolism in higher plants: Localization and regulation of activity of key enzymes. *Critical Reviews in Plant Sciences*, 19(1), 31–67.

Wishart, J., George, T. S., Brown, L. K., Ramsay, G., Bradshaw, J. E., White, P. J., and Gregory, P. J. 2013. Measuring variation in potato roots in both field and glasshouse: the search for useful yield predictors and a simple screen for root traits. *Plant and Soil*, 368(1–2), 231–49.

Wurr, D. C. E., Fellows, J. R., Akehurst, J. M., Hambidge, A. J., and Lynn, J. R. 2001. The effect of cultural and environmental factors on potato seed tuber morphology and subsequent sprout and stem development. *Journal of Agricultural Science*, 136(01), 55–63.

Wurr, D. C. E., Fellows, J. R., Lynn, J. R., and Allen, E. J. 1993. The impact of some agronomic factors on the variability of potato tuber size distribution. *Potato Research*, 36(3), 237–45.

Xing, Z., Chow, L., Rees, H. W., Meng, F., Monteith, J., and Stevens, L. 2011. A comparison of effects of one-pass and conventional potato hilling on water runoff and soil erosion under simulated rainfall. *Canadian Journal of Soil Science*, 91, 279–90.

Xu, X., van Lammeren A. A., Vermeer, E., and Vreugdenhil, D. 1998. The role of gibberellin, abscisic acid, and sucrose in the regulation of potato tuber formation in vitro. *Plant Physiology*, 117(2), 575–84.

Xu, X., Vreugdenhil, D. and van Lammeren, A. A. M., 1998. Cell division and cell enlargement during potato tuber formation. *Journal of Experimental Botany*, 49(320), 573–82.

Yamaguchi, J. and Tanaka, A., 1990. Quantitative observation on the root system of various crops growing in the field. *Soil Science and Plant Nutrition*, 36(3), 483–93.

Published by Burleigh Dodds Science Publishing Limited, 2018.

Yuan, B.-Z., Nishiyama, S. and Kang, Y., 2003. Effects of different irrigation regimes on the growth and yield of drip-irrigated potato. *Agricultural Water Management*, 63(3), 153–67.

Zebarth, B. J. and Rosen, C. J., 2007. Research perspective on nitrogen BMP development for potato. *American Journal of Potato Research*, 84(1), 3–18.

Zebarth, B. J., Tai, G., Tarn, R., De Jong, H., and Milburn, P. H. 2004a. Nitrogen use efficiency characteristics of commercial potato cultivars. *Canadian Journal of Plant Science*, 84(2), 589–98. © Canadian Science Publishing or its licensors.

Zebarth, B. J., Leclerc, Y., Moreau, G., and Botha, E. 2004b. Rate and timing of nitrogen fertilization of Russet Burbank potato: Yield and processing quality. *Canadian Journal of plant Science*, 84(3), 855–63.

Zommick, D. H., Knowles, L. O., Pavek, M. J., and Knowles, N. R. 2014. In-season heat stress compromises postharvest quality and low-temperature sweetening resistance in potato (*Solanum tuberosum* L.). *Planta*, 239(6), 1243–63.

Chapter 2

Understanding ageing processes in seed potatoes

Paul C. Struik, Wageningen University and Research, The Netherlands

1 Introduction: seed tubers as the main propagules in potato production

During its entire life cycle, a potato (*Solanum tuberosum* L.) tuber passes through several developmental phases. These include organogenesis, tuber bulking, dormancy, resource mobilization and sprouting, followed by the creation of a new plant that will produce the next generation of tubers. Although a potato tuber is a relatively simple organ consisting essentially of a swollen below-ground stem, its life cycle requires the orchestration of a series of complex physiological processes and metabolic pathways, many of which are most likely regulated by many different genes.

In high-tech, modern agriculture, potato crops are usually grown from seed tubers, that is, mother tubers harvested from seed crops specifically grown for the purpose of producing propagules for the next crop. These seed tubers can consist of whole seed or cut seed. Cutting seed has many consequences for the different quality aspects of

http://dx.doi.org/10.19103/AS.2017.0016.17

the propagules planted, including their physiological condition and their health status. Therefore, we only discuss whole seed tubers in this chapter.

The size of the seed tubers planted varies from country to country and may also vary with the market outlet for which the crop is grown. The tuber shape, specific to the cultivar, also plays a significant role in deciding how large the planted seed should be, and therefore how large the quantity of seed tubers planted per hectare will be. As a rule of thumb, for a good ware crop we could say that about 2.5 Mg of seed tubers (50 000 seed tubers weighing 50 g each) are planted per ha to reach a stand of 45 000–50 000 plants per ha (or 150 000 stems per ha), yielding about 500 000 marketable tubers with an average fresh weight of 120 g (total fresh yield: 60 Mg per ha; cf. Struik and Wiersema, 1999). That means a multiplication factor in number of ten and a multiplication factor in weight of 24. For seed crops, farmers will tend to aim for a higher stem number (up to 250 000 stems per ha) as they aim for a high number of relatively small and uniform-sized tubers, for example 750 000 marketable tubers of 50 g, resulting in a yield of 37.5 Mg per ha (multiplication factor in number of 15 and in weight of 15).

After haulm killing or natural senescence, tubers become independent of the mother plant and their developmental changes are no longer buffered by the presence of an active haulm. After harvesting, seed tubers are usually stored for several months before planting. Storage can be done either at low temperature (to prolong dormancy, suppress sprouting and prevent losses of water and storage carbohydrates and proteins) or in diffused-light storage facilities where some level of cooling is combined with the sprouting-inhibiting and ageing-delaying effect of light.

2 Quality characteristics of seed tubers

Crop growth, development and yield very much depend on the quality of the seed tubers (Haverkort and Struik, 2015). Important seed quality characteristics include:

1 physical quality (seed tuber weight, other physical characteristics such as shape, presence of wounds, cracks, etc.);
2 physiological quality (i.e. the physiological status of the seed tuber and its sprouts, but also dry matter content, absence of physiological disorders, such as heat sprouts, secondary growth, etc.);
3 genetic quality (variety, true-to-typeness); and
4 phytosanitary quality (i.e. seed tuber health).

The physical quality is influenced by the grower during cultivation, harvesting, storage and grading. The physiological quality, however, is influenced by the combined effect of all factors influencing the behaviour of the seed tuber, from its initiation until its final use as seed and the emergence of its sprouts. The influences are not always straightforward and there are large genotype × environment interactions during the different stages of the life cycle of the seed tuber, as well as all kinds of feedback and feedforward mechanisms; therefore, physiological quality is a complex trait. It becomes even more complex as it is also influenced by physical quality (mainly weight). The direct and indirect relationships between seed weight and physiological quality definitely deserve attention. They will also be discussed below.

In this book chapter, I will not discuss genetic and phytosanitary quality in detail. The genetic quality is the domain of the breeders and of the growers involved in the early stages of the seed multiplication scheme. The seed tuber health is monitored and managed by the inspection services in close collaboration with seed growers and traders. Seed systems in developed potato-growing countries heavily rely on inspection and flush-out systems to limit seed degeneration (i.e. accumulation of seed-borne pathogens and pests during subsequent multiplication in the field) in seed production schemes. In seed systems in developing countries, alternative approaches to limit seed degeneration are necessary (Thomas-Sharma et al., 2016). The same institutions could also be used to monitor physiological quality of seed tubers but this usually does not take place.

Until now, we have used the phrase *physiological quality*. For seed tubers, we can identify two major aspects: *dormancy* and *physiological age*. These aspects are partly overlapping but they need to be introduced separately first in order to sort out all the various details, including the periods without sprout growth, vigorous sprout growth and senility.

In this introduction, I first define and describe the main components of physiological quality: dormancy and physiological age to provide an early, albeit slightly abstract, conceptualization of these phenomena. I then briefly describe the relevance of physical and physiological quality as yield-limiting and quality-determining factors in the production of seed and ware to provide a sense of relevance of the issues at stake. Subsequently, I zoom in on the details of the ageing processes of seed potatoes: dormancy, bud activation and initial sprout growth, apical dominance and the differences between physiological ageing of mother tubers and sprouts. I will also provide a detailed analysis of the development of physiological age and its consequences for crop performance, the causes of variation in physiological age between and within seed lots, the options for manipulation of physiological ageing and the agronomic importance of physiological age. I will end with a brief outlook on future research and some indications where to look for further information.

The approach chosen unavoidably results in some overlap between different parts of the text, but this is done on purpose and hopefully will make it easier for the reader to digest this text on such a complex subject.

3 Dormancy and physiological age

3.1 Dormancy

The term 'ageing processes in seed potatoes' in the title of this chapter should not be considered too negatively. It is not merely an expression indicating an increasingly severe loss of function or performance because the seed tubers are getting older. Ageing also includes loss of characteristics that can be a nuisance (such as dormancy or apical dominance) and can also reflect a progress towards better performance of the seed tubers for crops grown for specific outlets.

A tuber starts its life cycle when the stolon tip on a potato stem starts to swell by a change from transverse to longitudinal cell division in pith and cortex (Xu et al., 1998) and begins to accumulate starch and other storage compounds (such as patatins) (Visser et al., 1994; Appeldoorn et al., 1999). While the accumulation of starch and storage proteins continues during tuber enlargement, the accumulation of storage compounds, especially starch, slows down during maturation (Viola et al., 2007). During its growth, a tuber has a certain level of natural dormancy (see definition below) which deepens when the tuber starts to mature (Van Ittersum, 1992a). Early during the growth, this dormancy can still be broken by conditions that are adverse for tuber induction or tuber bulking, such as alternating dry and wet periods, high air temperatures or sudden increases in availability of nitrogen (Struik and Wiersema, 1999).

After harvesting, the tubers are generally dormant, at least for the cultivars that are commonly grown in areas with only one growing season per calendar year. Dormancy can be defined as 'the absence of visible growth of any plant structure containing a meristem' (Lang et al., 1987). Dormancy means that the potato tubers cannot be induced to form sprouts without any treatment (e.g. applying a stress such as heat or cold shock or applying growth regulators such as gibberellic acid, carbon disulphide or Rindite; Struik and Wiersema, 1999). This dormancy during early stages of storage is defined as *endodormancy* (Lang et al., 1987; Teper-Bamnolker et al., 2012). This endodormancy is associated with the suppression of meristem growth mediated by some unknown endogenous signal(s) (Suttle, 2004a). The duration of this period of endodormancy depends on many different factors. One of the most significant factors is genotype (Bachem et al., 2000): there are genotypes with a very short or almost non-existent period of endodormancy. This might be useful in areas with several potato-growing seasons within one calendar year but is certainly a disadvantage when one wants to store seed potatoes or ware potatoes for a longer period of time. Most commonly grown cultivars have a period of endodormancy of several months. However, there are also many other factors influencing endodormancy. These include the growing conditions during tuber bulking, the timing of the haulm killing, the duration of the maturation and storage in soil, the time of and conditions during harvest, the storage conditions and any treatments during crop growth and/or storage (Turnball and Hanke, 1985; Van Ittersum, 1992a; Wiltshire and Cobb, 1996; Struik and Wiersema, 1999). Especially, temperatures (both average temperature and temperature fluctuations) during crop growth, skin set, wound healing and storage have a large and complicated impact on the duration of endodormancy (Van Ittersum, 1992a; Struik and Wiersema, 1999; Struik et al., 2006).

Once the period of endodormancy has passed (i.e. when endodormancy is lost), tubers become physiologically competent to sprout (Sonnewald, 2001), but internal factors (apical dominance) and environmental factors (e.g. low temperature) may still impede sprouting of some or all buds. The status of dormancy caused by internal factors (e.g. apical dominance) is called *paradormancy*, whereas the status of dormancy caused by inhibition of sprouting by non-conducive environmental conditions is called *ecodormancy* (Lang et al., 1987; Suttle, 2007; Sonnewald and Sonnewald, 2014).

Dormancy does not mean that the tuber does not change. Viola et al. (2007) claimed that in potato tubers, dormancy only occurs in the tuber buds (or eyes) containing the meristem; the rest of the tuber remains metabolically and physiologically active and can therefore progress towards a state in which bud activation can occur more readily. In other words, during dormancy the seed tuber progresses towards readiness to sprout. This has also been demonstrated, for example, by Van Ittersum (1992a) who showed that it takes increasingly less powerful chemical treatments (e.g. lower concentrations of growth regulators) to break dormancy.

Dormancy is maintained by endogenous hormones and their balance (Suttle, 2009; Hartmann et al., 2011; Teper-Bamnolker et al., 2012): ethylene and abscisic acid induce and maintain endodormancy, while cytokinins and gibberellins are associated with loss of dormancy and initiation of sprout growth. There are also indications that auxins are involved in the control of endodormancy of potato seed tubers (Sorce et al., 2000; Sorce et al., 2009).

Moreover, some role might also be played by strigolactones (Sonnewald and Sonnewald, 2014). However, it should be stressed that the hormonal regulation of dormancy breaking, bud activation and sprouting is still incompletely understood (Sonnewald and Sonnewald, 2014; Eshel, 2015). I will discuss this in more detail in Section 5.1.

3.2 Physiological age

A seed tuber ages by accumulating days or accumulating day-degrees (Struik et al., 2006), but physiological ageing is a process that is strongly influenced by genotype in interaction with conditions during the various stages of the life cycle of a seed tuber (Struik and Wiersema, 1999; Caldiz, 2009). Physiological age can be defined as 'the stage of development of a seed tuber, which changes progressively by increasing chronological age, depending on growth history, storage conditions and treatments during the different stages of the life cycle' (cf. Reust, 1986; Struik and Wiersema, 1999). This definition implies that physiological age includes other aspects than just chronological age. The definition also implies that seed tubers of exactly the same *chronological age* can differ greatly in *physiological age*. Caldiz (2009) stated that physiological age can be seen as the physiological status, determined by genotype, chronological age and environmental conditions from tuber initiation until new plant emergence. In other words, by definition, physiological age is determined by the interplay between chronological age, intrinsic characteristics, environmental factors and management factors.

Many attempts have been made to characterize this ageing process based on accumulated day-degrees from dormancy break (O'Brien and Allen, 1981; O'Brien et al., 1983); temperature sum during storage (Scholte, 1987, 1989; Struik and Wiersema, 1999; Struik et al., 2006. Caldiz, 2009); relative growth vigour indices that compare aged with less aged seed tubers (Bodlaender et al., 1987; Van der Zaag and Van Loon, 1987; Van Ittersum et al., 1990; Van Ittersum, 1992a); structural changes (Sonnewald and Sonnewald, 2014); or biophysical (e.g. electrolyte leakage across membranes; De Weerd et al., 1995), physiological (loss of apical dominance; Eshel and Teper-Bamnolker, 2012), biochemical (specific hormone balances; Caldiz et al., 2001; Sonnewald and Sonnewald, 2014), metabolic (changes in sugar metabolism; Viola et al., 2007) or even molecular (e.g. the activation of certain genes; Bachem et al., 2000; Eshel 2015) markers. For details, see also the overviews written by Coleman (2000), Caldiz et al. (2001), Struik et al. (2006), Struik (2007) and Delaplace et al. (2008).

Genetic factors (cultivar choice), environmental factors, management factors and their interactions during the entire life cycle of the crop have an influence on physiological age (Struik and Wiersema, 1999; Struik et al., 2006). This means that in order to account for the effects of physiological age, it is necessary to consider the conditions during the growth of the crop producing the seed, the period during haulm killing and harvest (often overlooked, but probably highly relevant due to the possibly large variation in soil temperature during this phase), the conditions during harvest, the conditions during the phase between harvest and storage (including the period of seed tuber curing), the conditions during storage, the conditions during any management activity during storage (such as grading), the period during preparation for planting and the pre-emergence phase immediately after planting (cf. Caldiz, 2009). The myriad of possible influences also makes it very difficult to do good research on physiological age. Proper evaluation of treatments and/or cultivars and their interactions is only possible if the seed source is standardized, the growing season is the same for all treatments and the post-harvest handling and treatment of harvested seed are entirely controlled.

Relevant factors during the production phase of the seed tubers that influence the physiological age include source of starting material, all factors influencing the timing of the onset of tuberization and the photoperiod, temperature, light intensity, water supply and nitrogen supply during tuber bulking, as well as the method and timing of haulm killing (Van Ittersum, 1992a; Struik and Wiersema, 1999). Factors between haulm killing and harvest mainly include soil temperature (with its large variation within the ridge depending on the position of the tubers) and soil moisture (Struik and Wiersema, 1999). Factors during storage include temperature, relative humidity, photoperiod and diffused light (Struik and Wiersema, 1999). I will discuss these factors later in more detail.

Temperature effects on physiological age are especially complicated because of possible feedback and feedforward mechanisms. Temperature during dormancy differs in effect compared with temperature during sprouting. Heat shocks, cold shocks and similar accumulated day-degrees, built up in different ways, all have very specific effects, very much depending on cultivar (Van Ittersum, 1992a; Struik and Wiersema, 1999; Struik et al., 2006), suggesting that simple predictions of seed performance based on accumulated day-degrees might not work (see also Scholte, 1987, 1989; Struik et al., 2006).

In later sections, I will discuss the morphological, physiological and agronomic aspects of physiological ageing of seed tubers, including bud activation, initial sprout growth, apical dominance, senility, incubation and the implications for the performance of crops grown from these seed tubers in more detail.

4 The importance of seed quality as a yield-limiting and quality-determining factor

A potato crop should be grown using the best-quality seed (Haverkort and Struik, 2015). That is to say, seed tubers of the best genetic quality (true-to-type and of a superior cultivar), the best physical quality (undamaged, proper size and shape), the right physiological age (i.e. seed that delivers the right number of stems per seed tuber each demonstrating maximum vigour while minimizing inter-stem competition, with proper timing of tuber set and the right number of marketable tubers) and the best possible seed health status, grown under conditions without aberrant abiotic and biotic stresses, and managed with the best combination of agronomic techniques, including the best possible crop protection, and wise use of inputs of nutrients and water in order to obtain a yield that is close to what is potentially feasible (Haverkort and Struik, 2015).

In general, seed quality can be considered a yield-limiting factor: when the seed is not of the best possible quality, the crop cannot fulfil the full potential of its yielding ability. However, in the case of potato the influence of the quality of the seed goes beyond affecting initial growth rate and initial vigour (Ewing and Struik, 1992; Struik and Wiersema, 1999).

Physiological age of seed potatoes has a strong and very diverse impact on the performance of the crop grown from these seed tubers. Physiological age affects: the emergence of the crop (both in terms of time of emergence and percentage of emerged plants), the number of stems per seed, the number of tubers formed per stem, the proportion of tuber-bearing stems, the vigour (and rates of growth and production) of each individual tuber-bearing stem, the onset of tuber bulking, the duration of tuber bulking, the number of tubers set, the tuber quality (e.g. the proportion of saleable tubers, tuber shape, proportion of physiological disorders, greening), the earliness of the crop and the process of maturation, the dry matter concentration, average tuber weight and tuber size distribution and tuber yield of the progeny crop (Reust, 1982; O'Brien et al., 1983; Van der Zaag and Van Loon, 1987; Van Loon, 1987; Van Ittersum, 1992a; Moll, 1994; Struik and Wiersema, 1999; Struik et al., 2006; Delaplace et al., 2008; Caldiz, 2009 and many other authors).

The seed lot planted needs to acquire the physiological age that is best suited to produce a canopy and a tuber system that allow tuber production for specific outlets (Struik et al., 1990, 1991). In many parts of the world, it is difficult to obtain seeds in the optimal stage. This is especially true for areas with double cropping and/or regions where cold storage facilities are not available.

However, when the physiological age is accurately and reliably manipulated, the farmers have a powerful tool to influence the earliness of their crop, independent of the maturity class of the cultivar grown. Moreover, they can influence, at least to some extent, the time of completion of the crop cycle, the average tuber size and tuber size distribution. For example, in seed tuber crops in Western Europe, often a slightly older seed is planted than in ware crops. Organic farmers might also want to plant slightly older seeds that are properly pre-sprouted to advance tuber bulking, thus escaping – at least partly – an early infection with late blight. To reliably manipulate physiological age of the seed tubers (i.e. to create seed that delivers the right number of stems per seed tuber of a maximum vigour, with proper timing of tuber set and the right number of marketable tubers), farmers can influence the conditions of the seed crop, the time and method of haulm killing and harvesting, the storage conditions and treatments during crop growth as well as during and after storage (see below). For a more elaborate overview of all factors involved in physiological age of seed tubers and the impact of physiological age on the performance of the crop grown from these seed tubers, see Reust (1982); Struik and Wiersema (1999); Caldiz et al. (2001) and Struik et al. (2006); and the section on 'Causes of variation in physiological age and options for manipulation'.

Table 1 Comparison of some characteristics of crops grown from physiologically young and physiologically old seed tubers (based on Caldiz, 2009)

Characteristic	Physiological age	
	Young	Old
Time of emergence	Later	Earlier
Rate of emergence	Slower	Faster
Number of stems per seed tuber	Lower	Higher
Time of tuber initiation	Later	Earlier
Number of tubers per stem	Higher	Lower
Chance of secondary growth	Lower	Higher
Canopy development	Exuberant	Poor
Tuber yield in short season	Lower	Higher
Tuber yield in long season	Higher	Lower
Maturity	Later	Earlier

A summary of a comparison between relatively young and relatively old seeds is given in Table 1. A more quantitative analysis will be provided in Section 7.

In the following sections, I will explain dormancy, bud activation and initial sprout growth, apical dominance, the different sites of ageing and stages of physiological age in more detail.

5 Understanding dormancy, bud activation, initial sprout growth and apical dominance

5.1 Understanding dormancy

The start of tuber growth is also the inception of dormancy (Fernie and Willmitzer, 2001). Eshel and Teper-Bamnolker (2012) indicated that it is unclear whether each individual bud on a single seed tuber demonstrates a certain level of autonomy regarding the onset and end of dormancy and the initiation of sprouting. During tuber growth, an increasing number of eyes (with the meristems of the sprouts being present) are formed, so the different eyes differ in age and conditions during development.

Endogenous hormone concentrations and their balances are responsible for the state of dormancy, its depth and the breaking of dormancy (Sonnewald and Sonnewald, 2014), in a cultivar-specific way and a seed tuber size-specific manner. Certainly, the end of the dormancy of the entire seed tuber and/or a specific bud is also influenced by endogenous hormones regulating bud activation and subsequent sprouting (Suttle, 2004a; Suttle, 2009; Sorce et al., 2009; Hartmann et al., 2011).

The onset of dormancy and its maintenance have been associated with changes in the concentrations of ethylene and, especially, abscisic acid (Suttle and Hultstrand, 1994; Sonnewald and Sonnewald, 2014). Based on molecular analyses, it has been concluded that the genes that are associated with the anabolic and catabolic metabolism of abscisic acid correlate with dormancy in potato tubers (Simko et al., 1997; Ewing et al., 2004; Campbell et al., 2008). According to Suttle (1995), abscisic acid content is highest immediately after harvest (i.e. at the time when the endodormancy is deepest), and gradually declines when dormancy weakens during storage. This suggests that the trend over time of the abscisic acid content reflects the progress of the seed tuber towards readiness to sprout. However, Suttle et al. (2012) indicated that the decline in abscisic acid content is not a prerequisite for dormancy release. Cytokinins and gibberellins play an important role in bud activation, sprouting and sprout elongation once dormancy is released (Suttle, 2004b; Eshel and Teper-Bamnolker, 2012; Sonnewald and Sonnewald, 2014), but gibberellin concentrations do not show a clear association with maintenance of dormancy (Eshel and Teper-Bamnolker, 2012). However, gibberellins are able to break dormancy, initiate active bud growth and stimulate sprouting (Salimi et al., 2010). Cytokinins seem to increase in concentration during dormancy, suggesting that they might play a role in dormancy break and bud activation as well (Suttle, 1998; Eshel and Teper-Bamnolker, 2012). It is likely that the stimulating effect of gibberellins on bud break and sprout growth requires a certain level of cytokinins (Eshel and Teper-Bamnolker, 2012). Finally, auxins may also play a role in the onset and end of potato tuber endodormancy: the concentration of indol acetic acid is at its highest at the beginning of tuber dormancy and later decreases during storage, at least in the buds (Sorce et al., 2000; Eshel and Teper-Bamnolker, 2012). Moreover, auxins do play a significant role in vascular development (Sonnewald and Sonnewald, 2014), thus releasing a specific bud from its symplastic isolation.

There are many compounds available to chemically break dormancy. Rindite (a mixture of ethylene chlorohydrin, ethylene dichloride and carbon tetrachloride) is widely used on a commercial scale (Rehman et al., 2001). Other compounds have been used as well, including gibberellic acid (GA$_3$; Rappaport et al., 1957; Salimi et al., 2010), carbon disulphide (Salimi et al., 2010), bromoethane (Coleman, 1984), thourea and benzyladenine (Struik and Wiersema, 1999). Chlorpropham (isopropyl N-chlorophenylcarbamate or CIPC) is widely used to prevent sprouting; its mode of action is to interfere with cell division (Campbell et al., 2010).

5.2 Understanding bud activation and initial sprout growth

During tuber growth and development, large quantities of reserves are stored in the parenchymatous tissue of the tubers, mainly in the form of starch and storage proteins (patatins). When buds are being activated, this storage material must be converted into components that can be rapidly transported to the developing sprouts (Eshel, 2015). The transport mainly occurs in the form of sucrose. Sucrose availability is a prerequisite for bud activation, and sucrose is probably active both as a nutrient and as a signal (Sonnewald and Sonnewald, 2014). This resource mobilization is essential for successful bud break, bud activation and sprouting, but requires a strong symplastic connection in actively growing buds (Viola et al., 2007). Resting buds show symplastic isolation and therefore do not obtain the resources necessary to sprout. Viola et al. (2007) therefore suggested that initial sprout growth might in fact be limited by the local availability of substrate, despite the fact that there is more than enough substrate in the seed tuber to support sprout growth. Using minitubers of different sizes and assessing losses over long periods of storage, Lommen (1993a,b) showed that the minimum size of a seed tuber required to supply enough substrate and nutrients to growing sprouts is relatively small, at least well below the sizes commonly used for seed. Sprouting therefore mainly depends on efficient remobilization and connectivity.

During later stages of sprout growth, sprout growth can also be impeded by local shortages of other nutrients, such as calcium, inducing subapical necrosis followed by subsequent branching (DeKock et al., 1975). Dyson and Digby (1975) even suggested that calcium is required to maintain apical dominance (see Section 5.3 for definition), but this effect is associated with its role in preventing subapical necrosis.

Sprout growth results in quality loss of stored potatoes, both seed and ware (see, e.g. Lommen, 1993a,b; Struik et al., 2006; Sonnewald and Sonnewald, 2014), due to remobilization (and use) of starch and proteins and due to increased water loss (Coleman, 1987; Sonnewald, 2001).

5.3 Understanding apical dominance

Krijthe (1962) already indicated that after the end of endodormancy, there are four different stages in seed tuber development which reflect the progress of physiological age of the seed tuber:

1 apical dominance, a stage during which only the top sprout of the seed tuber develops;
2 reduced apical dominance during which several buds are sprouting;
3 multiple sprouting associated with extensive branching of the sprouts;
4 incubation, that is, when seed tubers have become so senile that they no longer produce sprouts but little daughter tubers directly on the mother tuber.

Between Stages 3 and 4, it is also possible that aged or senile seed tubers produce multiple hairy sprouts that show very little vigour. These stages together with their consequences for growth of a plant from the seed tuber are demonstrated in Fig. 1.

It is important to note that prolonged storage at low temperatures results in a loss of apical dominance, resulting in more sprouts per seed tuber once the seed tuber starts sprouting (Hartmans and Van Loon, 1987; Struik and Wiersema, 1999; Struik, 2007). Warm storage, on the other hand, enhances apical dominance and reduces the number of sprouts that emerge per seed tuber. In other words, prolonged storage at low temperatures is required to enhance the number of stems produced per seed tuber (Struik, 2007).

Teper-Bamnolker et al. (2012) suggested that the potato tuber, being a swollen stem, 'exhibits stem-like behaviour, with various strengths of apical bud dominance over other buds'. They identified three different types of loss of apical dominance in stored potato seed tubers:

1 loss of dominance of the apical bud (i.e. the bud at the rose end) over the buds situated closer to the stolon end;
2 loss of dominance of the main bud in any given eye over the subtending axillary buds present within the same eye; and
3 loss of dominance of the developing sprouts over their own branching, meaning that branches grow out on the individual sprout.

The first type of loss of dominance will give an increase in the number of stems from a single seed tuber and may be highly desirable in common agricultural practice. These different stems can all show considerable vigour and become, once rooted, independent tuber-bearing main stems. The second type of loss of dominance will create more between-stem competition during early phases of stem growth. The third type of loss of apical dominance can result in a complex system of branches on one and the same main stem, which may not all bear tubers.

Based on the different types of loss of apical dominance, Eshel and Teper-Bamnolker (2012) asked the intriguing question whether the loss of apical dominance could serve as a marker of physiological age. To some extent, the loss of the different types of apical dominance indeed reflects progress towards senility (see also Fig. 1). However, the different types of apical dominance can be influenced by abiotic factors during storage and by chemical treatments of the seed tubers, suggesting that the programme of physiological ageing is not always the same. On the other hand, effects of abiotic and chemical treatments can be either on the rate of progress towards ageing, on the relative rate at which the different steps towards ageing are followed or even on the occurrence of the different steps. Cultivars also vary greatly in their rate of physiological ageing (Van Ittersum et al., 1990; Struik et al., 2006).

Teper-Bamnolker et al. (2012) and Esher (2015) suggested that the weakening of the apical dominance was associated with programmed cell death in the apical meristem of

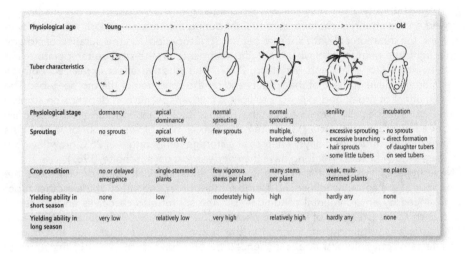

Physiological age	Young- - - - - - - - - - - ->- - - - - - - - - - - ->- - - - - - - - - - - - - ->- - - - - - - - - - - - ->- - - - - - - Old					
Tuber characteristics						
Physiological stage	dormancy	apical dominance	normal sprouting	normal sprouting	senility	incubation
Sprouting	no sprouts	apical sprouts only	few sprouts	multiple, branched sprouts	- excessive sprouting - excessive branching - hair sprouts - some little tubers	- no sprouts - direct formation of daughter tubers on seed tubers
Crop condition	no or delayed emergence	single-stemmed plants	few vigorous stems per plant	many stems per plant	weak, multi-stemmed plants	no plants
Yielding ability in short season	none	low	moderately high	high	hardly any	none
Yielding ability in long season	very low	relatively low	very high	relatively high	hardly any	none

Figure 1 Physiological age of tubers, the effect of physiological age on sprouting behaviour and the subsequent effects on the above-ground development of the crop and tuber yield with different lengths of the growing period. Changes from one stage of physiological age to another are gradual and not abrupt. The yield characteristics are only valid for normal seed rates and main ware crops. For example, seed tubers showing apical dominance may produce high yields with increased seed rates and may even produce a higher yield of marketable tubers than multi-sprouted seed tubers if grown for the early premium market. Based on Struik and Wiersema (1999).

the seed tuber. Auxins play a role in all three types of apical dominance; in fact, auxins may be the link between bud activation and its apical dominance (Teper-Bamnolker, 2012); see also Section 5.2.

6 Understanding ageing in sprouts and mother tubers

6.1 Understanding the differences between ageing of sprouts, ageing of mother tubers and interactions between the two

Both the physiological status of the mother tuber and the physiological status of the sprout affect the performance of a seed tuber. This combined influence was already demonstrated by Krijthe (1962) and Van Ittersum (1992a). Until breaking of the dormancy, changes in the physiological status of the seed are only reflected by biochemical and physiological changes in the seed tuber itself, as morphological changes have not occurred and the buds have not been activated yet. After dormancy breaking, the physiological age is still influenced by the age of the mother tuber, but at the same time also modified by the additional effects of conditions and treatments on the behaviour of the sprouts (Caldiz et al., 2001). However, experimentation to provide evidence for these separate effects is difficult and therefore evidence is still scarce. An interesting example is the interaction between photoperiod and temperature during storage under light, either at relatively low (Johansen and Mølmann, 2017) or at relatively high (McGee et al., 1987, 1988a,b; Scholte,

1989) light intensity. Light may prevent rapid ageing of seed tubers but the effects may depend on the timing of the exposure to light during the storage period, the intensity of the light, the duration of light exposure per day (photoperiod), on the duration of storage under light (the number of days under light treatment), on the storage temperature and on the cultivar (Johansen and Mølmann, 2017). This positive effect is realized both by effects on the development of the sprouts and on the condition of the mother tuber. The positive effect of prolonged exposure to light is cultivar-specific and depends on storage temperature (especially perceived by the mother tuber) and photoperiod (especially perceived by the sprout). At a storage temperature of 16°C growth vigour of seed tubers remains highest under long days, whereas at a storage temperature of 28°C growth vigour decreases much faster under long days than under short days (Scholte, 1989).

Nevertheless, it is relevant from a scientific and practical point of view that the behaviour of the sprout is influenced by the mother tuber, the effect of the mother tuber is influenced by environmental conditions, but the sprout also perceives environmental cues, such as the presence of light and photoperiod (Caldiz, 2009; Struik and Wiersema, 1999). It is an intriguing phenomenon in the studies of physiological age that part of the effects of physiological age is influenced by processes at the level of the tuber or the tissues in the tuber, whereas another part is associated with processes in the sprout itself. A clear example of the former is the metabolic change that needs to take place to initiate resource mobilization required to enable bud activation and sprout growth (Sonnewald and Sonnewald, 2014). A clear example of the latter is the removal of the symplastic isolation and the creation of symplastic connection in the awakening bud (Viola et al., 2007). Obviously, these different types of processes need to be orchestrated and synchronized for successful bud development. That requires signalling, signal perception and interpretation, communication and organization (for a detailed discussion on such phenomena, see Struik et al., 2008).

Nevertheless, it is possible to generate relatively old sprouts on relatively young seed tubers. This may happen when cold shocks trigger early break of dormancy (Van Ittersum, 1992a). It is also possible to generate relatively young sprouts on relatively old tubers. This may happen during diffused-light storage, when the mother tubers age rapidly because of the prolonged storage under relatively high temperatures, whereas the sprouts remain relatively young and vigorous because of the inhibitory and/or delaying effects of the diffused light. Young sprouts on old tubers may also appear upon de-sprouting, when the subsequent sprouts are relatively young but the de-sprouting (especially when repeated) accelerates the ageing of the seed tuber.

6.2 Understanding the relationship between mother tuber size and physiological age

The seed tuber weight determines how much resources (energy, calcium, other nutrients, etc.) an individual seed tuber carries to sustain the growth and development of vigorous sprouts, but also – together with seed shape – how many eyes an individual seed tuber will have and what proportion of eyes will produce a viable stem. Choosing the right number of seeds per unit area and the most suitable seed size are already important steps towards obtaining the right number of stems per hectare. But the story is more complicated because of the interrelationships between seed weight and physiological age in determining the number of stems per seed tuber (Fig. 2) and in determining the crop performance (Fig. 3). How many buds per seed tuber will develop viable sprouts, how many stems will be produced per individual sprout and how many of these stems will survive and finally will bear tubers (and how many tubers they will bear) will depend on cultivar, seed tuber size, seed tuber health, seed tuber physiological age and environmental conditions (Struik and Wiersema, 1999).

In this respect, it is important to distinguish between effects of seed tuber size and of seed physiological age but also acknowledge their interaction. The size of the seed tuber not only affects the number of emerging stems per seed tuber, but the vigour of growth of those stems as well and subsequently the number of tubers per plant and per individual stem, and their growth. Next to its influence on the number of stems per seed tuber, the physiological age of the seed tuber also affects the growth vigour and the development of each individual stem. This is, among others, expressed by the initial growth rate, the onset of tuberization, the number of tubers, the duration of the growing cycle of the crop and (partly because of this) the growth period of the tubers (Fig. 3). Therefore, farmers can influence the timing of the potato harvest (both in seed and ware crops) to a certain degree, partly independently of the earliness of the cultivar grown. This gives them, to a certain extent, the chance to manipulate the earliness of the crop, the tuber yield, the average tuber size and the tuber size distribution. In Western Europe, for instance, seed growers usually plant seed potatoes of older age than ware growers, because then the seed potatoes develop more stems per plant, more tubers per stem, tuberize earlier and the maturation of the crop can be advanced. However, the seed tubers should not become too old as a very advanced physiological age will be associated with an increase in physiological disorders, loss of vigour, a too short crop cycle and a low yield (see also Section 7 for a more quantitative analysis).

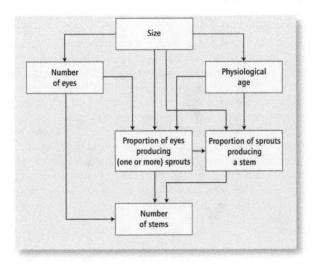

Figure 2 The effect of size and physiological age of the seed tuber on the number of main stems per seed tuber. Based on Struik and Wiersema (1999).

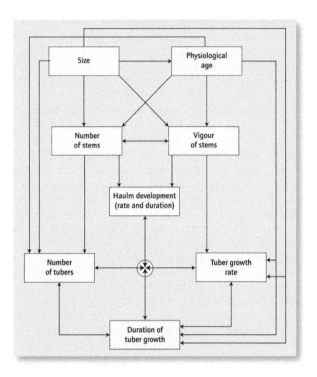

Figure 3 The different effects of the size of the seed tuber and its physiological age on crop performance and their interactions. Based on Struik and Wiersema (1999).

The seed tuber of the proper quality (both the right physical and physiological quality) produces the proper number of stems, with a maximum vigour of growth, the right moment of onset of tuber bulking, the proper number of marketable tubers of the right size and the right time of crop maturation. These characteristics depend to a large extent on the interactions between the seed tuber weight and shape and the physiological age of the seed tuber.

The interaction between seed tuber size and physiological age in determining the vigour of the seed tubers also depends on the cultivar and the type of propagule. Van Ittersum (1992b) demonstrated for normal seed tubers that the relation between duration of dormancy and seed tuber size was cultivar-specific, in the sense that both the size of the effect and the range of weights over which the largest change occurred differed between cultivars tested. Lommen (1993a) showed for minitubers that the dormant period was longer in minitubers with lower weights than in minitubers with higher weights across the entire range of measured minituber sizes.

7 Analysis of the dynamic development of physiological age and resulting crop performance

As indicated above, a seed tuber is dormant during a considerable proportion of its life cycle. Immediately after the seed tuber is initiated (i.e. during the early stages of tuber bulking of the seed crop) a tuber develops a certain degree of dormancy. The dormancy may be temporarily interrupted (e.g. when conditions become conducive for secondary growth or formation of heat sprouts), but under conditions favourable for tuber growth there is a gradual deepening of the dormancy, which may even continue after haulm killing and harvesting. During storage, the dormancy becomes gradually less deep until it is broken. The rate with which this occurs depends on the cultivar, the initial depth of the dormancy and the conditions during dormancy break.

For practical reasons, dormancy is defined as the physiological state of the seed tuber during which autonomous sprout growth will not occur within two weeks, even not when the tuber is kept in conditions ideal for sprout growth (Reust, 1986; Van Ittersum, 1992a; Struik and Wiersema, 1999). It is important to realize that biochemical and physiological processes do occur during dormancy; however, these processes do not immediately trigger morphological changes ('immediately' meaning not within two weeks under conditions favourable for sprout growth). Yet these processes are relevant for what will happen next: they do affect the number of sprouts produced after breaking of the dormancy, and they do determine the development of the growth vigour of the seed tuber during physiological ageing. Conditions during dormancy and thereafter affect the progress of the physiological ageing and therefore influence the performance of the seed tuber as reflected by the vigour of the plant developing from it after planting.

From the breaking of the dormancy onwards, physiological age influences the behaviour of each bud of the seed tuber (see also Section 5.3), thus affecting the number of sprouts per eye and their growth vigour. Moreover, physiological age also influences the physiological behaviour of the resulting stems, even well after emergence, that is, even during tuber initiation, tuber bulking and crop maturation. After breaking of dormancy the seed tuber goes through different phases: apical dominance (only one sprout), normal sprouting (only a few sprouts per seed tuber), normal, advanced sprouting (many sprouts per seed tuber which become increasingly branched with an increase in physiological age), senility (excessive sprouting with very weak sprouts) and incubation (little tuber formation, that is, formation of new, small tubers directly on the mother tuber). The rate with which the seed tubers go through these stages is very much dependent on the cultivar. Cultivars also differ in their sensitivity to the environment during the different stages. For example, the level of apical dominance and the duration of this phase of physiological ageing can vary greatly among cultivars. Crops grown from seed tubers of different physiological ages will differ in canopy structure (e.g. number of stems per plant, number of lower branches per stem and number of sympodial branches), number of tubers, above-ground biomass, tuber yield, tuber size distribution and tuber quality (Ewing and Struik, 1992; Struik and Wiersema, 1999).

De-sprouting advances ageing of the seed tuber, especially when it is carried out repeatedly (Struik and Wiersema, 1999).

The successive stages of the physiological age of seed tubers, the consequences for sprouting behaviour and the subsequent effects on the above-ground development of the crop and tuber yield with different lengths of the growing period were already illustrated in general terms in Fig. 1 (Section 5.3). Figure 4a illustrates in detail how physiological age affects the ability to sprout (in number of sprouts as well as in weight of sprouts per tuber) and the growth vigour of a crop with different durations of the growing season. The corresponding patterns of above-ground growth and tuber growth for a crop grown from physiologically relatively young seed tubers and a crop from physiologically relatively old seed tubers are also indicated schematically (Fig. 4b). Figure 4b demonstrates that

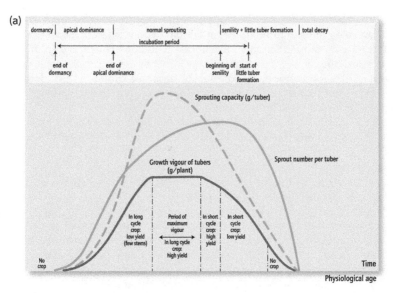

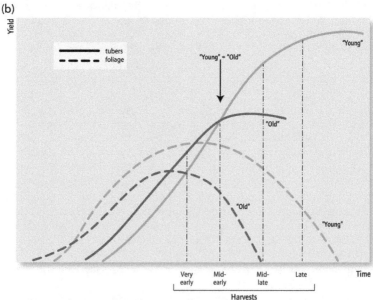

Figure 4 The effect of physiological age of the seed tuber on the sprouting capacity (expressed in grams sprouts per seed tuber), number of sprouts per seed tuber and its growth vigour (expressed in grams biomass per plant) (a). The growth vigour curve illustrates the possible effect of different levels of growth vigour on productivity with different lengths of the growing cycle. The different stages of physiological ageing are indicated at the top. (b) Illustrates the yield of foliage and tubers from physiologically relatively old and relatively young seed tubers in the course of time. The figure clearly shows that old seed tubers are to be preferred above young ones for short cycle crops, but when the growth cycle gets longer young seed tubers should be chosen. Based on Struik and Wiersema (1999).

the best physiological age depends on how long the crop can maintain its foliage. Note that this depends on the physiological age of the seed tubers and also on variety, crop management and environmental conditions. Crops grown for seed tuber production are short cycle crops as it is often recommended to kill the haulm of such crops early to prevent infection with viruses and other pathogens, and therefore such crops should have an early start of tuber bulking, and should not invest in a long period of green canopy. Therefore, they should be grown from older seed tubers than crops, which are left to mature without haulm killing. The physiological age interferes with maturity type: using older seed advances the development of late cultivars, thus making them tuberize and mature earlier, while using younger seed can postpone the development of early cultivars, thus making them tuberize and mature later than normal.

Physiologically young and old seed tubers affect the size distribution of tubers as well: older seed tubers produce more stems and sometimes more tubers per stem and thus produce a crop with a smaller average tuber weight and a smaller relative variability of tuber size.

Caldiz et al. (2001) developed a so-called physiological age index and demonstrated that this index could be used to assess age differences caused by differences in growing conditions, seed origin, haulm killing date, seed storage conditions and cultivar. They also demonstrated that the physiological age index showed a good correlation with performance of the crop grown from such seed. Their physiological age index was calculated as (the sampling date minus the haulm killing date) divided by (the date of incubation minus the haulm killing date). The index ranges from zero for very young seed tubers evaluated immediately after haulm killing to 1 for old seed tubers assessed at incubation. Delaplace et al. (2008) later confirmed the value of this index for Belgian conditions. However, Johansen et al. (2008) found that the physiological age index did not reflect the physiological age of seed tubers produced in different environments in Norway very well.

8 Causes of variation in physiological age and options for manipulation

The factors influencing physiological age are manifold. First and foremost, as has already been stressed several times above, the genotype has a strong influence on all stages of the development of the tuber, on the rate with which a seed lot passes through the different stages and on the responses of the seed lot to environmental factors and management practices during these different stages. Genotypic variation adds a lot of complexity to research on physiological age. It also means that standardized treatments (e.g. standard storage regimes) are often the causes of suboptimal usage of the potential of seed lots. In fact, there should be a separate storage recipe for each cultivar or even each seed lot to maximize its growth vigour depending on when, where and for what purpose it is going to be planted.

There are many other reasons why seed lots vary in dormancy and physiological age. In addition to origin and chronological age, physiological age depends on (Struik and Wiersema, 1999):

1 the size of the individual seed tuber planted (Van Ittersum, 1992b; Struik and Wiersema, 1999);
2 intrinsic factors within the plant (e.g. tuber position on the stem, tuber position within the ridge (Van Ittersum and Struik, 1992; Struik and Wiersema, 1999);

3 the growth history of the seed crop from which the seed tuber has been obtained (e.g. temperatures during tuber bulking, photoperiod during tuber bulking; Van Ittersum 1992a);

4 management of the seed crop (e.g. nitrogen fertilizer) and possible treatments applied to the seed crop (e.g. hormonal sprays on the canopy, one or two weeks before haulm killing) (Van Ittersum, 1992a);

5 the timing and method of haulm killing (Struik and Wiersema, 1999; Caldiz et al., 2001);

6 the conditions between haulm killing of the seed crop and the harvesting of the seed tuber (unfortunately, there are no references for this item, but this is definitely an area for research);

7 the size of the seed tuber harvested (Van Ittersum, 1992b; Struik and Wiersema, 1999);

8 the conditions during harvesting (note that handling of seed tubers induces additional respiration and reduces the depth of the dormancy; unpublished data from the author's laboratory);

9 the conditions during storage (temperature, light, relative humidity, composition of the atmosphere (concentrations of ethylene, CO_2, O_2, etc.)) (Struik and Wiersema, 1999);

10 possible treatments during harvesting or storage (e.g. cold or heat shocks, sprout inhibitors, dormancy breaking chemicals) (Struik and Wiersema, 1999);

11 possible practices during storage (such as grading, by which the variation within the seed lot but also the average value for dormancy and physiological age might alter);

12 possible treatments between storage until planting (warming up, pre-sprouting, de-sprouting, etc.) (Struik and Wiersema, 1999).

9 Summary and future trends

One of the most challenging aspects of future research on physiological age is the task to find a reliable, rapid and cheap diagnostic tool that is capable of assessing the physiological age at any stage during the development of the seed tuber (i.e. from its initiation to its planting), predicting the progress of the further ageing process of this seed tuber, preferably as a function of some kind of temperature sum and quantifying the variation in progress of the physiological age present in the seed lot (see also Van Ittersum, 1992a; Van Ittersum and Struik, 1992). Many research groups have taken up this challenge in the past, see the overview of different markers provided in the introduction on physiological age. However, it proved to be very difficult to identify characteristics, traits or markers that could assess or proxy the status of the seed tuber in a way that is consistent across cultivars and environments (either during seed growth, storage or the pre-planting phase), let alone predict future trends of physiological ageing of the seed tuber. However, there are hopes that with modern molecular tools, and especially with metabolomics, it will be possible to assess physiological age more precisely, to predict its behaviour in relation to storage conditions much more accurately, and to allow characterization of genotypes regarding their behaviour and specific responses to environmental cues. Such a tool will allow the optimization of the seed handling in all its stages of production and storage, will make it possible to create the best possible seed vigour as dictated by the market outlet for which the crop is grown and the site at which it is grown, and to optimize crop management based on the knowledge of the physiological age of the seed (and therefore the expected behaviour of the crop). Moreover, such a tool will also be highly relevant for breeding: it will allow breeding of cultivars that show slow ageing at a given, preferred, relatively long duration of dormancy as desired for conditions in environments with one growing season per calendar year, or cultivars that show rapid progress towards optimal growth vigour for environments with multiple potato crops per calendar year.

10 Where to look for further information

Research on dormancy, physiological age and growth vigour is highly relevant for countries with several growing seasons per calendar year (such as Tunisia, Israel and Ethiopia) or those with areas in which the growing seasons fall in different periods of the year (such as Argentina and Australia) and in countries that are exporting large quantities of seed potatoes to other countries with different growing seasons (such as the Netherlands, exporting to North Africa, Asia and many other regions, or Australia, exporting to Indonesia). It is also very important for countries with very short growing seasons, for example, Nordic countries (Sweden, Finland), countries with a short rainy season (certain parts of Africa), countries with only short periods without heat stress (countries in Northern Africa and Western Asia) or countries with crop production systems that only allow a short cycle potato crop (e.g. in Vietnam, early-maturing potato is grown between two rice crops).

In all these cases, crop management, storage of seed potatoes and preparing seed potatoes for planting become crucial for getting the seed tubers in the condition in which they perform best. In some situations, advancing physiological age is required, especially when the period between harvesting and planting is very short (e.g. in Ethiopia), in some cases slowing down the ageing is necessary, especially when the storage period is very long (e.g. in Tunisia when producing seed potatoes in certain growing seasons for use in certain growing seasons) or when temperatures during storage cannot be controlled very well. There is also wide genetic variation in duration of dormancy, but there can be a trade-off between the need to have a short dormancy for seed tubers and the need for a long dormancy for ware potatoes.

The International Potato Center (CIP; headquarters in Lima, Peru) has performed outstanding research on developing, analysing and promoting diffused light storage in warmer areas where potato can be grown but where cold storage is not feasible. This technology is only suitable for seed potatoes, as storage in light will induce the formation of glycoalkaloids and thus make the potatoes inedible. Diffused light storage creates an environment of lower temperatures than ambient temperatures and delays and reduces sprout growth, and results in tubers with short, sturdy sprouts that are suitable for planting. The physiological ageing of the seed tubers stored in diffused light storages, and of their sprouts, is slowed down and the seed tubers can be stored for a long period without losing their vigour entirely. CIP is also breeding for optimal dormancy and physiological age; note that these characteristics are not related to maturity type.

Most of the work on ageing of seed tubers is strategic or applied, and rightfully so, given the need to optimize storage recipes for very diverse cultivars and for very specific outlets. Fundamental research on the genetics, (molecular) physiology and metabolomics of dormancy and physiological age has been carried out or is going on in institutions in the Netherlands (e.g. Wageningen University and Research; breeding and genomics), Germany (e.g. the different Max Planck Institutes; breeding, molecular genetics, molecular physiology of dormancy and physiological age), Israel (e.g. the Volcani Center; physiology of apical dominance) and elsewhere in the world.

11 References

Appeldoorn, N. J. G., S. M. de Bruijn, E. A. M. Koot-Gronsveld, R. G. F. Visser, D. Vreugdenhil and L. H. W. van der Plas, 1999. Developmental changes in enzymes involved in the conversion of hexose phosphate and its subsequent metabolites during early tuberization. *Plant, Cell & Environment* 22: 1085–96.

Bachem, C. W. E. B., R. G. F. Visser and P. C. Struik (2000). Tuber dormancy and sprouting. Special issue of potato research. *Potato Research* 43, 297–454.

Bodlaender, K. B. A., H. M. Dekhuijzen, J. Marinus, A. van Es, K. J. Hartmans, L. J. P. Kupers, C. D. van Loon and D. E. van der Zaag, 1987. *Effect of Physiological Age on Growth Vigour of Seed Potatoes. A Study with Seed Tubers of Two Cultivars Stored at Two Different Temperatures.* Rapport 555, Instituut voor Bewaring en Verwerking van Landbouwprodukten (IBVL), Wageningen, the Netherlands.

Caldiz, D. O., 2009. Physiological age research during the second half of the twentieth century. *Potato Research* 52: 295–304.

Caldiz, D. O., L. V. Fernandez and P. C. Struik, 2001. Physiological age index: A new, simple and reliable index to assess the physiological age of seed potato tubers based on haulm killing date and length of the incubation period. *Field Crops Research* 69: 69–79.

Campbell, M., E. Segear, L. Beers, D. Knauber and J. Suttle, 2008. Dormancy in potato tuber meristems: Chemically induced cessation in dormancy matches the natural process based on transcript profiles. *Functional & Integrative Genomics* 8(4): 317–28.

Campbell, M. A., A. Gleichsner, R. Alsbury, D. Horvath and J. Suttle, 2010. The sprout inhibitors chlorpropham and 1,4-dimethylnaphthalene elicit different transcriptional profiles and do not suppress growth through a proliferation of the dormant state. *Plant Molecular Biology* 73: 181–9.

Coleman, W. K., 1984. Large scale application of bromoethane for breaking potato tuber dormancy. *American Potato Journal* 61(9): 587–9.

Coleman, W. K., 2000. Physiological ageing of potato tubers: A review. *Annals of Applied Biology* 137: 189–99.

DeKock, P. C., P. W. Dyson, A. Hall and F. B. Grabowska, 1975. Metabolic changes associated with calcium deficiency in potato sprouts. *Potato Research* 18: 573–81.

Delaplace, P., Y. Brostaux, M.-L. Fauconnier and P. du Jardin, 2008. Potato (*Solanum tuberosum* L.) tuber physiological age index is a valid reference frame in postharvest ageing studies. *Postharvest Biology and Technology* 50: 103–6.

De Weerd, J. W., L. K. Hiller and R. E. Thornton, 1995. Electrolyte leakage of aging potato tubers and its relationship with sprouting capacity. *Potato Research* 38: 257–70.

Dyson, P. and J. Digby, 1975. Effects of calcium on sprout growth and sub-apical necrosis in Majestic potatoes. *Potato Research* 18(2): 290–305.

Eshel, D., 2015. Chapter 11. Bridging dormancy release and apical dominance in potato tuber. In: J. V. Anderson (Ed.), *Advances in Plant Dormancy*. Springer International Publishing, Switzerland, pp. 187–96.

Eshel, D. and P. Teper-Bamnolker, 2012. Can loss of apical dominance in potato tuber serve as a marker of physiological age? *Plant Signaling & Behavior* 7 (9): 1158–62.

Ewing, E. E. and P. C. Struik, 1992. Tuber formation in potato: Induction, initiation, and growth. *Horticultural Reviews* 14: 89–198.

Ewing, E. E., I. Simko, E. A. Omer and P. J. Davies, 2004. Polygene mapping as a tool to study the physiology of potato tuberization and dormancy. *American Journal of Potato Research* 81(4): 282–9.

Fernie, A. R. and L. Willmitzer, 2001. Molecular and biochemical triggers of potato tuber development. *Plant Physiology* 127: 1459–65.

Hartmann, A., M. Senning, P. Hedden, U. Sonnewald and S. Sonnewald, 2011. Reactivation of meristem activity and sprout growth in potato tubers require both cytokinin and gibberellin. *Plant Physiology* 155: 776–96.

Hartmans, K. J. and C. D. van Loon, 1987. Effect of physiological age on growth vigour of seed potatoes of two cultivars. I. Influence of storage period and temperature on sprouting characteristics. *Potato Research* 30: 397–410.

Haverkort, A. J. and P. C. Struik, 2015. Yield levels of potato crops: Recent achievements and future prospects. *Field Crops Research* 182: 76–85.

Johansen, T. J. and J. A. B. Mølmann, 2017. Green-sprouting of potato seed tubers (*Solanum tuberosum* L.) – Influence of daily light exposure. *Potato Research* 60: 159–70.

Johansen, T. J., P. Møllerhagen and E. Haugland, 2008. Yield potential of seed potatoes grown at different latitudes in Norway. *Acta Agriculturae Scandinavica Section B – Soil and Plant Science* 58: 132–8.

Krijthe, N. (1962). Observations on the sprouting of seed potatoes. *European Potato Journal* 5: 316–33.

Lang, G. A., J. D. Early, G. C. Martin and R. L. Darnell, 1987. Endo-, para-, and ecodormancy: Physiological terminology and classification for dormancy research. *Horticultural Science* 22: 371–7.

Lommen, W. J. M., 1993a. Post-harvest characteristics of potato minitubers with different fresh weight and from different harvests. I. Dry-matter concentration and dormancy. *Potato Research* 36: 265–72.

Lommen, W. J. M., 1993b. Post-harvest characteristics of potato minitubers with different fresh weight and from different harvests. II. Losses during storage. *Potato Research* 36: 273–82.

McGee, E., R. H. Booth, M. C. Jarvis and H. J. Duncan, 1987. The inhibition of potato sprout growth by light. I. Effects of light on dormancy and subsequent sprout growth. *Annals of Applied Biology* 110: 339–404.

McGee, E., R. H. Booth, M. C. Jarvis and H. J. Duncan, 1988a. The inhibition of potato sprout growth by light. II. Effects of temperature and light intensity. *Annals of Applied Biology* 113: 137–47.

McGee, E., R. H. Booth, M. C. Jarvis and H. J. Duncan, 1988b. The inhibition of potato sprout growth by light. III. Effects on subsequent growth in the field. *Annals of Applied Biology* 113: 149–57.

Moll, A., 1994. Einfluss der physiologischen Alterung bei Pflanzknollen verschiedener Sorten auf Wachstumscharakteristika in Laborversuch. *Potato Research* 37: 11–20.

O'Brien, P. J. and E. J. Allen, 1981. The concept and measurement of physiological age. In: *Abstracts of Conference Papers, 8th Triennial Conference EAPR*, 30 August–4 September 1981, München, Germany. European Association for Potato Research, Wageningen, the Netherlands, pp. 64–6.

O'Brien, P. J., E. J. Allen, J. N. Bean, R. J. Griffith, S. A. Jones and J. L. Jones, 1983. Accumulated day degrees as a measure of physiological age and the relationships with growth and yield in early potato varieties. *Journal of Agricultural Science, Cambridge* 101: 613–31.

Rappaport, L., L. F. Lippert and H. Timm, 1957. Sprouting, plant growth, and tuber production as affected by chemical treatment of white potato seed pieces. *American Potato Journal* 34(9): 254–60.

Rehman, F., S. K. Lee, H. S. Kim, J. H. Jeon, J. Park and H. Joung, 2001. Dormancy breaking and effects on tuber yield of potato subjected to various chemicals and growth regulators under greenhouse conditions. *Journal of Biological Science* 1(9): 818–20.

Reust, W., 1982. *Contribution à l'appreciation de l'âge physiologique des tubercules de pomme de terre (Solanum tuberosum L.) et étude de son importance sur le rendement*. Thèse no. 7046, Ecole Polytechnique Fédérale Zürich, Suisse.

Reust, W., 1986. EAPR Working group 'Physiological age of the potato'. *Potato Research* 29: 268–71.

Salimi, K., R. Tavakkol Afshari, M. B. Hosseini and P. C. Struik, 2010. Effects of gibberellic acid and carbon disulphide on sprouting of potato minitubers. *Scientia Horticulturae* 124: 14–18.

Scholte, K., 1987. Relation between storage T sum and vigour of seed potatoes. In: *Abstracts of Conference Papers, 10th Triennial Conference EAPR*, Aalborg, Denmark. European Association for Potato Research, Wageningen, the Netherlands, pp. 28–9.

Scholte, K., 1989. Effect of daylength and temperature during storage in light on growth vigour of seed potatoes. Report of the meeting of the section Physiology of the EAPR, Kiryat Anavim (Israel), May 29-June 4, 1988. *Potato Research* 214–15.

Simko, I., S. McMurry, H. M. Yang, A. Manschot, P. J. Davies and E. E. Ewing, 1997. Evidence from polygene mapping for a causal relationship between potato tuber dormancy and abscisic acid content. *Plant Physiology* 115(4): 1453–9.

Sonnewald, U., 2011. Control of potato tuber sprouting. *Trends in Plant Science* 6: 333–5.

Sonnewald, S. and U. Sonnewald, 2014. Regulation of potato tuber sprouting. *Planta* 239:27–38.

Sorce, C., L. Lombardi, L. Giorgetti, B. Parisi, P. Ranalli and R. Lorenzi, 2009. Indolacetic acid concentration and metabolism changes during bud development in tubers of two potato (*Solanum tuberosum*) cultivars. *Journal of Plant Physiology* 16: 1023–33.

Sorce, C., R. Lorenzi, N. Ceccarelli and P. Ranalli, 2000. Changes in free and conjugated IAA during dormancy and sprouting of potato tubers. *Functional Plant Biology* 27: 373–7.

Struik, P. C., 2007. The canon of potato science: 40. Physiological age of seed tubers. *Potato Research* 50: 375–7.

Struik, P. C., A. J. Haverkort, D. Vreugdenhil, C. B. Bus and R. Dankert, 1990. Manipulation of tuber-size distribution of a potato crop. *Potato Research* 33: 417–432.

Struik, P. C., D. Vreugdenhil, A. J. Haverkort, C. B. Bus and R. Dankert, 1991. Possible mechanisms of size hierarchy among tubers on one stem of a potato (*Solanum tuberosum* L.) plant. *Potato Research* 34, 187–203.

Struik, P. C. and Wiersema, S. G. (1999). *Seed Potato Technology*. Wageningen Pers, Wageningen, the Netherlands.

Struik, P. C., P. E. L. van der Putten, D. O. Caldiz and K. Scholte, 2006. Response of stored potato seed tubers from contrasting cultivars to accumulated day-degrees. *Crop Science* 46: 1156–68.

Struik, P. C., X. Yin and H. Meinke, 2008. Plant neurobiology and green plant intelligence: Science, metaphors and nonsense. *Journal of the Science of Food and Agriculture* 88: 363–70.

Suttle, J. C., 1995. Postharvest changes in endogenous ABA levels and ABA metabolism in relation to dormancy in potato tubers. *Physiologia Plantarum* 95: 233–40.

Suttle, J. C., 1998. Involvement of ethylene in potato microtuber dormancy. *Plant Physiology* 118: 843–8.

Suttle, J. C., 2004a. Physiological regulation of potato tuber dormancy. *American Journal of Potato Research* 81: 253–62.

Suttle, J. C., 2004b. Involvement of endogenous gibberellines in potato tuber dormancy and early sprout growth: A critical assessment. *Journal of Plant Physiology* 161: 157–64.

Suttle, J. C., 2007. Dormancy and sprouting. Advances and perspectives. In: D. Vreugdenhil (Ed.), *Potato Biology and Biotechnology*. Elsevier, Amsterdam, pp. 287–309.

Suttle, J. C., 2009. Ethylene is not involved in hormone- and bromoethane-induced dormancy break in Russet Burbank minitubers. *American Journal of Potato Research* 86: 278–85.

Suttle, J. C., S. R. Abrams, L. De Stefano-Beltrán and L. L. Huckle, 2012. Chemical inhibition of potato ABA-8′-hydroxylase activity alters in vitro and in vivo ABA metabolism and endogenous ABA levels but does not affect potato microtuber dormancy duration. *Journal of Experimental Botany* 63(15): 5717–25.

Suttle, J. C. and J. F. Hultstrand, 1994. Role of endogenous abscisic acid in potato microtuber dormancy. *Plant Physiology* 105: 891–6.

Teper-Bamnolker, P., Y. Buskila, Y. Lopesco. S. Ben-Dor, I. Saad, V. Holdengreber, E. Belausov, H. Zemach, N. Ori, A. Lers and D. Eshel, 2012. Release of apical dominance in potato tuber is accompanied by programmed cell death in apical bud meristem. *Plant Physiology* 158: 2053–67.

Thomas-Sharma, S., A. Abdurahman, S. Ali, J. L. Andrade-Piedra, S. Bao, A. O. Charkowski, D. Crook, M. Kadian, P. Kromann, P. C. Struik, L. Torrance, K. A. Garrett and G. A. Forbes, 2016. Seed tuber degeneration in potato: The need for a new research and development paradigm to mitigate the problem in developing countries. *Plant Pathology* 65: 3–16.

Turnball, C. G. N. and D. E. Hanke, 1985. The control of bud dormancy in potato tubers. *Planta* 165: 359–65.

Van der Zaag, D. E. and C. D. van Loon, 1987. Effect of physiological age on growth vigour of seed potatoes of two cultivars. 5. Review of literature and integration of some experimental results. *Potato Research* 30: 451–72.

Van Ittersum, M. K., 1992a. *Dormancy and Vigour of Seed Potatoes*. PhD Thesis, Wageningen Agricultural University, Wageningen, the Netherlands.

Van Ittersum, M. K., 1992b. Variation in the duration of tuber dormancy within a seed potato lot. *Potato Research* 35(3): 261–9.

Van Ittersum, M. K., K. Scholte and L. J. P. Kupers, 1990. A method to assess cultivar differences in rate of physiological ageing of seed tubers. *American Potato Journal* 67: 603–13.

Van Ittersum, M. K. and P. C. Struik, 1992. Relation between stolon and tuber characteristics and the duration of tuber dormancy in potato. *Netherlands Journal of Agricultural Science* 40: 159–72.

Van Loon, C. D., 1987. Effect of physiological age on growth vigour of seed potatoes of two cultivars. 4. Influence of storage period and storage temperature on growth and yield in the field. *Potato Research* 30(3): 441–50.

Viola, R., J. Pelloux, A. van der Ploeg, T. Gillespie, N. Marquis, A. G. Roberts and R. D. Hancock, 2007. Symplastic connection is required for bud outgrowth following dormancy in potato (*Solanum tuberosum* L.) tubers. *Plant, Cell & Environment* 30(8): 973–83.

Visser, R. G. F., D. Vreugdenhil, T. Hendriks and E. Jacobsen, 1994. Gene expression and carbohydrate content during tuber to stolon transition in potatoes (*Solanum tuberosum*). *Physiologia Plantarum* 90: 285–92.

Wiltshire, J. J. J. and A. H. Cobb, 1996. A review of the physiology of potato tuber dormancy. *Annals of Applied Biology* 129: 553–69.

Xu, X., D. Vreugdenhil and A. A. M. van Lammeren, 1998. Cell division and cell enlargement during potato formation. *Journal of Experimental Botany* 49: 573–82.

Ensuring the genetic diversity of potatoes

John Bamberg and Shelley Jansky, USDA-ARS, USA; Alfonso del Rio, University of Wisconsin-Madison, USA; and Dave Ellis, International Potato Center (CIP), Peru

1 Introduction

1.1 Overview

In this chapter, we aim to provide a brief assessment of the contribution of genebanks to potato science and industry.

Significant improvements in the cropping system can, of course, be accomplished through better production and management practices, machinery, agrichemicals and marketing. In this chapter, we focus only on better genetics – although genetic components do also potentially interact with most of these other means of advancing crop production. For example, while we might improve irrigation management, the inherent water use efficiency of the cultivar will most likely also influence the final formula for optimal production.

Current potato varieties already provide a crop that stands out for productivity, affordability, palatability and versatility with a widespread positive effect on society (DeJong, 2016). However, production constraints and consumer and farmer needs do change over time. Thus, there is value to society in continually finding ways to use genetics to make potato crop better in every way possible – if not, then public funding of a genebank has little justification. Thus, a primary and overarching concept is that genebank staff need to have a corresponding attitude of innovation in all they do in order to maximize the genebank's support of innovation in the crop.

http://dx.doi.org/10.19103/AS.2017.0016.02
Published by Burleigh Dodds Science Publishing Limited, 2018.

Primitive cropping provided a genetic buffer in which diversity, by planting of different landraces, was broad over space but relatively narrow over time. Thus, at least some individuals in the diverse mix of landraces grown across the landscape every year were likely to be productive in any one season, while others would be productive in different seasons, making total crop failures unlikely. In modern agriculture, in contrast, breeders seek to fine-tune a very limited number of clones to specific environments and optimize cropping systems for maximum production. Thus, diversity is now narrow over space. But because of disease/stress/pest challenges and with production technology and consumer demands varying over time, germplasm must be preserved in a condition that is ready to be quickly deployed – so genetic diversity can be broad over time, as needed.

The sports model is a good parallel. We should not condemn the strategy of depending almost wholly on a few star players, since this often results in winning the game. But then one also needs to hedge by having backups ready to take the field very quickly if the stars are injured. Thus, 'monocultures' of crops are not bad as long as we have genebanks, and breeding programmes using them, which can quickly deploy alternatives.

1.2 Basic genebank mission

We have established that genebanks serve the basic function of holding useful genetic variation in reserve. The method of accomplishing this is usually presented under the five general headings of acquire, classify, preserve, evaluate and distribute. Many concepts about these objectives are basic and do not change over time. Here, we provide general concepts, relying on the reader interested in more details to refer to references contained therein. For example, the 2014 Potato Crop Germplasm Committee Vulnerability Report (http://www.ars-grin.gov/npgs/cgc_reports/potatovuln2014.pdf) and a previous review (Bamberg and del Rio, 2005) provide good overviews. Other issues are evolving rapidly, so we provide suggested key words with which the reader can access the latest information on the internet. The basic challenges and opportunities under each of the five objectives will be reviewed, particularly noting more recent developments and their impact.

Here again, a parallel from common experience – a tool store – provides a useful illustration of the basic concepts surrounding the genebank mission:

Acquisition is stocking the store with a diversity of tools to maximize the breadth of tasks the customer is able to do. These should include familiar tools to address well-known problems, but also unfamiliar tools that address the unknown or unexpected. Thus, the genebank should acquire stocks with, for example, known superior resistance to a common disease, but also stocks with uncharacterized general sequence diversity and breadth of phenotypic resistance mechanisms which might provide a source of resistance to a disease or pest or environmental condition yet unknown. **Classification** of germplasm is needed, for the same reason that a store must organize and clearly label related tools so that staff can easily manage the inventory and customers can do the most efficient shopping. **Preservation** is needed for the same reasons that the tool store maintains the functionality and sufficient inventory of the product. This includes monitoring viability with germination testing and having plenty of propagules available. Having a genebank with germplasm that is dead, diseased or with too few propagules to distribute is like a store with only a display model of a tool that is actually non-functional. **Evaluation** has its importance in the principle that, to maximize service to the customer, the tool store should maximize available information about the tools' functions. Thus, staff should provide instruction manuals and, ideally, do additional tests on the tools themselves as

Published by Burleigh Dodds Science Publishing Limited, 2018.

time and resources permit. Genebank staff should be involved in the kind of evaluation that examines – and thereby improves – the quality of genebank service. Examples include studies that evaluate which accessions deserve particular emphasis because they are exceptionally rich in genetic diversity and which genebank techniques minimize the risk of losing diversity. **Distribution** applies primarily to *working* genebanks described in this chapter. In contrast, a 'doomsday' vault like the Svalbard Global Seed Vault in Norway, or the United States base collection at the National Laboratory for Germplasm Resources Preservation (Keyword 'NLGRP') at Ft. Collins, Colorado emphasizes long-term backup preservation, not ongoing rapid distribution for research and breeding. Finally, distribution of useful germplasm has unintended negative consequences without a programme to exclude systemic plant pathogens. In the tool store analogy, this is like making sure tools meet all global safety standards and are free of dangerous defects.

The five categories above provide a good working framework, but it is easy to imagine cases in which they overlap and interact with each other. For example, evaluation informs classification and both have implications for preservation and distribution. If a DNA-based evaluation reveals unexpected close similarity of two accessions, it implies they ought to be classified into the same taxon, and it also implies some degree of genetic redundancy that makes preserving and distributing both of them a lower priority.

A sixth consideration is often avoided because it seems negative: Even if we have the resources needed to expand the store's capacity, it must be true that some tools have become obsolete and no longer merit a place on the store's shelves. In the case of germplasm, we know that broken genes that make the plant unthrifty are routinely purged in nature (Schoen et al., 1998). Since it takes extraordinary effort to maintain those traits or individuals that are, for whatever reason, very unsuited to genebank cultivation and reproduction, good management demands some attention to discarding and archiving obsolete germplasm.

1.3 Special considerations for potato

Overlaid on the general nature of genebanks discussed above are considerations particularly relevant to a potato genebank.

Many wild relatives. Potato currently has about 100 wild related species (Spooner et al., 2014, 2016), many of which are fairly easily accessible to the crop breeding pool.

Two forms of reproduction. The crop comprises highly heterozygous clonal cultivars, but wild species typically reproduce sexually. Botanical seed storage has obvious advantages in longevity, being less vulnerable to temperature and moisture extremes for shipping, and requiring less storage and shipping space. Thus, potato genebanks need technology for efficient clonal maintenance and distribution of disease-free plantlets, usually accomplished *in vitro* (Bamberg et al., 2016c), and botanical seed, usually dried, sealed and stored at −20°C or colder. Botanical seed populations are typically multiplied by hand pollination in a greenhouse or screenhouse that excludes natural pollinators (bumblebees). The latest and best techniques are often best accessed through genebank websites or personal contact with genebank staff. Natural species of potato are diploid, tetraploid or hexaploid, with both inbreeding and outcrossing breeding systems.

Latin American origin. Except for two wild species in the United States, primitive cultivars and wild species originated from Latin America.

Systemic diseases. Viruses and other diseases require special care and quarantine testing for international germplasm import, routine within-country distributions from the genebank and during in-house germplasm handling in the genebank.

Published by Burleigh Dodds Science Publishing Limited, 2018.

Easy manipulation. Potato seeds are not 'recalcitrant', and often maintain good germination for decades in a simple household freezer. The majority of taxa grow and reproduce well in standard greenhouse conditions. With refinements of standard tissue culture techniques, organs like anthers can often be cultured to regenerate haploids; protoplasts can be made, fused and regenerated.

Multiple world genebanks. Other collections provide the opportunity for meshing data associated with sites of natural origin, evaluation data for accessions held in common, sharing technology and germplasm backups for each other.

1.4 New potato germplasm considerations

Some developments made, or will make, fundamental changes in how potato genebanks operate. Three fundamental problems of wild germplasm are crossing barriers to the cultivated forms, weedy traits and lack of efficient methods to evaluate a large amount of diversity. Hence, any advance in technology that mitigates these limitations increases the practical value of the genebank without adding any new stocks.

Information storage, access and management. Those involved professionally for less than 25 years may not be able to fully appreciate the tremendous advantage that computerization brought to germplasm management. In particular, staff gained the ability to quickly sort and search multiple information fields associated with any accession. The evolution of the Germplasm Resources Information Network over the past 30 years (keywords 'GRIN germplasm') was the culmination of these advances for germplasm in the United States, providing standardized online storage and access for virtually all genebank-related information, including germplasm ordering. An updated version, GRIN-Global, is now being adopted by genebanks worldwide which will further enhance interoperability, data sharing and facilitate users to precisely identify the best germplasm for their needs.

Electronic maps and GIS (geographic information systems). For collecting data, GPS devices based on GIS provided precise documentation of the natural origins of wild populations. Coupled with mapping programmes and aerial images overlaid with roads, other landmarks and climate data, it is possible to plot the relative natural origins of accessions. One can then plan the best route to visit sites targeted for collection (e.g. cited on herbarium specimens), and even predict likely habitats for exploration such as searching for commonly associated vegetation or slopes with the commonly favoured exposure (http://www.ars-grin.gov/nr6/coll_trips/SW2015.pdf). These resources are often available online for free.

DNA markers to assess general genetic diversity. Differences in random neutral markers provide a much more abundant and objective resource for stratifying genetic relationships and heterogeneity than by other assumed diversity-associated traits like variation in phenotypes, geographic origins or habitats. DNA markers became the standard for determining taxonomic relationships, whose acquisitions would be most valuable to the genebank, their accessions within a species encompassing the most diversity, and basic genebank 'housekeeping' functions like detecting mixes, mislabelling or other mistakes. Examples of the widespread application of DNA markers in the genebank are emphasized in other sections of this chapter.

Small modifications to elite clones. Although inbreeding is common in potato wild species, current cultivated potatoes are all heterozygotes fixed by clonal reproduction. Selecting elite potato cultivars takes much effort and time, so it would be highly desirable to make only small improvements to existing cultivars and not recombine the entire genome

Published by Burleigh Dodds Science Publishing Limited, 2018.

in a breeding cross. The single-trait substitution into existing elite lines accomplished by backcrossing in inbreeding crops could only previously be attempted in potato by mutation breeding or selecting spontaneous clonal variants. However, the potential for small changes has expanded in the form of intentional regeneration of selected somaclones (Nassar et al., 2014) and transformations with foreign genes or targeted modifications of existing genes (Butler and Douches, 2016). While these new technologies hold great promise, the traditional method of mining germplasm by screening for desirable phenotypes and crossing is not obsolete (Miller, 2016). A related strategy that is getting increasing attention these days is reinventing potato cultivar breeding in the form of inbred diploids (Jansky et al., 2016).

Restriction of germplasm exchange. Raw germplasm is self-reproducing, and has no immediate commercial value, so was historically widely regarded as free. But the country of origin of germplasm resulting from potato collection and exchange is now recognized as having a sovereign right over the current access, as well as benefit sharing from commercial use of the germplasm. Thus access to germplasm is becoming highly regulated, in many cases germplasm has become virtually inaccessible. Expanded discussion of this topic is provided in Section 6.

1.5 Extended genebank services

The optimal tool store is more than a source of goods, and so it is with a genebank.

Administration and outreach. The successful tool store would ideally increase its commercial impact by advertising and extending its expertise to the public. Similarly, genebank staff often have specialized expertise that allows them to serve as reviewers or editors for manuscripts and grant applications, and on technical and policy committees both nationally and internationally, as well as *ad hoc* hosting of genebank visits, seasonal internships or graduate students. These contributions have the reciprocal benefit of keeping genebank staff in touch with the needs of the germplasm users, and crop producers and consumers. Staff can contribute to popular articles on genetic resources and otherwise help inform the public on the role of crop relatives in supporting the efficiency and safety of the food industry, while conserving natural resources and protecting the environment.

Custom service. We have already outlined genebank service in the form of optimized acquisition, classification, preservation, evaluation and distribution. However, both a store and a genebank can increase their impact through *custom* service. This happens when genebank staff are familiar with the research literature, build rapport with customers and direct some of their resources and expertise towards supplying custom information, propagules or samples to cooperative research projects with multiple germplasm users from an array of specialties.

On-site research. Staff research was introduced previously under the heading of *evaluation.* It not only increases information about the genebank's genetic tools, but makes staff able to give advice on the best tool and application from personal experience. Staff research can have particular impact if directed towards germplasm handling technology and exploration for new traits. For example, discovering more efficient techniques for making hybrids is a good topic for in-house genebank research, since it has broad payoffs both inside and outside the genebank, but is a problem that few specialists like breeders have time to intently pursue. Evaluation for novel traits is also a particularly apt research area, since genebank staff have the special opportunity to notice new phenotypes among the broad spectrum of exotic germplasm they routinely grow for other purposes.

Published by Burleigh Dodds Science Publishing Limited, 2018.

1.6 Achieving a comprehensive genebank service

Putting all the objectives, services and research together, the optimal genebank would contribute to every step in the germplasm deployment process: it would assess the needs for acquisition both from the wild and elsewhere, organize expeditions, collect germplasm, research the nexus of *in situ* versus *ex situ* diversity, guide stocks through import quarantine, optimize organization of stocks by taxonomic and other classification, identify core subsets of accessions within species, develop and apply better technology, keep the maximum diversity in the raw germplasm alive through efficient seed increase and clonal *in vitro* maintenance and readily available to cooperators through germplasm orders, enhance raw germplasm, log data in GRIN or another publicly available database (such as Genesys), publish results in research journals and release hybrids or selections with value-added beyond that of the raw germplasm. This complete spectrum of germplasm support would be impractical for any researcher or breeder to provide for himself.

2 Acquisition of potato genetic material

There are three basic forms of acquisition: (1) direct collecting, (2) sharing, and (3) germplasm development.

2.1 Direct collecting

This is when novel germplasm is accessed from the wild environment (wild species) or farmers' fields or markets (cultivated species). Potato (*Solanum*) species occur from the southwest United States to southern Chile, mostly at higher elevations. There is a concentration of species diversity in Mexico and in Peru/Bolivia (Hijmans et al., 2002, p. 120). Genebanks may arrange and fund collecting expeditions to cover gaps in geography, taxonomy or particular needed traits. A parallel discipline to germplasm collecting exists with botanists who document the natural locations of plants in floras and herbaria, thereby becoming very familiar with particular geographic regions of interest. Collaboration with such professionals is of great use for planning germplasm collecting expeditions. A relatively recent development is the ability to use DNA markers to assess factors that influence the efficiency of collecting germplasm. Since such collecting research has been of particular emphasis at the US Potato Genebank (USPG), we highlight some relevant findings here.

Comparing ex situ *and* in situ *diversity.* Is genetic diversity in the wild changing through time? If so, how representative are the *ex situ* preserved samples when compared to the original *in situ* samples? del Rio et al. (1997a) and Cadima-Fuentes et al. (2016) provided DNA marker evidence that re-collections of natural populations from the same sites but at different years of collection are significantly genetically different. The unfortunate implication is that *ex situ* sites cannot be relied upon as backups of populations in the genebank, but the fortunate implication is that they are worth re-collecting for novel diversity.

Determining associations between eco-geographic and genetic patterns. Traditionally, collecting samples from diverse geographical origins was recommended to increase diversity (Frankel and Soule, 1981; Perry and McIntosh, 1991). However, it would be even more effective if areas of high or distinctive genetic diversity were predicted by specific

Published by Burleigh Dodds Science Publishing Limited, 2018.

eco-geographical parameters. A series of studies in geographically diverse populations of potato species *S. jamesii, S. fendleri, S. sucrense* and *S. verrucosum* tested whether genetic diversity was correlated with different ecological, geographical and reproductive variables (del Rio et al., 2001; del Rio and Bamberg, 2002; del Rio and Bamberg, 2004). In all cases, genetic associations among populations were efficiently detected with DNA markers. However, associations with eco-geographical variables were rarely detected. Hamrick (1987) warned that habitats may be quite heterogeneous, even within small areas, which can explain why genetic differentiation among populations was generally independent of spatial separation. Only for the inbred species *S. verrucosum* was spatial separation predictive of genetic distance, perhaps because inbreeding exposes all the population's genetics to adaptation to the characteristic environment at locations.

Easy versus remote. It has been shown that collecting is biased towards sites with easy access (i.e. places close to roads and villages). Hijmans et al. (2000; 2002) reported this as 'infrastructure bias', noting that about 60% of the wild potato collections in Bolivia were made within 2 km of a road. Bamberg et al. (2010) tested whether the extra time and effort needed to get to 'remote' sites pays off in the capture of more diversity than at easy sites. In paired sites at three mountain ranges in southeast Arizona, more diversity in *S. fendleri* was sometimes captured at the more remote site, but sometimes at the easy site. Thus, both types of locations need to be sampled and assessed in the lab.

Predicting and assessing genetic diversity hot spots and mega-populations. One 'easy' site, the Santa Catalina Mountains, northeast of Tucson, was identified as having particular genetic richness, despite relatively few samples. Hence, a more intensive collecting strategy was deployed at this site in September 2009. AFLP markers confirmed that this was a rich site (Bamberg et al., 2011) having 24 marker alleles unique to the entire region. This demonstrated that preliminary DNA evidence of genetic richness can identify sites that deserve a higher priority for more thorough collecting.

The ultimate in efficiency for collecting and other *in situ* work would be to identify 'mega-populations': easily accessible, very large and robust, very localized single populations that possess the maximum possible share of known diversity for that species. Such populations would be the wild equivalent of top-ranked members of core collections in the genebank. Bamberg et al. (2016) reported such a single natural population at Mesa Verde, Colorado, that captured >80% of the known genetic diversity for *S. jamesii*. In addition to its value for collecting, this area could be considered a site for habitat protection and an *in situ* reserve, as continuity of the evolutionary response of a species is also an important objective in conservation. Areas with such high genetic diversity could have evolutionary potential critical for adapting to climate change.

Impact of agrichemicals on reproductive traits and genetic diversity. Agricultural practices might be narrowing the genetic diversity at natural habitats, even if those activities are not extinguishing the populations. For example, indirect pesticide contamination may occur in natural populations growing near agricultural fields. If so, then populations growing distant from agriculture would be predicted to be richer in genetic diversity. Two research projects evaluated 15 different species known to grow within or very close to fields in the Andes, and assessed two aspects: the impact of pesticides on traits related to plant's reproductive capacity (del Rio et al., 2012a) and the effect on genetic diversity (del Rio et al., 2012b). The results revealed that the pesticide affected reproductive traits, in particular reducing the production of viable pollen and the duration of flowering. However, DNA markers used to measure changes in allele frequencies at different loci did not detect genetic drift caused by the hypothesized pesticide-mediated selection.

Published by Burleigh Dodds Science Publishing Limited, 2018.

How many populations are enough? Many of the approximately 100 potato species are represented by only a few populations, but we have hundreds of some species. An obvious question is whether there is a standard pattern for the general accumulation of diversity. We used DNA marker data on three model species to examine the actual rate of accumulation of polymorphic loci as populations were added over time, or if they had been randomly added on the basis of the empirical frequencies of polymorphic loci (Bamberg and del Rio, 2016). Addition of new polymorphic loci greatly slows after one has about 100 populations. Of course, it is impossible to know if valuable new alleles and traits exist in a population in the wild but not yet in the genebank, or if these species are representative of all potato species. But this model suggests that if the current total number of populations in USPG could be doubled (all species had 100 populations), one could consider it 'enough' since additional populations would net greatly diminishing returns in allelic diversity.

2.2 Sharing

This is the exchange of germplasm already sequestered in a formal genebank or in a public or private collection. Potato is a crop grown from tuber 'seed' pieces, so there is an extensive industry to produce vigorous propagules with low pathogen levels. This necessitates multiple localized *de facto* clonal genebanks for numerous cultivars in the form of state 'seed certification' organizations. There are also private organizations of citizens who preserve and share germplasm, motivated by a common enthusiasm for expanding the palette of available food vegetables. With varying degrees of formality, they collect, grow, evaluate for culinary quality and exchange exotic germplasm. If the national genebank builds a rapport with state seed certification and NGO collections such as Seed Savers Exchange, an efficient network can result, whereby the genebank staff know where particular germplasm items are available. If the genebank does not maintain those items itself, it can refer germplasm requesters to those alternative sources.

Unless labelling mistakes have occurred, clonally propagated stocks with the same name should be genetically equal among genebanks. In contrast, botanical seed populations with the same origin data might vary due to sampling differences. Often samples from a single seedlot are split and then separately regenerated in different genebanks for many years. Differences in propagation and seed multiplication techniques could theoretically introduce selection or drift that altered one or both of the samples. We tested this by comparing pairs of populations of different potato species which were reputed duplicates between USPG and the Vavilov Institute Potato Genebank (VIR) in Russia (Bamberg et al., 2001) and the International Potato Center Genebank (CIP) in Peru (del Rio et al., 2006). Very few genetic differences were detected between these reputed duplicates from different genebanks. One important implication is that all evaluation data from one genebank can be applied with confidence to the corresponding germplasm sample in an alternative genebank.

2.3 Germplasm development

Germplasm may be 'acquired' in the sense that novel subsets are developed from existing raw accessions. This is not an unnecessary burden on genebank capacity, but, on the contrary, is very encouraging since it demonstrates that value is present and being mined

Published by Burleigh Dodds Science Publishing Limited, 2018.

in the raw germplasm. For example, a rare useful mutant in a population may become a new acquisition as a clonal isolate, or fixed in a true-breeding seedlot. Populations of different interspecific hybrids or stocks with different ploidies that promote crossability may become distinct genebank holdings. Such items are usually indexed separately as 'genetic stocks' or 'breeding stocks'. Useful traits that are enhanced can cover the entire continuum from basic anatomical and physiological mutants, those aimed at benefitting farmers (e.g. disease and pest resistances), or those aimed at benefitting consumers (e.g. nutritional improvements). A relatively recent challenge to genebanks is the increasing opportunity to be a central source for large *families* characterized for both genetic markers and an array of useful traits. Clonal maintenance and distribution of hundreds of lines for a single germplasm accession requires a great deal of genebank resources. One possible solution on the horizon: fixing segregated genetic variants in the form of inbred botanical seedlots – that is, recombinant inbred lines. More details on germplasm development are presented in Section 5.

2.4 Impediments

A discussion of acquisition reasonably includes impediments. Collecting in the wild has inherent limitations since wild potatoes grow across vast areas, often accessible only by primitive roads or trails. Unfortunately, the habitats where potatoes are found, and the associated flora, are not consistent or obvious, making it difficult to predict exactly where to look. Plants in the wild are often not showy and in some cases only available for collecting at certain times in some seasons (e.g. not collectable at all in a dry year). Even when previous reports are available, as in herbarium records, the locations given may not be sufficiently precise or accurate to allow the germplasm collector to find the exact small area where potatoes are growing. In some cases plants may be found, but viable propagules for the genebank are not available for collection.

Two other factors impeding acquisition are restricted exchange or collecting due to germplasm ownership, or due to quarantine regulations aimed at preventing import of pathogens. These topics are discussed in detail in Section 6.

3 Classification of potato genetic material

Classification may be divided into three somewhat overlapping types of germplasm categorization.

3.1 Taxonomic classification

Species is one of the primary identifiers used to categorize and document accessions in potato genebanks. But potato has a particularly large number of recognizable distinct forms, often including intermediates, so there has been much difference in opinion as to how to best set species' limits. Since potato is a major commercial food crop, its genebanks have often had the advantage of a professional taxonomist devoted to the classification of their stocks. Since 1987, Dr D. M. Spooner has served as the taxonomist for USPG. The reader interested in additional details is directed to his publications (Keywords 'Spooner publications') which offer a wealth of information on his own novel research on potato systematics, extensive reviews of the works of others, the details of the latest taxonomic treatment (Spooner et al., 2014, 2016) and, particularly, his personal 30-year retrospective as a potato taxonomist (Spooner, 2016).

Splitting. If two distinct forms exist with different traits for agriculture, one might be motivated to taxonomically 'split', differentiating those by species name for the sake of utility, even if intermediate forms exist. Potato sub-taxa or even species have been concluded to lack predictivity (Khiutti et al., 2015; Jansky et al., 2015), but if genetics is the basis of both taxonomy and traits, the challenge is finding the connection that must exist (Bamberg et al., 2016). With splitting, intermediate accessions will have the awkward denomination of 'hybrid', potentially obscuring how they fit into the overall pattern of species relationships. Or, if ambiguous accessions are more aggressively assigned a species name, they may flip in subsequent determinations, creating instability and confusion in genebank documentation. Splitting creates a need for higher-level names to show which species are related. This has often been accomplished in potato by assigning species to 'Series', 'Groups' or 'Clades'.

Lumping. Alternatively, in a 'lumping' scheme, distinct forms may be combined with their intermediates into a single species. Unless the two original species were not suspected of being relatives, lumping inexorably reduces information, making the species name less significant in the utilitarian sense. For example, a single species name for a collie, bloodhound and poodle *does* provide the useful information that they are the same at a general level, *but only for those who did not already know that.* Such lumping in dog types automatically shifts emphasis to the lower-order names for use by those working on a less general level, where all dog breeds are definitely not interchangeable. The same is true for the genebank – if species is the primary identifier, lumping increases the need to designate and more routinely use sub-specific names or other differentiating descriptors like ploidy. This is true whenever these lower rank names are not simply arbitrary, but really do represent objective significant genetic or phenotypic differences – which is sometimes the case with potato (Hardigan et al., 2015).

Is an objective baseline for species boundaries in potato possible? One cannot appeal to the empirically determined biological species concept, since no potato treatment proposes combining all of the many taxa known or assumed to be able to produce fertile F2 progeny, except as per utility as described in Bradeen and Haynes (2011, p. 7). A less extreme objective baseline would be to set a single minimum quantity of general genetic difference between species. Or, even better, to require that lumping creates species in which all accessions are more closely related to each other than to any accession in a different species. Accessions usually, but not always, group together by species (Hardigan et al., 2015; Jacobs et al., 2011). Even the latest treatment (Spooner et al., 2014) can have separate species for which all descriptor metrics overlap, suggesting that a hypothetical individual or accession could exist that would properly key to two species names. Spooner (2016) opines that a simple 'formula' for species limits will 'forever be an elusive goal'. But there is hope in noting that the advance of technological and analytical tools has often clarified what previously seemed unresolvable.

In practice, taxonomic authority is often codified by centralized computerization. Genebank and database administrators decide on a certain recognized taxonomy and organize the database to reflect it, as occurs with the USA GRIN system. Thus an 'official' taxonomy may be reinforced by virtue of being the *only* option presented to users who access the database for germplasm and information.

Published by Burleigh Dodds Science Publishing Limited, 2018.

3.2 Core collections

As already mentioned, there is an increased need to 'classify' below the species level when lumping broadens the diversity within a species or when, for any reason, a species has so many variable accessions that the typical germplasm user needs a smaller representative set to make evaluation practical. One approach is designating core collections, where the goal is a minimum number of accessions (e.g. 10%) that capture the maximum diversity (e.g. 90%) within species (Bamberg and del Rio, 2014 and references therein). The simplest decision in core selection is when exact duplicates are identified and one is eliminated. Core collection members can be designated and ranked by their density of diversity, so that the germplasm user is able to balance genetic coverage against the cost of testing more samples (see Bamberg et al., 2016a and references therein).

Although core collections can be composed to capture diversity for various parameters (e.g. country of origin), the modern trend has been to avoid stratifying by such traits that are not necessarily linked to general genetic diversity. Instead, DNA markers based on sequence differences are used. These are presumed to be mostly neutral – have little or no phenotype impact that could result in a genotype by environment (GxE) skew. If a DNA marker technique generates many independent polymorphic loci and is consistent and free of bias, it can provide a very objective approach for comparing germplasm. Practicality of such markers has also improved with greatly reduced cost per data point over time. However, we still lack the optimal marker system that can do everything needed: (1) detect more than two alleles per locus; (2) test a bulk of many seedlings, thereby characterizing the net diversity of a heterogeneous population in a single sample; and (3) use mapped markers so that results of various experiments at different times and places can be readily combined and compared.

The core collection concept may also be applied to the whole genebank, where one designates a reduced array of representative species intended particularly for initial screening surveys when there is no a priori information on the trait. The USPG has created and validated such a 'mini-core' of 25 potato species (Bamberg et al., 2016a).

3.3 Cogs

One might set 100 populations per species as a reasonable maximum, since gain in diversity was observed to slow markedly after acquiring that many (Bamberg and del Rio, 2016). These populations may be heterogeneous, indicating there is much potential diversity to classify within. With such a large amount of potential variation to classify, a quick, simple and low-cost approach is attractive. We recently successfully tested whether the remarkable power of human visual subjective categorization could be systematically merged with the scientific method to classify within species. The term 'cog' (short for cognate = born together) was coined for a group within species composed by rapid visual impression and then validated as distinctive by replication and empirical genetic and phenotypic tests (Bamberg et al., 2016b).

4 Preservation of potato genetic material

Managing the technical aspects of seed and clonal multiplication, storage and viability takes a major share of genebank's time and resources, but a detailed discussion on this topic is beyond the scope of this chapter. The reader interested in specifics of the latest recommended techniques may access genebank websites or contact genebank staff, since advice on how to propagate the germplasm is part of genebank service. A recent treatment of *in vitro* methods at the genebank is available (Bamberg et al., 2016c).

In the following section, we highlight a few results of studies we have done using DNA markers to examine the impact of various techniques on the preservation of the genetics of genebank populations. This is important, because if weaknesses are found, the genebank would need to direct limited resources to stop attrition of diversity.

4.1 DNA markers for assessing diversity

Effects of seed increase. Most wild potato species are propagated as botanical seed resulting from growing and intermating multiple parents of a seedlot that has a low number of seeds or low germination. Seeds are light, small and long-lived, thus convenient for shipping and storage. However, in contrast to clonal propagation, sampling seedlots introduces the possibility of genetic drift among generations. When del Rio et al. (1997b) examined the impact of genebank seed increase on different generations of *S. fendleri* and *S. jamesii* accessions as models, significant differences were not found, suggesting that seed increase does not result in large genetic shifts.

Seedling selection. One specific way that genetic shifts could take place within populations during seed multiplication is inadvertent exclusion of some genotypes at transplanting. One plants extra seeds to ensure that there are enough plants (typically >20) to intermate and produce a new generation of seeds. Consequently, some extra seedlings are always discarded at transplanting. Bamberg and del Rio (2006) examined the risk of losing diversity at the seedling transplant stage if small seedlings are discarded. In most cases, seedlings within populations actually looked identical. But when smaller seedlings were present in a population, no genetic differences were observed between them and their normal sibs. The occasional difference seen in seedling size is due to either random environmental effects or genetic differences that are too small to be detected by the marker system used.

Use of balanced bulks for seed multiplication. Similar to the threat of mis-sampling the population at transplanting, loss of alleles could occur if some seed increase parents make an unbalanced contribution of offspring to the next generation (Breese, 1989). When bulk intermating, one does not know if some plants contributed more or less as fathers, but bulking an equal number of seeds from each plant logically ought to balance maternal representation in the next generation. This, therefore, is a widely recommended alternative to a general bulk of seeds (Rao et al., 2006). However, balanced bulks require a significant investment, essentially doubling the number of samples that need to be prepared, stored and documented – is it worthwhile? When we investigated (Bamberg and del Rio, 2009), for the great majority of populations, plants naturally produced similar numbers of berries, making an intentional balanced bulk moot. However, we also examined two populations identified as having the most variable seed production among parents. Even in this extreme case, over 12 replicate seed increase generations, only about 4% of RAPD loci had alleles practically vulnerable to loss. Thus, we conclude that balanced maternal bulks have small genetic benefits in the next seed increase generation.

Heterogeneity within populations. Since drift cannot affect loci with fixed alleles, heterogeneity dictates the appropriate strategy for germplasm protection, preservation and evaluation. Potato species would be expected to differ, since some are clearly natural

Published by Burleigh Dodds Science Publishing Limited, 2018.

selfers and others are self-incompatible. We used RAPDs (Bamberg and del Rio, 2004) and SNPs (Bamberg et al., 2015), with four species representing different ploidies and breeding systems as models to estimate population heterogeneity. Diploid and tetraploid outcrossers had the most marker heterogeneity. A single plant of these species does not represent the population very well, but bulk samples of about 20 plants have a high probability of uniformly capturing almost all the variation.

General status of vulnerability of alleles in the genebank. As described above, sampling opportunities like seedling selection and non-balanced seed bulking do not appear to be major threats to maintaining alleles in the genebank across generations of sexual reproduction. Random neutral DNA markers suggest that the reason for this is threefold: (1) low-frequency alleles are not very common in populations, in part because some species are inbreeders, (2) to confidently maintain low-frequency alleles in a population, one needs population sizes that are practically impossible to attain in the genebank, so it is a rather hopeless cause to try to keep truly vulnerable alleles unless they are isolated and maintained clonally, and, most importantly, (3) low-frequency alleles vulnerable in any given population are almost always *fixed* in another population, making them invulnerable to loss in the context of the whole genebank (Bamberg and del Rio, 2003). The most vulnerable situation is when an allele is found only in one population of its species, and it also has a low frequency within its population. But the evidence suggests that alleles that are rare within a population are usually common or fixed in another population (Bamberg and del Rio, 2009) and alleles that occur in only one population are usually present in a relatively high frequency there (Bamberg and del Rio, 2009).

5 Evaluation and enhancement of potato genetic material

5.1 Evaluation

A daunting challenge. Evaluation to detect phenotypes useful for improving the crop is as very broad a topic as the many disciplines of potato study – genetics, pathology, physiology, entomology, horticulture, nutrition, biochemistry, production and management, and more – and cannot be addressed in detail in this chapter. As previously mentioned, the time and resources needed to evaluate a large number of germplasm in several different ways is a major limitation to its use, and any advances in technology for phenotyping will increase the practical value of exotic germplasm.

Basic evaluation that identifies the germplasm. DNA markers have great advantages since they provide many data points and are stable in any environment and from any of the plant's tissues at any stage of development. But the link between DNA markers and useful traits will be weak without accurate and relevant phenotype data. For example, we know that strong GxE exists such that some germplasm only exhibit desirable high performance for certain traits when evaluated as field tubers, not tubers grown in greenhouse pots (Hale et al., 2008). Similarly, we also need accurate descriptive and natural origin information, because provenance and taxonomic identity may also be associated with practical phenotypic traits. For example, certain species originating from certain locations are associated with greater frost tolerance (Hijmans et al., 2003). The greatest inefficiency and potential blurring of such patterns occurs when erroneous descriptive data make two identical germplasm items appear to be different. In del Rio and Bamberg (2000) we investigated a population of *S. sucrense* with implausible origin data. We could not, of course, determine its precise origin with DNA markers, but *could* show it was not an exact duplicate of another population in the genebank and thus was worthy of separate maintenance and evaluation. On the other hand, we also showed that AFLP and SNP markers could solve the mystery of why a certain population of *S. okadae* reputedly from Bolivia had characteristics found only in populations from Argentina – it quite clearly was a mislabelled identical duplicate sample of another *okadae* population from Argentina already in the genebank (Bamberg et al., 2016b).

5.2 Enhancement

Interest in using exotics for enhancement is what motivated establishing centralized genebanks as an alternative to an inefficient *ad hoc* approach by multiple unorganized separate research and breeding programmes. The use of exotic germplasm for potato improvement dates back to the mid-nineteenth century, when South American land races were imported to Europe and North America after the late blight epidemics. The common belief was that centuries of asexual reproduction had made potato crop weak and susceptible to diseases (Goodrich, 1863). The introduction of both 'new blood' and sexual propagation was expected to restore the vigour of the crop. Of course, sexual propagation eliminated many of the pathogens that had led to the decline in vigour over time. The addition of new germplasm probably also benefitted the crop, although most of the new seedlings were inferior to existing cultivars, as modern breeders would predict. Notable successes were Garnet Chile and the Chilean cultivar Daber, which figured prominently in the pedigrees of many North American and European cultivars (Glendinning, 1983; Bradshaw et al., 2006).

The next major effort to introgress wild germplasm was nearly a century later, when breeders in both North America and Europe began using the hexaploid wild species *S. demissum* as a source of late blight resistance genes (Stevenson and Clark, 1937; Ross, 1986). However, breeders found this germplasm challenging to work with and, when resistant cultivars were developed, strains of the pathogen were able to overcome the *S. demissum* R genes. Heterosis was noted, though, when adapted hybrids between cultivated and wild potatoes were generated (Ross, 1986).

The *S. demissum* R gene breeding efforts ushered in a new era of germplasm enhancement in potato. *Solanum chacoense* was used as a source of late blight and common scab resistance in the development of the cultivar Lenape (Akeley et al., 1968). As an unexpected bonus, the wild germplasm appears to also have contributed exceptional processing quality. Lenape is in the pedigrees of many major potato cultivars developed for chip production (Love et al., 1998). However, an unexpected negative contribution of *S. chacoense* was high tuber glycoalkaloid content (Zitnak and Johnston, 1970).

In 1963, Sherret Chase suggested that an analytic breeding scheme might be appropriate for potato. In this scheme, cultivated potato would be reduced to the diploid level, where crossing and selections would be carried out. Then, the germplasm would be returned to the tetraploid level for the production of cultivars (Chase, 1963a,b). This scheme was adopted by several programmes worldwide and has been described by many (Mendiburu and Peloquin, 1977; Peloquin et al., 1989; Jansky et al., 1990; Carputo and Barone, 2005; Thieme et al., 2008; Ortiz et al., 2009). Basically, parthenogenesis is used to create potato dihaploids, which are crossed to diploid wild relatives. The resulting hybrids are selected for agronomic traits and the new traits introduced by the wild relatives. Then, sexual polyploidization is employed to return the germplasm to the tetraploid level.

The analytic breeding scheme has contributed to the development of several successful cultivars. Yukon Gold is the product of a cross between a tetraploid cultivar and a 2n pollen-producing Phureja x dihaploid hybrid (Johnston and Rowberry, 1981). Dakota Russet resulted from crossing a tetraploid cultivar to a diploid hybrid containing *S. raphanifolium*. Other cultivars are more indirect products of analytic breeding. For example, the selections S438 and S440 are products of sexual polyploidization between a tetraploid *S. tuberosum* clone and a dihaploid x *S. tarijense* hybrid. S438 and S440 are parents of the chip processing cultivars Accumulator, Kalkaska, Lelah, Nicolet, Tundra and White Pearl, while S440 is the grandparent of the chip processing cultivar Pinnacle.

Even in the absence of a systematic breeding effort such as the analytic breeding scheme, wild relatives have been incorporated into the pedigrees of many cultivars. Some of these introgressions date back to the early efforts to bring in late blight resistance genes from *S. demissum*. Other breeding endeavours have focused on pest and pathogen resistance from a handful of wild species, especially *S. acaule, S. chacoense, S. fendleri, S. maglia, S. microdontum, S. spegazzinii, S. stoloniferum, S. toralapanum* and *S. vernei* (reviewed by Plaisted and Hoopes, 1989; Bradshaw et al., 2006). Typically, after a cross is made to a wild species, several generations of backcrossing to cultivated potato are necessary to recover commercially acceptable phenotypes (Black, 1949; Rudorf, 1958; Lauer, 1959; Bradshaw et al., 2006).

For all of these efforts, genebanks provided support, making stocks more secure, accessible, healthy and better documented with catalogues, and eventually online access, providing enhancers with information they would otherwise have to glean from the literature.

Genebanks also facilitated enhancement by providing a platform for evaluation and through genebank-sponsored and in-house evaluation. They have provided, preserved and distributed model stocks for technical research. Staff and associated programmes developed technology and related information. Examples are discovering how to induce haploids by simple crossing, making interploidy hybrids through use of gametes with the sporophytic chromosome number (2n gametes), establishing crossability groups (Endosperm Balance Number, EBN), creating monoploids that isolate a single allele per locus to facilitate sequencing and employing protoplast fusion to combine species that do not naturally hybridize. Details of these methods have been reviewed elsewhere (Hanneman, 1999; Bradeen and Haynes, 2011).

6 Legal custody and access to potato genetic material

Undisputedly the largest single event effecting the collection, acquisition, conservation, exchange, distribution and ownership of plant genetic resources in the past twenty-five years is the Convention of Biological Diversity (CBD; https://www.cbd.int/doc/legal/cbd-en.pdf). First opened for signatures at the United Nations Conference on Environment and Development (known as the Rio 'Earth Summit'; https://www.un.org/geninfo/bp/enviro.html), CBD came into force in December 1993, with 168 countries ratifying it. The foundation of CBD was built on three principles: *the conservation of biological diversity, the sustainable use of its components* and *the fair and equitable sharing of the benefits arising out of the utilization of genetic resources* (Article 1 CBD). The definition of genetic material, *any material of plant, animal, microbial or other origin containing functional units of heredity*, is sufficiently broad to allow interpretations for inclusion of virtually all forms of materials distributed from genebanks, including DNA. Finally, CBD reaffirms that countries have sovereign rights over their genetic resources, which includes the right to equitable sharing of benefits arising from their use. This also gives countries the *de facto* right to control and regulate access.

Although CBD was conceived as an instrument to enhance the conservation and use of genetic resources, two factors, sovereign rights and the economic value, that such rights might entail, halted much of the sharing of germplasm at the national level in the developing world. The rationale is complicated and multifaceted, yet a lack of infrastructure, policy and legislative and economic mechanisms to finance and carry out conservation in many countries resulted in a general lack of conservation measures and very limited access (Chandra and Idrisova, 2011). If a country does not have the capacity to regulate and set a value on its genetic resources, and most countries do not, the result is not to allow access until such time as a value can be determined. Since this is challenging, access to genetic resources in the developing world is limited. In addition, the much-needed conservation efforts expounded by CBD are unfortunately slow to become a reality as plant species continue to be threatened with extinction. As an example, outside of efforts by the USDA in the United States, there are virtually no publicly available collections of wild potato species in the past 17 years.

In response to the difficulty in the implementation of CBD at the national level, as well as the lack of benefit sharing from the use of plant genetic resources, a second instrument, the International Treaty for Plant Genetic Resources for Food and Agriculture (ITPGRFA; http://www.fao.org/plant-treaty), came into force in June 2004. The ITPGRFA covers only 64 crops (Annex 1 crops); however, the Annex 1 list includes all cultivated and wild potato species section tuberosa *except S. phureja. S. phureja* is not the only crop notably missing from Annex 1 (sugar cane, wild *Manihot* spp., peanut and soya bean are also excluded from Annex 1), yet it does uniquely pose a fascinating dilemma. The recent taxonomic revision of cultivated potato (Spooner et al., 2014) eliminates the species designation *S. phureja* and lumps *S. phureja* with four other Annex 1 species into *S. tuberosum*. How this will be dealt with in the future by the South American countries which negotiated the exclusion of *S. phureja* in the Annex 1 listing will be interesting.

While both CBD (and its Nagoya Protocol) and ITPGRFA defined conservation, access and benefit sharing as key components, as mentioned earlier, for potato they have fallen short of actually implementing new programmes for conservation or access. Although the key principles for implementing access have been outlined for prior informed consent (PIC) and mutually agreed terms (MAT), uniform standards defining what is needed for compliance of PIC and MAT do not exist. Progress has been made, however, with the ITPGRFA defining the terms for transfer of Annex 1 germplasm with the standard material transfer agreement (SMTA). Unfortunately, far too few countries, and even most party nations, have yet to implement the regulations and infrastructure for the transfer of germplasm with the SMTA. In fact, the vast majority of germplasm transferred with an SMTA is from non-national collections (IT/GB-6/15/20; *Report from Institutions that have signed Article 15 Agreements*, Sixth session of the Governing Body of the ITPGRFA). Few, if any, national plant germplasm collections from developing countries have been made broadly available with easily assessable and searchable websites as a result of the ITPGRFA.

Despite significant issues such as use restricted to breeding, training and research, no time limit to the terms of the SMTA and set payment of benefits (sales) with no adjustment

Published by Burleigh Dodds Science Publishing Limited, 2018.

for crop or % incorporation of covered material, it is supported by a multilateral system with defined terms. This greatly facilitates the transfer of germplasm as terms for transfer and use are already defined and do not need to be negotiated on a one-to-one basis as is the case with CBD. The potato genetic resources community is, therefore, fortunate that potato is included in Annex 1 of the ITPGRFA, but more needs to be done by the national programmes to allow access under the terms of the ITPGRFA to potato germplasm under national control. The ratification of the ITPGRFA by the United States (ratification consented by Senate vote, 28 August 2016; https://www.congress.gov/treaty-document/110th-congress/19/resolution-text) should contribute to building confidence on the ITPGRFA and hopefully enable needed capacity building for greater collaboration with national programmes for the collection, sharing and conservation of potato genetic resources.

A final note supporting the need to enhance capacity for the access to potato germplasm from national programmes is that approximately 94% of all material transferred through the multilateral system with an SMTA was distributed by the eleven genebanks in the Consultative Group on International Agricultural Research (CGIAR; http://www.fao.org/3/a-mo439e.pdf). This clearly illustrates the points made above that access to germplasm from party countries to the ITPGRFA is not happening and one overreaching reason for this is the lack of capacity and infrastructure to do so. In the case of potato, however, there are many positive changes in the Andean countries. Most notable is the announcement that Bolivia became a contracting party to the ITPGRFA on 4 December 2016 (http://www.fao.org/plant-treaty/news/detail-events/en/c/448725/). Additionally, the Instituto Nacional de Innovacion Agraria (INIA, the Peruvian National Program for Agriculture) hopes to initiate collections of wild potato in 2017 (Ellis, pers. comm.).

7 Conclusion and future trends

Legal custody of germplasm is one prerequisite for access and another is the distribution across international boundaries and the assurance that such distribution does not spread diseases or pests. The widespread distribution of clonal plant material poses the risk of simultaneous distribution of plant diseases and pests (Brasier, 2008). National agricultural pest and disease agencies are charged with keeping unwanted pests and pathogens out of their country and hence there are strict certification requirements for all countries that imported plant material is free of harmful pests and pathogens. Although this prevents their spread, it also prevents the exchange, access and use of genetic resources such as potato. If the exporting country does not have the infrastructure, expertise, technology or ability to perform the tests needed for phytosanitary certification, the genetic resources cannot be moved outside the country of origin. This is too often the case in developing countries and thus constitutes a major limitation to global access and the use of genetic resources for crops such as potato.

Potato crop wild relatives and cultivars/varieties from some potato genebanks (James Hutton Institute, UK and USDA, USA) are distributed as seed which is generally easier to certify as phytosanitary clean and thus ship internationally. At the International Potato Center in Lima, Peru (CIP), the presence of the following viruses, shown to be seed transmitted, are tested for parental material pre-flowering: arracacha virus B -oca strain (ABV – O), alfalfa mosaic virus, Andean potato latent virus, potato yellowing virus, tobacco mosaic virus, potato virus T (PVT) and potato spindle tuber viroid (PSTVd). If any of the viruses are present in the parental material, that plant is removed. Remaining plants are monitored throughout the growing season for symptoms and the screenhouses used for seed regeneration are maintained free of insects which transmit disease. Seeds regenerated under clean, insect-free conditions from disease-free parents can be certified as disease free and hence distributed internationally.

In the case of clonal material, not only does the genebank material have to be certified clean of phytopathogens, it also has to be shipped in a way that the material cannot become contaminated during shipping. As mentioned above, the shipping of disease-free *in vitro* material has become the standard for international shipment of clonal material from genebanks. The phytosanitary cleaning process of clonal potato accessions involves the introduction of the plants into *in vitro* culture, the treatment of the *in vitro* plantlets with thermotherapy (36–40°C for three weeks), followed by the aseptic isolation of a 0.3 mm meristem and the regeneration of *in vitro* plants from this meristem. Although the cleaning process is straightforward and highly effective with a success rate of 96% in potato at CIP (data unpublished), the diagnostic tests to confirm a disease-free status lag behind the physical phytosanitary cleaning of accessions and thus can delay the process for distribution of material for 1–2 years!

The time it takes to certify material as clean of diseases of import importance is a limiting factor for potato improvement programmes needing germplasm under a short or limited funding cycle. Next-generation DNA sequencing and other developments are targeting the limitation in the certification of disease-free status of potato germplasm. One such technology is small RNA sequencing and reassembly (sRSA: Kreuze et al., 2009), where the whole sequencing of small RNA from plants is proving to be a very sensitive method for detection of virus infection and could reduce the time for the confirmation of disease-free status of potato genebank accessions from a year or more to a month. Other developing technologies which could significantly facilitate the movement of non-*in vitro* clonal germplasm or advanced breeding lines/cultivars include the use of loop-mediated isothermal amplification assay-based systems (Liljander et al., 2015), which offer battery-operated portable cartridges able to detect the presence of viruses in less than an hour in the field or at point of entry of genetic resources. In the future, such portable assay systems could be powered by cell phones.

Another important factor in the access and benefit sharing equation is the role of genebanks in providing benefits to the communities whose ancestors 10 000 years ago began the process of the selection of potato landraces from which all existing cultivars originated. In the Andes, families traditionally planted twenty or more landraces of potatoes as an insurance policy. The bitter potatoes (*S. juzepczukii*, *S. ajanhuiri*, *S. curtilobum*) tend to be very hardy and, except for extreme years, always offer some production, and are hence critical for the sustenance of farming communities. The landraces of the non-bitter potatoes (collectively classified as *S. tuberosum* by Spooner et al. (2014)) tend to be more environmentally, and thus annually, unpredictable. This centuries-old insurance of planting large numbers of varieties ensures that, when one desired cultivar does not produce well one year, chances are another will. Unfortunately, this insurance in planting large numbers of cultivars is breaking down in some Andean communities. A recent report from the communities of Santa Cruz de Pichiu in Peru notes:

> In the last 20 years the number of native varieties have dramatically declined from 60 to 80 varieties to 6 or 8 varieties of potatoes. (translation by D. Ellis; p. 2. Aguilar, A.M. 2016)

Published by Burleigh Dodds Science Publishing Limited, 2018.

In an effort to help restore the natural balance in sustenance potato farming communities in the Peruvian Andes, the CIP genebank has been involved in a programme since the late 1998 focusing on the repatriation of native potato landraces back to the communities whose forefathers preserved them. Starting with returning native potato landraces collected and conserved by CIP over the past 45 years to a few select communities, the CIP genebank repatriation programme has now benefitted over 90 indigenous communities in Peru. The repatriation programme returns disease-free seeds of landrace potatoes to communities which were collected near or were known to grow in the localities in which these communities reside. To date, CIP has given back (repatriated) to native Peruvian farmers over one-third of its potato landrace collection.

Programmes such as the repatriation project provide a small effort towards the total need for support of the conservation of habitats, wild relatives and communities whose relatives were critical for the domestication of our crop and whose livelihoods still depend on farming landraces by traditional means. These very communities and environments are key to sustaining the diversity we need for continued improvements to maintain potato yields. A changing climate is having a major impact in the Andes and thus the diversity of potato and its wild relatives. The *ex situ* conservation of this genetic diversity in genebanks is the only guarantee that future generations will have the same opportunity we have today to use this diversity for sustained potato productivity.

8 Where to look for further information

In addition to the references cited below, the authors invite readers to consult CIP and USPG websites for the most current information on genebank issues.

9 References

Aguilar, A. M. (2016). Informe Final de siembra de papas nativas repatriadas en la Comunidad Campesina de Santa Cruz de Pichiu, Chana, Huari, Ancash. Report by Comunidad Campesina de Santa Cruz de Pichiu. 51 p.

Akeley, R. V., Mills, W. R., Cunningham, C. E. and Watts, J. (1968). Lenape: A new potato variety high in solids and chipping quality. *American Potato Journal* 45:142–5.

Bamberg, J. B. and del Rio, A. H. (2003). Vulnerability of alleles in the US Potato Genebank Extrapolated from RAPDs. *American Journal of Potato Research* 80:79–85.

Bamberg, J. B. and del Rio, A. H. (2004). Genetic heterogeneity estimated by RAPD polymorphism of four tuber-bearing potato species differing by breeding structure. *American Journal of Potato Research* 81:377–83.

Bamberg, J. B. and del Rio, A. H. (2005). Conservation of Genetic Resources. In: Maharaj, K. Razdan and Autar, K. Mattoo (Eds), *Genetic Improvement of Solanaceous Crops* Vol.1: Potato. Science Publishers, Inc., Enfield, USA, 451p.

Bamberg, J. B. and del Rio, A. H. (2006). Seedling transplant selection does not cause genetic shifts in potato genebank populations. *Crop Science* 46:424–7.

Bamberg, J. B. and del Rio, A. H. (2009). Unbalanced bulk of parents' seed does not cause significant drift in germplasm regeneration of two model potato (*Solanum*) species populations. *American Journal of Potato Research* 86:391–7.

Bamberg, J. B. and del Rio, A. H. (2014). Selection and validation of an AFLP marker core collection for the wild potato *Solanum microdontum*. *American Journal of Potato Research* 91:368–75.

Bamberg, J. B. and del Rio, A. H. (2016). Accumulation of genetic diversity in the US Potato Genebank. *American Journal of Potato Research* 93:430–5.

Bamberg, J. B., del Rio, A. H. and Navarre, R. (2016b). Intuitive visual impressions (cogs) for identifying clusters of diversity within potato species. *American Journal of Potato Research* 93:350–9.

Bamberg, J. B., del Rio, A. H. and Penafiel, J. (2011). Predicting genetic richness at wild potato collection sites in Southeastern Arizona. *American Journal of Potato Research* 88:398–402.

Bamberg, J. B., del Rio, A. H., Coombs, J. and Douches, D. (2015). Assessing SNPs versus RAPDs for predicting heterogeneity in wild potato species. *American Journal of Potato Research* 92:276–83.

Bamberg, J. B., del Rio, A. H., Fernandez, C., Salas, A., Vega, S., Zorrilla, C., Roca, W. and Tay, D. (2010). Comparison of 'Remote' versus 'Easy' *in situ* collection locations for USA wild *Solanum* (potato) germplasm. *American Journal of Potato Research* 87:277–84.

Bamberg, J. B., del Rio, A. H., Kinder, D., Louderback, L., Pavlik, B. and Fernandez, C. (2016a). Core collections of Potato (*Solanum*) Species Native to the USA. *American Journal of Potato Research*. In Press.

Bamberg, J. B., Kiru, S. and del Rio, A. H. (2001). Comparison of reputed duplicate populations in the Russian and US Potato Genebanks using RAPD markers. *American Journal of Potato Research* 78:365–9.

Bamberg, J. B., Martin, M. W., Abad, J., Jenderek, M. M., Tanner, J., Donnelly, D. J., Nassar, A. M. K., Veilleux, R. E. and Novy, R. G. (2016c). *In vitro* technology at the US Potato Genebank. *In Vitro Cellular and Developmental Biology – Plants* 52:213–25.

Black, W. (1949). Breeding for disease-resistance in potatoes. *Empire Journal of Experimental Agriculture* 17:116–24.

Bradeen, J. and Haynes, K. (2011). Introduction to potato. In: Bradeen, J. and Kole, C. (Eds), *Genetics, Genomics and Breeding of Potato*. Science Publishers, P. O. Box 699, Enfield, NH, USA.

Bradshaw, J. E., Bryan, G. J. and Ramsay, G. (2006). Genetic resources (including wild and cultivated Solanum species) and progress in their utilization in potato breeding. *Potato Research* 49:49–65.

Brasier, C. M. (2008). The Biosecurity threat the UK and global environment from international trade in plants. *Plant Pathology* 57:792–808.

Breese, E. L. (1989). Regeneration and multiplication of germplasm in seed genebanks: The scientific background. International Board for Plant Genetic Resources, Rome.

Brown, G. M., Jr. (1990). Valuation of Genetic Resources. In: Orians, G. H. et al. (Eds), *The Preservation and Valuation of Biological Resources*. University of Washington Press, Washington, pp. 203–28.

Butler, N. and Douches, D. (2016). Sequence-specific nucleases for genetic improvement of potato. *American Journal of Potato Research* 93:303–20.

Cadima-Fuentes, X., van Treuren, R., Hoekstra, R., van den Berg, R. G. and Sosef, M. S. M. (2016). Genetic diversity of Bolivian wild potato germplasm: Changes during *ex situ* conservation management and comparisons with resampled *in situ* populations. *Genetic Resources and Crop Evolution*, DOI 10.1007/s10722–015-0357-9.

Carputo, D. and Barone, A. (2005). Ploidy level manipulations in potato through sexual hybridization. *Annals of Applied Biology* 146:71–9.

Chandra, A. and Idrisova, A. (2011). Convention on biological diversity: A review of national challenges and opportunities for implementation. *Biodiversity and Conservation* 20:3295–316.

Charlesworth, D. and Charlesworth, B. (1995). Quantitative genetics in plants: The effect of the breeding system on genetic variability. *Evolution* 49:911–20.

Chase, S. S. (1963a). Analytic breeding of amphipolyploid plant varieties. *Crop Science* 4:334–7.

Chase, S. S. (1963b). Analytic breeding in Solanum tuberosum L. – A scheme utilizing parthenotes and other diploid stocks. *Canadian Journal of Genetics and Cytology* 5:389–95.

Crossa, J. and R. Vencovsky (1999). Sample size and variance effective population size for genetic resources conservation. *Plant Genetic Resources. Newsl.* 119:15–25.

Crossa, J., Hernandez, C. M., Bretting, P., Eberhart, S. A. and Taba, S. (1993) Statistical genetic considerations for maintaining germplasm collections. *Theoretical and Applied Genetics* 86:673–8.

Published by Burleigh Dodds Science Publishing Limited, 2018.

DeJong, H. (2016). Impact of potato on society. *American Journal of Potato Research* 93:415–29.

del Rio, A. H., Bamberg, J. B. and Huaman, Z. (2006). Genetic equivalence of putative duplicate germplasm collections held at CIP and US Potato Genebanks. *American Journal of Potato Research* 83:279–85.

del Rio, A. H., Bamberg, J. B., Huaman, Z., Salas, A. and Vega, S. E. (1997b). Assessing changes in the genetic diversity of potato gene banks. 2. *In situ vs. ex situ Theoretical and Applied Genetics* 95:199–204.

del Rio, A. H. and Bamberg, J. B. (2000). RAPD markers efficiently distinguish heterogeneous populations of wild potato (*Solanum*). *Genetic Resources and Crop Evolution* 47:115–21.

del Rio, A. H. and Bamberg, J. B. (2002). Lack of association between genetic and geographic origin characteristics for the wild potato *Solanum sucrense* Hawkes. *American Journal of Potato Research* 79:335–8.

del Rio, A. H. and Bamberg, J. B. (2004). Geographical parameters and proximity to related species predict genetic variation in the inbred potato species *Solanum verrucosum* Schlechtd *Crop Science* 44:1170–7.

del Rio, A. H., Bamberg, J. B. and Huaman, Z. (1997a). Assessing changes in the genetic diversity of potato gene banks. 1. Effects of seed increase. *Theoretical and Applied Genetics* 95:191–8.

del Rio, A. H., Bamberg, J. B., Centeno-Diaz, R., Salas, A., Roca, W. and Tay, D. (2012a). Effects of the pesticide *Furadan* on traits associated with reproduction in wild potato species. *American Journal of Plant Sciences* 3:1608–12.

del Rio, A. H., Bamberg, J. B., Centeno-Diaz, R., Soto, J., Salas, A., Roca, W. and Tay, D. (2012b). Pesticide contamination does not affect the genetic diversity of potato species. *American Journal of Potato Research* 89:384–91.

del Rio, A. H., Bamberg, J. B., Huaman, Z., Salas, A. and Vega, S. E. (2001). Association of eco-geographical variables and RAPD marker variation in wild potato populations of the USA. *Crop Science* 41(3):870–8.

Delbaere, B. (2005). European policy review: biodiversity research to support European policy. *Journal of Nature Conservation* 13(2):213–14.

FAO (2010). The Second Report on the State of the World's Plant Genetic Resources. Rome, Italy: FAO; 370 p.

Foll, M. and Gaggiotti, O. (2008). A genome scan method to identify selected loci appropriate for both dominant and codominant markers: A Bayesian perspective. *Genetics* 180:977–93.

Franco, J., Crossa, J., Villaseñor, J., Taba, S. and Eberhart, S. A. (1997). Classifying Mexican maize accessions using hierarchical and density search methods. *Crop Science* 37:972–80.

Franco, J., Crossa, J., Villaseñor, J., Taba, S. and Eberhart, S. A. (2005). A sampling strategy for conserving genetic diversity when forming core subsets. *Crop Science* 45:1035–44.

Franco, J., Crossa, J., Warburton, M. L., Taba, S. and Eberhart, S. A. (2006). Sampling strategies for conserving maize diversity when forming core subsets using genetic markers. *Crop Science* 46:854–64.

Frankel, O. and Soule, M. (1981). *Conservation and Evolution*. Cambridge University Press, Cambridge, UK.

Frankel, O. H. (1984). Genetic perspectives of germplasm conservation. In: Arber, W. K. et al. (eds), *Genetic Manipulation: Impact on Man and Society*. Cambridge University Press, Cambridge, UK, pp. 161–70.

Frankham, R., Ballou, J. D. and Briscoe, D. A. (2002). *Introduction to Conservation Genetics*. Cambridge University Press, Cambridge, UK.

Glendinning, D. R. (1983). Potato introductions and breeding up to the early 20th century. *New Phytologist* 94:479–505.

Goodrich, C. E. (1863). The origination and test culture of seedling potatoes. *Transactions of the New York State Agricultural Society* 23:89–102.

Hale, A. L., Reddivari, L., Nzaramba, M. N., Bamberg, J. B. and Miller, J. C., Jr. (2008). Interspecific variability for antioxidant activity and phenolic content among *Solanum* species. *American Journal of Potato Research* 85:332–41.

Hanneman, R. E., Jr. (1999). The reproductive biology of the potato and its implications for breeding. *Potato Research* 42:283–312.

Hardigan, M., Bamberg, J., Buell, C. Robin and Douches, D. (2015). Taxonomy and genetic differentiation among wild and cultivated germplasm of Solanum sect. Petota. *The Plant Genome* 8:1–16.

Hereford, J. (2010). Does selfing or outcrossing promote local adaptation? *American Journal of Botany* 97(2):298–302.

Hermsen, J. G. T. (1994). Introgression of genes from wild species, including molecular and cellular approaches. In Bradshaw, J. E. and Mackay, G. R. (Eds), *Potato Genetics*. CAB International, Wallingford, Oxon, UK, pp. 515–38.

Hijmans, R. J., Garrett, K. A., Huaman, Z., Zhang, D. P., Schreuder, M. and Bonierbale, M. (2000). Assessing the geographic representativeness of genebank collections: the case of Bolivian wild potatoes. *Conservation Biology* 14:1755–65.

Hijmans, R. J., Jacobs, M., Bamberg, J. B. and Spooner, D. M. (2003). Frost tolerance in wild potato species: Assessing the predictivity of taxonomic, geographic, and ecological factors. *Euphytica* 130:47–59.

Hijmans, R. J., Spooner, D. M., Salas, A. R., Guarino, L. and de la Cruz, J. (2002). *Atlas of Wild Potatoes*. International Plant Genetic Resources Institute, Rome, Italy.

Jacobs, M. M. J., Smulders, M. J. M., van den Berg, R. G. and Vosman, B. (2011). What's in a name; Genetic structure in *Solanum* section Petota studied using population-genetic tools. *BMC Evolutionary Biology* 11:42.

Jansky, S., Dawson, J. and Spooner, D. (2015). How do we address the disconnect between genetic and morphological diversity in germplasm collections? *American Journal of Botany* 102:1–3.

Jansky, S., Yerk, G. and Peloquin, S. (1990). The use of potato haploids to put 2x wild species germplasm into a usable form. *Plant Breeding* 104:290–4.

Jansky, S. H. and Bamberg, J. B. (2014). Potato Crop Germplasm Committee Vulnerability Report (http://www.ars-grin.gov/npgs/cgc_reports/potatovuln2014.pdf).

Jansky, S. H., Charkowski, A. O., Douches, D. S., Gusmini, G., Richael, C., Bethke, P. C., Spooner, D. M., Novy, R. G., De Jong, H., De Jong, W. S., Bamberg, J. B., Thompson, A. L., Bizimungu, B., Holm, D. G., Brown, C. R., Haynes, K. G., Sathuvalli, V. R. et al. (2016). Reinventing potato as a diploid inbred line-based crop. *Crop Science* 56:1–11.

Johnston, G. and Rowberry, R. (1981). Yukon Gold: A new yellow-fleshed, medium-early, high quality table and french-fry cultivar. *American Potato Journal* 58:241–4.

Khiutti, A., Spooner, D. M., Jansky, S. H. and Halterman, D. A. (2015). Testing taxonomic predictivity of foliar and tuber resistance to *Phytophthora infestans* in wild relatives of potato. *Phytopathology* 105:1198–12015.

Kreuze, J. F., Perez, A., Untiveros, M., Quispe, D., Fuentes, S., Barker, I. and Simon, R. (2009). Complete viral genome sequence and discovery of novel viruses by deep sequencing of small RNAs: A generic method for diagnosis, discovery and sequencing of viruses. *Virology* 388: 1–7.

Lauer, F. I. (1959). Recovery of recurrent parent characters from crosses of Solanum demissum x S. tuberosum in successive backcrosses. *American Potato Journal* 36:345–57.

Lenné, J. M. and Wood, D. (1999). Optimizing biodiversity for production agriculture. In: Wood, D. and Lenné, J. M. (Eds), *Agrobiodiversity: Characterization, Utilization, and Management*. CAB International, Wallingford, UK, pp. 447–70.

Liljander, A., Yu, M., O'Brien, E., Heller, M., Nepper, J. F., Weibel, D. B., Gluecks, I., Younan, M., Frey, J., Falquet, L. and Jores, J. (2015). Field-applicable recombinase polymerase amplification assay for rapid detection of *Mycoplasma capricolum* subsp. *capripneumoniae*. *Journal of Clinical Microbiology* 53:2810–15.

Love, S. L., Pavek, J. J., Thompson-Johns, A. and Bohl, W. (1998). Breeding progress for potato chip quality in North American cultivars. *American Journal of Potato Research* 75:27–36.

Mendiburu, A. and Peloquin, S. (1977). The significance of 2n gametes in potato breeding. *Theoretical and Applied Genetics* 49:53–61.

Miller, Nicole. Potato, Interrupted. On Wisconsin magazine 117(4) p. 42.

Published by Burleigh Dodds Science Publishing Limited, 2018.

Myers, N. (1983). *A Wealth of Wild Species: Storehouse for Human Welfare*. Westview Press, Boulder, CL.

Myers, N. (1990). The biodiversity challenge: Expanded hot spots analysis. *The Environmentalist* 10:243–56.

Myers, N., Mittermeier, R. A., Mittermeier, C. G., da Fonseca, G. A. B. and Kent, J. (2000). Biodiversity hotspots for conservation priorities. *Nature* 403:853–8.

Nassar, A. M. K., Kubow, S., Leclerc, Y. N. and Donnelly, D. (2014). Somatic mining for phytonutrient improvement of 'Russet Burbank' potato. *American Journal of Potato Research* 91:89–100.

Ortiz, R., Simon, P., Jansky, S. and Stelly, D. (2009). Ploidy manipulation of the gametophyte, endosperm and sporophyte in nature and for crop improvement: a tribute to Professor Stanley J. Peloquin (1921–2008). *Annals of Botany* 104:795–807.

Peeters, J. P. and Martinelli, J. A. (1989). Hierarchical cluster analysis as a tool to manage variation in germplasm collections. *Theoretical and Applied Genetics* 78:42–8.

Peloquin, S. J., Jansky, S. H. and Yerk, G. L. (1989). Potato cytogenetics and germplasm utilization. *American Journal of Potato Research* 66:629–38.

Perry, M. C. and McIntosh, M. S. (1991). Geographical patterns of variation in USDA soybean germplasm collection: I. Morphological traits. *Crop Science* 31:1350–5.

Plaisted, R. and Hoopes, R. (1989). The past record and future prospects for the use of exotic potato germplasm. *American Journal of Potato Research* 66:603–27.

Rao, N. K., Hanson, J., Dulloo, M. E., Ghosh, K., Nowell, D. and Larinde, M. (2006). *Manual of Seed Handling in Genebanks*. Handbooks for Genebanks 8. Rome, Italy: Bioversity International.

Ross, H. (1986). *Potato Breeding – Problems and Perspectives*. Verlag Paul Parey, Berlin and Hamburg.

Rudorf, W. (1958). The significance of wild species for potato breeding. *Potato Research* 1:10–20.

Schoen, D. J., David, J. L. and Bataillon, T. M. (1998). Deleterious mutation accumulation and regeneration of genetic resources. *Proceedings of the National Academy of Sciences USA* 95:349–99.

Spooner, D. M. (2016). Species delimitations in plants: lessons learned from potato taxonomy by a practicing taxonomist. *Journal of Systematics and Evolution* 54:191–203.

Spooner, D. M. and van den Berg, R. G. (1992). An analysis of recent taxonomic concepts in wild potatoes (Solanum sect. Petota). *Genetic Resources and Crop Evolution* 39:23–37.

Spooner, D. M., Alvarez, N., Peralta, I.E., and Clausen, A.M. (2016). Taxonomy of wild potatoes and their relatives in southern South America (Solanum sects. Petota and Etuberosum). *Syst. Bot. Monogr.* 100:1–240.

Spooner, D. M., Ghislain, M, Simon, R., Jansky, S. H. and Gavrilenko, T. (2014). Systematics, diversity, genetics, and evolution of wild and cultivated potatoes. *Botanical Review* 80:283–383.

Stevenson, F. J. and Clark, C. F. (1937). Breeding and genetics in potato improvement. Yearbook of Agriculture, USDA pp. 405–44.

Thébault, E. and Loreau, M. (2003). Food-web constraints on biodiversity-ecosystem functioning relationships. *Proceedings of the National Academy of Sciences of the USA* 100:14949–54.

Thieme, R., Rakosy-Tican, E, Gavrilenko, T., Antonova, O., Schubert, J., Nachtigall, M., Heimbach, U. and Thieme, T. (2008). Novel somatic hybrids (*Solanum tuberosum* L. + *Solanum tarnii*) and their fertile BC(1) progenies express extreme resistance to potato virus Y and late blight. *Theoretical and Applied Genetics* 116:691–700.

van Hintum, T. J. and van Treuren, R. (2002). Molecular markers: tools to improve genebank efficiency. *Cellular & Molecular Biology Letters* 7:737–44.

Waycott, W. and Fort, S. B. (1994). Differentiation of nearly identical germplasm accessions by a combination of molecular and morphological analyses. *Genome* 37:577–83.

Zitnak, A. and Johnston, G. (1970). Glycoalkaloid content of B5141-6 potatoes. *American Potato Journal* 47:256–60.

Chapter 4

Advances in conventional potato-breeding techniques

Jai Gopal, ICAR-Central Potato Research Institute, India

1 Introduction

2 Parental line selection

3 Progeny selection

4 Improving the speed and success rate of selection

5 Summary

6 References

1 Introduction

Plant breeders have the endless task of developing new crop varieties with increased yield and quality, possessing durable resistance to fast-evolving pathogens and pests, and adapted to changing environmental conditions. Potato breeding can be considered as old as its domestication. Archaeological evidence indicates cultivation of potato in the Andean highlands (the centre of origin of potato) as early as 200 A.D. (Hougas 1956). No doubt, many Andean cultivars were produced by farmer selection from naturally occurring variation. Potato breeding in the modern sense began in 1807 in England when Thomas Andrew Knight made deliberate hybridizations between varieties by artificial pollination (Knight 1807). However, potato (*Solanum tuberosum*) proved to be a difficult species for genetical research because of its complex inheritance patterns, and it was not until the end of the 1930s that geneticists recognized that it was a tetraploid (2n = 4x = 48), which displays tetrasomic inheritance. Furthermore, most traits of economic importance displayed continuous variation for which Mendelian analysis was not possible because distinct classes could not be discerned. As a consequence, potato breeding has remained empirical and genetically unsophisticated until now. There has been a lack of major genetic improvement in potato as unfavourable alleles easily remain hidden in the tetraploid genome and manifest at each breeding cycle (Lindhout et al. 2011). Century-old potato varieties such as Russet Burbank and Bintje are still cultivated, although more than 4000 potato varieties have been released to date.

Since potato is highly heterozygous, in order to maintain its productivity, improved potato varieties are developed by inter-mating desired parental lines and selecting superior clones from the progeny. The success of any potato-breeding programme thus depends on the selection of superior parents and crosses, and selection of superior clones from

http://dx.doi.org/10.19103/AS.2017.0016.03

the progeny. Potato being vegetatively propagated, any selected genotype can be fixed with all its intra- and inter-locus interactions responsible for phenotypic expression, and multiplied for commercial cultivation if desired. Although recent advances in molecular breeding provide opportunities for rapid genetic gain (Slater et al. 2014a), phenotypic selection remains the common practice in conventional potato-breeding programmes. Nearly all new varieties of potato still emerge from the process free from use of molecular technologies. This chapter reviews the progress and advances made in phenotypic selection techniques of conventional potato breeding. The role of molecular approaches in improving phenotypic selection is also described briefly.

2 Parental line selection

2.1 Combining ability

Combining ability of parents is studied to identify the parental lines that transmit superior performance for a characteristic to their offspring when crossed with a wide variety of other clones. Such parental lines are said to have good general combining ability (GCA) for the concerned characteristic. The deviation from what is expected on the basis of the GCA of the parents is called specific combining ability (SCA) of the cross. For many economically important traits it has been possible to partition genetical variation into components due to GCA and SCA.

Several methods are available for combining ability estimation, which are chosen as per the genetic structure and reproductive behaviour of the species concerned. In potato, Killick (1977) used half–diallel crossing designs, while North Carolina II design was employed by Plaisted et al. (1962) and Killick and Malcomson (1973). Gopal (1998a,b) used line × tester analysis on a pattern similar to North Carolina II designs for combining ability studies. In these test cross methods a large set of crosses are required to be made in which selected females are crossed with a random sample of testers in a specific pattern based on the design used. In order to simplify GCA estimation in potato, Gopal et al. (2008b) studied the effectiveness of top cross vs. poly-cross as an alternative to the test cross method. Matings based on selected individual single testers (top cross) resulted in GCA estimates similar to those based on a number of testers. The top cross method thus could substantially reduce the number of crosses required for estimating the GCA.

Several studies of combining abilities are available for a number of characteristics in potato (reviewed by Gopal 2015). Combining ability is an important feature of the expression of certain disease resistances also (Tung et al. 1990; Gopal 1998b; Mihovilovich et al. 2007; Gopal et al. 2008b). However, contradictory results have been obtained with regard to the relative importance of GCA and SCA for various characteristics. GCA has only moderate repeatability over generations for various characteristics (Gopal 1998a). Further, the relative importance of GCA or SCA in a population depends on the kind of material, experimental design and/or environmental conditions. Thus, the information on combining ability needs to be sought afresh for identifying promising parents.

2.2 Progeny test

Progeny test is an effective tool to evaluate the breeding value of parents and specific cross combinations (Bradshaw and Mackay 1994; Bradsaw et al. 1995; Gopal 1997), particularly for traits that display continuous variation and for which, in general, non-additive inheritance is more important. De Jong and Tai (1991) reported that in 4x–2x crosses due to a relatively low degree of determination (R2) of both 4x and 2x parents on the performance of their 4x–2x hybrid progenies, test crossing and subsequent progeny analysis was necessary to identify parents that can be used in a breeding programme.

Moderate to strong correlation coefficients between progeny means in different generations of potato breeding were reported by many workers (Brown and Caligari 1986, 1989; Maris 1969, 1988). However, Caligari et al. (1986) reported weak correlations. Gopal (1997) reported that moderately effective selection based on progeny means could be conducted for plant vigour; tuber yield; tuber number; average tuber weight; breeders' preference score; tuber colour; tuber shape; and uniformity in tuber colour, tuber shape and tuber size. Correlation coefficients between progeny means in different generations for tuber dry matter were reported to be strong (Maris 1962, 1969; Neele and Louwes 1989). Hayes and Thill (2002b) reported that progeny test is effective in family selection for chip colour. Cross prediction based on progeny mean has also been practiced for tuber blight and foliage blight (Caligari et al. 1983) as well as potato cyst nematodes, *Globodera pallida* (Phillips 1981).

Brown et al. (1988) and Brown and Caligari (1989) reported that progeny means for breeders' preference score in the seedlings stage grown in the glasshouse correlated well with progeny mean preference scores made in the first and second clonal generations at both seed and ware sites. Such a univariate cross prediction method had been tested by Caligari and Brown (1986) also. They reported that actual numbers of clones falling in desired category was also a good cross predictor as a reasonably good agreement between predicted and observed numbers was seen over sites and years for breeders' preference score. Based on similar studies, Brown et al. (1987a,b) and Bradshaw et al. (1998) reported that seedling progeny evaluation by breeders' preference score could be used to reject entire crosses on the ground that they were less likely than others to contain clones of commercial worth. Brown and Caligari (1989) found this method to be effective for total tuber weight (yield) also, but not for mean tuber weight and number of tubers. Total tuber weight has been reported to be an important component of breeders' preference score in early generations (Brown and Caligari 1986; Neele et al. 1991; Bradshaw et al. 1998). Tai (1975) reported that out of five characteristics recorded, tuber appearance and average weight of marketable tubers made an important contribution to the mean visual scores (breeders' preference score).

Three methods of cross prediction namely multivariate probabilities, the sum of ranks and the frequency of genotypes in a sample that transgress set target values were tested by Brown and Caligari (1988). The characteristics tuber weight, mean tuber weight, number of tubers and regularity of tuber shape were examined. The best estimates were obtained using multivariate probabilities based on the means, within-progeny variation and the phenotypic correlation between variates. It was found that a sample as small as 25 clones provided good predictions as the rankings of the crosses according to these multivariate probabilities provided good indications of the number of clones which survived selection in an actual breeding scheme. Brown et al. (1988) found that the progeny mean was a slightly better predictor of superior crosses over generations than the mean and within-progeny phenotypic variance. Caligari and Brown (1986) and Gopal (2001b) had also found that within-progeny variation was not a major component in prediction. Gopal (2001b) reported that family selection based on progeny mean is almost as effective as the one based on both progeny mean and within-family standard deviation.

Progeny tests based on observed and expected proportions of clones that transgress a given target value have also been tried. Brown et al. (1988) reported that predictions based on the expected proportions of clones that would transgress a given target value was better than the prediction based on the observed frequency of desirable clones in a progeny sample. However, Gopal (1997) did not find much difference in the effectiveness of these two types of proportions with regard to effectiveness of prediction. From a practical standpoint, breeders would be mainly interested in knowing whether the proportion of clones expected to transgress a particular target value in later generations can be predicted from the observed or predicted proportions transgressing the target value in an earlier generation. Further, all these studies (Caligari and Brown 1986; Brown et al. 1988; Gopal 1997) had reported that progeny means were better than the proportions for cross prediction.

The effectiveness of progeny tests in early generations to identify parents that produce broadly adapted clones was studied by Haynes et al. (2012). For this, seeds of all second field generation clones were distributed to five locations for selection. Based on common selections made at all the locations, they concluded that early generation selection can be used to identify parents that produce broadly adapted progeny. Because potato displays such strong G × E interactions, the selection based on multi-location progeny tests in early generations would be particularly beneficial for selecting potato cultivars with broad adaptation.

2.3 Estimated breeding values

Best linear unbiased prediction (BLUP) is an advanced biometric technique that permits the analysis of a large data set derived from pedigree for the calculations of estimated breeding values (EBV) of parents. Slater et al. (2014b) used BLUP in potato breeding after modifying it to suit autoteraploid potato as suggested by Kerr et al. (2012). Since BLUP considers information from all relatives in the analysis, it improves the accuracy of analysis, particularly for low heritability traits. Slater et al. (2014b) showed that BLUP analysis based on a large pedigree provided EBVs for the families, and all cultivars in the pedigree and the individual genotypes themselves, potentially enabling breeding from the best progeny. EBVs also improved cross-generational predictions, and superior genotypes could be identified for retention that would otherwise have been lost by the rejection of the entire low scoring families thereby demonstrating the advantage of using EBVs over progeny means. Paget et al. (2015b) used eight years of early stage trial data on potato tuber yield to identify parents with high empirical breeding values. They found that simple variance models may be preferable for early stage genetic evaluation of potato yield.

2.4 Mid-parent and mid-self values

Mid-parent values can be used as a preliminary guide at the beginning of a breeding programme to predict the mean performance of crosses (Bradshaw et al. 2003). However, this would be effective only for those traits which are predominantly controlled by additive gene action. As early as 1929, Krantz and Hutchin reported that average maturity of F1 potato progenies was closely associated with the mid-parent maturity value, and later Killick (1977) found that foliage maturity is mainly governed by additive gene action. But most of the quantitative characteristics in potato are non-additive in nature, and in such cases mid-parent values can be used just to complement the selection based on progeny means which are required to be estimated later. For tuber yield, correlation coefficients between progeny means and mid-parent values, or parental values, were found to be moderate or low (Brown and Caligari 1989; McHale and Lauer 1981; Maris 1989; Schroeder and Peloquin 1983; Veilleux and Lauer 1981b). For tuber dry matter, however, mid-parent values were moderately to strongly associated with progeny means (Maris 1989; Neele and Lowes 1989). Gopal (1998c) reported that there were, in general, no significant correlation coefficients between mid-parent values and progeny means for most of the characteristics. For fry colour, however, Bradshaw et al. (2003) reported that when an outlier was excluded from the data, a moderate correlation of 0.71 was found between the mid-parent values and offspring means, even when mid-parent values and offspring means were assessed in different years. A higher correlation of 0.86 was found in a diallel set of crosses in which parents and progenies were assessed in the same trial (Bradshaw et al. 2000). Neele (1990) reported that time of tuber initiation of the progenies was associated with the time of tuber initiation of the earliest tuberizing parent. The mean increase rate of the harvest index of the progenies was equal to the mean values of parents. He thus suggested that differences in harvest index between progenies can be predicted with a high degree of precision by using the harvest index parameters of the parents. This estimate of harvest index was further found to have a fairly good correlation to tuber yield.

Mid-self values have also been tested for cross prediction. Brown and Caligari (1989) reported that mid-self values provided better prediction than mid-parent values for mean tuber weight and number of tubers. Gopal (1998c) too reported that in contrast to mid-parent values, mid-self values were, in general, better for cross prediction. Better predictive value of mid-self value than mid-parent value may be because mid-self values reflect the potential of the parents based on the performance of new genotypic combinations produced after selfing, whereas parents' per se performance/mid-parent values are based on the phenotypic performance of the parents, which may or may not be reflected in the progeny. Compared to test crosses, this method has the advantage that it requires only one offspring population per evaluated parent, but it can only be applied to male fertile parental clones. As compared to mid-parent value, evaluation for mid-self values, however, is more time- and labour-consuming than simply evaluating the parents. Neele et al. (1991), however, reported that mid-self value did not improve the prediction of progeny means compared to the mid-parent value. They further reported that predictions based on the test crosses were the best and surpassed those of the mid-parent value.

3 Progeny selection

3.1 Genetic divergence

Crosses among genetically diverse parents are reported to result in heterotic progenies. Such parents–offspring relationship for tuber yield in potato has been confirmed by many authors, including Rowe (1967), Mendoza and Haynes (1974b), Mok and Peloquin (1975), Cubillos and Plaisted (1976), De Jong et al. (1981), Veilleux and Lauer (1981a), Landeo and Hanneman (1982), Schmiediche et al. (1982), Estrada-Ramos (1984), Gopal and Minocha (1997a) and Luthra et al. (2001, 2005). Negative heterotic effects (inbreeding depression) were observed for yield and other characteristics in progenies of Gp. Tuberosum crosses (Maris 1969, 1989; Tarn and Tai 1983; Tai 1994). New varieties did not result in a major impact on yield improvement of potatoes (Van der Plank 1946;

Seiffert 1957; Howard 1963a,b; Gopal 2006) in most of the countries. This is attributed to the relatively narrow genetic base of this subspecies (Mendoza and Haynes 1974a); most tuberosum varieties, then, are more or less closely related to each other (Gopal and Oyama 2005) being descended from a small number of introductions of South American potatoes (Simmonds 1962). Gp. Andigena also has narrow genetic base (Maris 1989). Thus, it has been suggested that high inbreeding coefficients in the progenies should be avoided. However, Loiselle et al. (1989) reported that in a multi-trait type of breeding programme, inbreeding coefficient was of little use in predicting a priori the selection pattern of a cross. They attributed this to several factors. One is the large influence of non-heritable variation of potato clones in visual selection of early breeding generations (Tai and Young 1984; Gopal et al. 1992). Another is the changing emphasis on different traits as the selection process progresses, which results in no significant correlation between selections at succeeding stages (Tai and Young 1984; Gopal et al. 1992). Selection primarily retains genotypes of a 'commercial' type. The emphasis is placed on good tuber size and a high marketable yield, which are negatively associated with inbreeding coefficient. Thus, heterosis breeding is essential for individual traits like tuber yield, although inbreeding coefficient may not be of direct use in selection of parents for a multi-trait-like breeding programme. Hayes and Thill (2002a) suggested inter-mating large numbers of unrelated but adapted parents for improving breeding efficiency for cold chipping in potato.

Evaluation of genetic diversity among parents based on morphological characteristics is not only time consuming but also cumbersome. Paz and Veilleux (1997) and Chimote et al. (2008) reported that RAPD markers may facilitate the identification of diverse parents to maximize the expression of heterosis. Gopal and Minocha (1997a) reported the importance of G × E interaction on the use of apparent genetic diversity determined on phenotypic basis of parents for predicting the progeny's performance (Gopal 2006). It was concluded that genetic divergence, though of use in identifying parents for exploiting heterosis in progeny, has only moderate effectiveness and that too when G × E interaction is taken care of.

3.2 Diploid progenies and use of 4x × 2x test crosses

Neele et al. (1991) used diploid progeny of a tetraploid clone for predicting the breeding value of tetraploid clone, but the results were not encouraging. A possible reason for the poor values of this predictor was that in the diploid progenies many poor growing plants occurred that seemed to be unrelated to the breeding value of the tetraploid parent. Neele et al. (1991) hypothesized that during the production of diploids after pollination of tetraploid with S. phureja recessive deleterious alleles might had been present in a homozygous state. In a tetraploid most such recessive alleles would go unnoticed. Use of diploid progenies is otherwise also not practicable as these are difficult to produce and maintain over subsequent clonal generations.

Neele et al. (1991) also tried 4x × 2x test crosses as predictors of progeny performance. These provided good estimates for the actual values of the progenies. For many characteristics the correlation coefficients exceeded 0.8. These were almost of the same value as mid-parent value in their study. The high predictive value of these parameters in this study was due to the preponderance of the additive gene action in this population. However, a drawback of 4x × 2x crosses is the investment in the production of true seeds through 2n-FDR pollen producing males.

3.3 Individual clone selection

Selection of promising clones from the segregating progeny is a crucial task as a large number of seedlings are grown in most potato-breeding programmes and these must be quickly reduced to a manageable number before statistically accurate and consistent tests can be performed. It is difficult to achieve this as many studies (Maris 1962, 1966, 1988; Swiezynski 1968; Kameke 1975, 1978; Tai 1975; Anderson 1977; Brown et al. 1984, 1987b; Brown and Caligari 1986; Caligari et al. 1986; Neele et al. 1988, 1989; Gopal et al. 1992; Bae et al. 2008) have shown early generation selection to be ineffective in identifying superior genotypes for subjective characteristics such as tuber yield and its components. The size of tuber produced in seedling generation affects the performance of a clone for these characteristics in first and later clonal generations non-genetically, thereby adversely influencing efficiency of selection (Tai 1975; Brown et al. 1984, 1987a; Brown 1988; Gopal et al. 1992). Comparing clones selected or discarded in one year with clones selected or discarded in previous years indicated poor effectiveness of selection. Neele et al. (1988) reported that use of two-plant plots instead of one-plant ones in first clonal generation helped reduce the environmental variance, but it had only slightly positive effect on the selection efficiency. These studies thus did not suggest stringent selection in early generations.

Different strategies have been suggested to improve the efficiency of selection in early generations. Growing the seedlings in field or in large pots has been suggested to improve the efficiency of selection for tuber characteristics as this results in more and bigger tubers (Zadinna 1971; Swiezynski 1978; Gopal et al. 1992). Love et al. (1997) suggested separation of breeder's preference, which has large G × E interaction, into its component traits – some of which may have a small G × E variance and thus better selection efficiency. This strategy was also earlier tried by Gopal et al. (1992), and was found that positive selection for tuber colour, tuber shape, skin smoothness and non-cracking can be done in seedling generation whereas no rejection should be done in this generation for tuber yield and its components. They further reported that seedlings of poor vigour may be rejected prior to transplanting in the field as they are likely to have low yields in later generations. Ortiz and Golmirzaie (2003) too had found significant and positive correlations between plant vigour after transplanting the seedlings and their tuber yield. In order to define the selectability of visible traits, Love et al. (1997) reported that positive selection in early generations was possible for degree of skin russetting, and the incidence and severity of skin defects. Negative selection was found to be appropriate for incidence of secondary growth, growth cracks, tuber malformations, tuber curvature, pointed ends, the incidence and severity of russet patchiness, eye depth, tuber shape, yield, tuber number, average tuber weight and flatness. Traits too inconsistent for use in selection in early generations included uniformity of skin russetting, uniformity in tuber shape and uniformity in tuber size. Hayes and Thill (2003) reported that selection for cold chipping was successful in seedling generation. Swiezynski (1984) reported that selection in seedlings for potato viruses PVX, PVY, PVA and PVS was effective, but not for potato leafroll virus (PLRV) before the C1. Levy et al. (1991) reported that selection for heat tolerance was possible at seedling stage. The heat tolerance was, however, associated with earliness (Levy 1986).

Gopal (1997) recommended that progeny selection should be practiced prior to clonal selection in early generations. As described above, progeny selection can be exploited for making initial choices in identifying the crosses which have a high probability of producing clones with desirable phenotypes. Those progenies which have low probability can be discarded in seedling generation itself. This can be adopted in large breeding programmes: a large number (may be hundreds of crosses) may be evaluated in seedling stage by raising as few as 50–60 genotypes per cross. Negative selection (rejection of poor phenotype) may be practiced for the characteristics under evaluation, and a larger population of the selected crosses can be raised to practice within-progeny individual clonal selection (Gopal 1997). Gopal et al. (1992, 1994) and Gopal (2015) suggested the following selection procedure for individual clonal selection in early generations.

1 Seedlings of poor vigour may be rejected prior to transplanting in the field as they represent inferior genotypes.
2 Those clones with undesirable tuber colour, tuber shape, eye depth and tuber cracking may be rejected from the seedling stage onwards as these characteristics have high repeatability.
3 No rejection should be done on the basis of tuber yield, average tuber weight or number of tubers in the seedling stage.
4 Negative selection for tuber yield and tuber weight can be initiated from the first clonal generation onwards, whereas number of tubers can be considered for the rejection of undesirable types from the second clonal generation onwards.
5 Both number of tubers and tuber size contribute to tuber yield, but they are negatively associated with each other. Hence it is suggested that a standard be fixed for the minimum number of tubers required in the selected types, before employing average tuber weight as a selection parameter for tuber yield; otherwise selection may result in retention of genotypes with oversized and fewer tubers.

3.4 *In vitro* selection

Gopal and Mincoha (1997b,c) explored the possibility of using micropropagation technique (Gopal et al. 1998) for *in vitro* selection of agronomic characteristics in potato. They compared the performance of 22 potato genotypes under *in vitro* and *in vivo* conditions. A considerable similarity was observed between *in vitro* and *in vivo* systems with regard to genetic parameters and characteristics associations in potato (Gopal 2001a). *In vitro* selection was found to be highly effective for tuber colour and number of eyes, and moderately effective for average tuber weight, plant vigour and foliage senescence under specific conditions (Gopal and Minocha 1997b). Morpurgo (1991) found *in vitro* selection effective for salt tolerance as highly significant correlation was found between *in vitro* growth parameters and field performance of ten potato clones. Kim et al. (2012) used *in vitro* microtuber production by TPS families at high temperature (20–30ºC) as a screen for identifying heat-tolerant progenies. Khan et al. (2015) too reported that both family and clonal level *in vitro* assays identified families and clones as per their expected tuberization capacity under heat stress.

Use of polyethylene glycol- or sorbitol-induced osmotic stress in tissue culture media was found to be effective in discriminating the genotypes for drought tolerance, and *in vitro* performance of genotypes correlated well with their drought tolerance under field conditions (Gopal and Iwama 2007). Even agar was found to be effective for this purpose (Gopal et al. 2008a). Since then, polyethylene glycol or mannitol/sorbitol have been used by many workers in tissue culture media for *in vitro* screening for drought tolerance (Junior et al. 2011; Gorji et al. 2012; Barra et al. 2013), as this is a quick method for determining initial response of a large number of genotypes.

4 Improving the speed and success rate of selection

4.1 Microtubers for accelerating selection

The limitation of few tubers in early generations can be overcome by rapidly multiplying the potato clones in tissue culture and producing microtubers in culture vessels or minitubers in greenhouse by transplanting tissue culture-propagated plantlets. The production of micro/minitubers, however, will involve cost of tissue culture and labour. In order to reduce this, micro/minituber production can be started after the harvest of seedlings with the clones selected following the procedure described here. The number of clones required to be multiplied this way can be further reduced if breeding for tuber yield and its components is combined with resistance to diseases such as late blight. Susceptible genotypes can be discarded in the seedling stage itself (Caligari et al. 1983; Stewart et al. 1983). Further, commercial-scale tissue culture devices such as fermenters and bioreactors can be used to reduce the time and cost of rapid multiplication (Donnellly et al. 2003). The benefit:cost analysis will depend on the corresponding direct and indirect costs involved in micro/minituber production and conducting of field trials for extra years due to limited tubers in early generations when micro/minitubers are not produced. In the crop raised from micro/minitubers, selection can be done for general impression (Gopal and Minocha 1997c), giving less weight to tuber yield. In crop raised from mintitubers, selection can be done even for tuber yield (Gopal et al. 2002), provided minitubers are of minimum 20–30 g each. This procedure of individual clonal selection would help in an early and safe elimination of unproductive genotypes. The production of microtubers in sufficient numbers, and the subsequent production of normal seed tubers from them, will result in earlier large-scale trials thereby substantially reducing the period taken to develop a variety. The suggested selection procedures based on only *in vivo* propagation and also combined with *in vitro* propagation is illustrated in Fig. 1.

4.2 Multi-location testing

Comstock and Moll (1963) reported that genotype by environment interaction is in some way involved in most problems of quantitative genetics and many problems of plant breeding. Genotypes by environment interactions have been recognized by plant breeders for many years as reducing the efficiency of selection (Hill 1975). In practical potato-breeding programmes, limited seed supply and logistical problems involved in the handling of large number of progeny usually restrict the selection of early generation clones to one location. Further, selection in later generations is also usually made at one location on each year's results, and only those genotypes that are selected in one year are grown in subsequent years. It is only the advanced few selections that are evaluated in multi-location trials over years. McCann et al. (2012) suggested that more precise evaluations of the genetic potential of individual clones would be achieved through use of small plots evaluated over several locations and/or years rather than increased replications at one location. Haynes et al. (2012) too suggested the evaluations of early generations over locations to identify broadly adapted genotypes before these are rejected. The

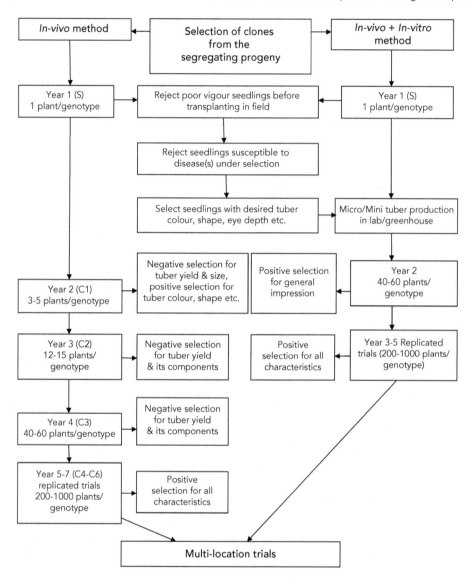

Figure 1 Step-wise procedure for selection of superior clones for agronomic characteristics in potato breeding, indicating the time advantage of combining *in vitro* and *in vivo* methods (Source: modified from Gopal 2015).

selected types from the early generations can be further evaluated in bigger and multi-locational trials for identifying the types widely adapted over environments. The number of environments and replications can be decided based on the number of genotypes under evaluation and the available number of tubers per genotype. Of course, the best performing genotypes are selected for release as cultivars based on performance over years and locations, but still new high-yielding cultivars did not necessarily show better yield stability than the old ones (Paget et al. 2015a).

4.3 Molecular markers in aid of phenotypic selection

Molecular genetic markers provide significant opportunities to improve the breeding efficiency and reduce time to develop improved potato varieties. The association of a genotype with a phenotype of interest allows genetically elite plants to be identified early in plant growth before the phenotype can be observed. The ability of these markers to identify elite plants and discard non-elite plants saves both time and money in the process of potato breeding as it reduces the number of clones to be evaluated in field trials. Slater et al. (2014a) proposed the combined use of marker-assisted selection (MAS) and EBVs to reduce the breeding cycle from over 10 to as few as four years. However, calculation of EBVs themselves requires large pedigree data (Slatter et al. 2014b) or many years' evaluation data (Paget et al. 2015a).

A substantial number of genetic markers linked to genes for qualitative and quantitative traits (QTLs) have been identified over the past 20 years (review by Barrell et al. 2013; Slater et al. 2014a; Ramakrishnan et al. 2015). Early examples of molecular markers include use of restriction fragment length polymorphism for linkage map construction (Bonierbale et al. 1988). This was followed by use of randomly amplified polymorphic DNA (RAPD) markers (Williams et al. 1990) and amplified fragment length polymorphism (AFLP) markers (Vos et al. 1995). However, RAPD markers have a low reproducibility and have been superseded by more reliable markers such as simple sequence repeats (SSRs, or microsatellites) (Provan et al. 1996). The most widely used markers in current plant breeding are SSRs and single-nucleotide polymorphisms (SNPs) because these are generally highly polymorphic, co-dominant, reliable, relatively simple and cheap to use, and they can also be multiplexed (Collard and Mackill 2008). MAS also helps in pyramiding of resistant loci in a single genotype (Dalton et al. 2013; Sudha et al. 2016). Now that the potato genome has been released (http://www.potatogenome.net/images/2/2e/PGSC_Press_Release_0909.pdf), genomic selection may become feasible in future (D'hoop et al. 2014).

Although MAS has the potential to select for traits several years earlier than would be the case in conventional breeding, yet its application in potato has not yet been to the extent that is being used in other crops. Tiwari et al. (2013) reported that limited progress has been made in the use of MAS even for late blight, the most devastating disease of potatoes worldwide. Pardo et al. (2013) observed considerable discrepancies in effectiveness of sequence-characterized amplified region and cleaved amplified polymorphic sequence markers in analysing progeny of controlled crosses involving resistance to PVY and cyst nematodes. Similar discrepancies were also observed by Li et al. (2013) as MAS with six markers for tuber quality traits did not give expected results in F1 progeny of a single-cross combination. This is mainly due to the genetic complexities caused by the autotetraploid and highly heterozygous status of cultivated potatoes, where allelic variation goes up to 20 alleles per locus (Graveland 2014). Further, potato shows a super high-SNP density, a SNP every 15–20 base pairs (Uitdewilligen 2012). Thus, use of molecular markers for identifying desired phenotype for characteristic(s) under selection requires validation and demonstration of their cost-effectiveness in order to be adopted by a practical potato breeder. Ortega and Lopez-Vizcon (2012) reported that although markers allowed selection of breeding clones with resistance to PVY, *Globodera rostochiensis* and/ or *G. pallida*, yet use of MAS in seedling generation was too costly due to a large number of DNA extractions and PCR assays required. More studies on usefulness and constraints found in applying MAS in actual practical potato-breeding programmes are needed.

5 Summary

Conventional potato breeding has led to tremendous genetic improvement in terms of number of improved varieties released, which are now more than 4000. Varieties not only having high yield but also possessing resistance to diseases and pests, and suitable for processing, are now available. However, some of the very old varieties are still in cultivation due to their stable and wider adaptability. Development of stable potato varieties with multiple desirable characteristics needs precise methods for identification of superior parents, crosses and clones from the segregating progeny. Although efforts made in this direction have led to identification of a number of selection parameters and procedures for effective and accelerated phenotypic selection, these are not robust enough to overcome the inherent problems of this tetraploid and highly heterozygous crop. Combining ability, EBV and progeny means-based tests are better than others, but their estimation needs considerable time and resources. Combining *in vitro* and *in vivo* approaches can not only reduce the time needed to develop a new variety, but also help in more reliable phenotypic selection. The use of MAS, though helpful in selecting phenotype based on genotype, in practical potato breeding has been limited for want of its validation and cost-effectiveness. Further studies are thus needed to make conventional potato breeding not only precise in phenotype selection, but also practically and economically viable.

6 References

Anderson, J. A. D (1977). Potato breeding. Selection for agronomic characters and resistance to cyst nematodes. MSc Thesis, Univ. Camb. p. 217.

Bae, J., Jansky, S. H. and Rouse, D. I. (2008). The potential for early generation selection to identify potato clones with resistance to *Verticillium* wilt. *Euphytica* 164:385–93.

Barra, M., Correa, J., Salazar, E., and Sagredo, B. (2013). Response of potato (*Solanum tuberosum* L.) germplasm to water stress under in vitro conditions. *Am. J. Potato Res.* 90:591–606.

Barrell, P., Meiyalaghan, S., Jacobs, J. M. E. and Conner, A. J. (2013). Applications of biotechnology and genomics in potato improvement. *Plant Biotechnol. J.* 11:907–20.

Bonierbale, M. W., Plaisted, R. L. and Tanksley, S. D. (1988). Construction of a genetic map of potato based on molecular markers from tomato. *Am. Potato J.* 65:471–2.

Bradshaw, J. E. and Mackay, G. R. (1994). Breeding strategies for propagated potatoes. In: Bradshaw, J. E. and Mackay, G. R. (eds), *Potato Genetics*, CABI Publishing, Wallingford, UK, pp. 467–97.

Bradshaw, J. E., Stewart, H. E., Wastie, R. L., Dale, M. F. B. and Phillips, M. S. (1995). Use of seedling progeny tests for genetical studies as part of a potato (*Solanum tuberosum* subsp. *tuberosum*) breeding programme. *Theor. Appl. Genet.* 90:899–905.

Bradshaw, J. E., Dale, M. F. B., Swan, G. E. L., Todd, D. and Wilson, R. N. (1998). Early-generation selection between and within pair crosses in a potato (*Solanum tuberosum* subsp. *tuberosum*) breeding programme. *Theor. Appl. Genet.* 97:1331–9.

Bradshaw, J. E., Todd, D. and Wilson, R. N. (2000). Use of tuber progeny tests for genetical studies as part of a potato (*Solanum tuberosum* subsp. *tuberosum*) breeding programme. *Theor. Appl. Genet.* 100: 772–81.

Bradshaw, J. E., Dale, M. F. B. and Mackay, G. R. (2003). Use of mid-parent values and progeny tests to increase the efficiency of potato breeding for combined processing quality and disease and pest resistance. *Theor. Appl. Genet.* 107:36–42.

Brown, J. (1988). The effect of the weight of the seedling-derived tuber on subsequent clonal generations in a potato breeding programme. *Ann. Appl. Biol.* 113:69–78.

Brown, J. and Caligari, P. D. S. (1986). The efficiency of seedling selection for yield and yield components in a potato breeding programme. *Z Pflanzenzüchtung* 96:53–62.

Brown, J. and Caligari, P. D. S. (1988). The use of multivariate cross prediction methods in the breeding of a clonally reproduced crop (*Solanum tuberosum*). *Heredity* 60:147–53.

Brown, J. and Caligari, P. D. S. (1989). Cross prediction in a potato breeding programme by evaluation of parental material. *Theor. Appl. Genet.* 77:246–52.

Brown, J., Caligari, P. D. S., Mackay, G. R. and Swan, G. E. L. (1984). The efficiency of seedling selection by visual preference in a potato breeding programme. *J. Agric. Sci. Camb.* 103:239–346.

Brown, J., Caligari, P. D. S., Mackay, G. R. and Swan, G. E. L. (1987a). The efficiency of visual selection in early generation of a potato breeding programme. *Ann. Appl. Biol.* 110:357–63.

Brown, J., Caligari, P. D. S. and Mackay, G. R. (1987b). The repeatability of progeny means in the early generations of a potato breeding programme. *Ann. Appl. Biol.* 110:365–70.

Brown, J., Caligari, P. D. S., Dale, M. F. B., Swan, G. E. L. and Mackay, G. R. (1988). The use of cross prediction methods in a practical potato breeding programme. *Theor. Appl. Genet.* 76:33–8.

Caligari, P. D. S. and Brown, J. (1986). The use of univariate cross prediction methods in the breeding of a clonal reproduced crop (*Solanum tuberosum*). *Heredity* 57:396–401.

Caligari, P. D. S., Mackay, G. R., Stewart, H. E. and Waistie, R. L. (1983). A seedling progeny test for resistance to potato foliage blight *Phytophthora infestans* (Mont.) de Bary). *Potato Res.* 27:43–50.

Caligari, P. D. S., Brown, J. and Abbott, R. J. (1986). Selection for yield and yield components in the early generations of a potato breeding programme. *Theor. Appl. Genet.* 73:218–22.

Chimote, V. P., Kumar, R., Sharma, N. and Kamal, S. (2008). Nuclear-cytoplasmic diversity in parental genotypes used for Indian early bulking potato (*Solanum tuberosum* ssp. *tuberosum*) breeding programme. *Indian J. Genet. Plant Breed.* 68:171–6.

Collard, B. C. and Mackill, D. J. (2008). Marker-assisted selection: an approach for precision plant breeding in the twenty-first century. *Philos. Trans. R. Soc. Lond.* B 363:557–72.

Comstock, R. E. and Moll, R. H. (1963). Genotype-environment interactions. In: Hanson, W. D., Robinson, H. F. (eds), Statistical genetics and plant breeding. *Nat. Acad. Sci. Nat. Res. Pub.* 982:164–96.

Cubillos, A. G. and Plaisted, R. L. (1976). Heterosis for yield in hybrids between *S. Tuberosum* ssp. *Tuberosum* and *tuberosum* ssp. *Andigena*. *Am. Potato J.* 53:143–50.

Dalton, E. de Griffin, D., Gallagher, T. F., Vetten, N. and Milbourne, D. (2013). The effect of pyramiding two potato cyst nematode resistance loci to *Globodera pallida* pa2/3 in potato. *Mol. Breed.* 31:921–30.

De Jong, H. and Tai, G. C. C. (1991). Evaluation of potato hybrids obtained from tetrapoid-diploid crosses. I. Parent-offspring relationships. *Plant Br: New Strat Pl Imp* 31:4:921–30; 107:177–82.

De Jong, H., Tai, G. C. C., Russel, W. A., Jonston, G. R. and Proudfoot, K. G. (1981). Yield potential and genotype-environment interactions of tetraploid-diploid (4x-2x) potato hybrid. *Am. Potato J.* 58:191–9.

D'hoop, B. B., Keizer, P. L. C., Paulo, M. J., Visser, R. G. F., van Eeuwijk, F. A. and van Eck, H. J. (2014). Identification of agronomically important Q. T. L. in tetraploid potato cultivars using a marker-trait association analysis. *Theor. Appl. Genet.* 127:731–48.

Donnelly, D. J., Coleman, W. K. and Coleman, S. E. (2003). Potato microtubers: production and performance: A review. *Am. J. Potato Res.* 80:103–15.

Estrada-Ramos, N. (1984). Acaphu: Tetraploid fertile breeding line selected from a *Solanum acaule* x *S. phureja* cross. *Am. Potato J.* 61:1–8.

Gopal, J. (1997). Progeny selection for agronomic characters in early generations of a potato breeding programme. *Theor. Appl. Genet.* 95:307–11.

Gopal, J. (1998a). General combining ability and its repeatability in early generations of potato breeding programmes. *Potato Res.* 41:21–8.

Gopal, J. (1998b). Heterosis and combining ability analysis for resistance to early blight (*Alternaria solani*) in potato. *Potato Res.* 41:311–17.

Gopal, J. (1998c). Identification of superior parents and crosses in potato breeding programmes. *Theor. Appl. Genet.* 96:287–93.

Gopal, J. (2001a). In vitro and in vivo genetic parameters and characters associations in potato. *Euphytica* 118:145–51.

Gopal, J. (2001b). Between and within families variation and family selection in potato breeding programmes. *J. Genet. Breed.* 55:313–18.

Gopal, J. (2006). Considerations for successful breeding. In: Gopal, J. and Paul Khurana, S. M. (eds), *Handbook of Potato Production, Improvement and Postharvest Management*. Food Product Press, New York, USA, pp. 77–108.

Gopal, J. (2015). Challenges and way-forward in selection of superior parents, crosses and clones in potato breeding. *Potato Res.* 58:165–8.

Gopal, J. and Minocha, J. L. (1997a). Genetic divergence for cross prediction in potato. *Euphytica* 97:269–75.

Gopal, J. and Minocha, J. L. (1997b). Effectiveness of in vitro selection for agronomic characters in potato. *Euphytica* 103:67–74.

Gopal, J. and Minocha, J. L. (1997c). Effectiveness of selection at microtuber crop level in potato. *Plant Breed.* 116:293–5.

Gopal, J. and Oyama, K. (2005). Genetic base of Indian potato selections as revealed by pedigree analysis. *Euphytica* 142:23–31.

Gopal, J. and Iwama, K (2007). In vitro screening of potato against water stress mediated through sorbitol and polyethylene glycol. *Plant Cell Rep.* 26:693–700.

Gopal, J., Gaur, P. C. and Rana, M. S. (1992). Early generation selection for agronomic characters in a potato breeding programme. *Theor. Appl. Genet.* 84:709–13.

Gopal, J., Gaur, P. C. and Rana, M. S. (1994). Heritability, and inter-generation associations between tuber yield and its components in potato (*Solanum tuberosum* L.). *Plant Breed.* 112:80–3.

Gopal, J., Minocha, J. L. and Dhaliwal, H. S. (1998). Microtuberization in potato (*Solanum tuberosum* L.). *Plant Cell Rep.* 17:794–8.

Gopal, J., Kumar, R. and Kang, G. S. (2002). The effectiveness of using a minituber crop for selection of agronomic characters in potato breeding programmes. *Potato Res.* 45:145–51.

Gopal, J., Iwama, K. and Jitsuyama, Y. (2008a). Effect of water stress mediated through agar on in vitro growth of potato. *In Vitro Cell. Dev. Biol. Plant* 44:221–8.

Gopal, J., Kumar, V. and Luthra, S. K. (2008b). Top-cross vs. poly-cross as alternative to test-cross for estimating the general combining ability in potato. *Plant Breed.* 127:441–5.

Gorji, A. M., Matyas, K. K., Dublecz, Z., Desci, K., Cernak, I., Hoffmann, B., Taller, J. and Polgar, Z. (2012). In vitro osmotic stress tolerance in potato and identification of major QTLs. *Am. Potato J.* 89:6:453–64.

Graveland, R. P. (2014). A potato breeding landscape. *Potato Res.* 57:339–41.

Hayes, R. J. and Thill, C. A. (2002a). Selection for potato genotypes from diverse progenies that combine 4°C chipping with acceptable yields, specific gravity, and tuber appearance. *Crop Sci.* 42:1343–9.

Hayes, R. J. and Thill, C. A. (2002b). Co-current introgression of economically important traits in a potato breeding program. *Am. Potato J.* 79:173–81.

Hayes, R. J. and Thill, C. A. (2003). Genetic gain from early generation selection for cold chipping genotypes in potato. *Plant Breed.* 122:158–63.

Haynes, K. G., Gergela, D. M., Hutchinson, C. M., Yencho, G. C., Clough, M. E., Henninger, M. R., Halseth, D. E., Sandsted, E., Porter, G. A. and Ocaya, P. C. (2012). Early generation selection at multiple locations may identify potato parents that produce more widely adapted progeny. *Euphytica* 186:573–83.

Hill, J. (1975). Genotype-environment interactions – a challenge for plant breeding. *J. Agric. Sci.* 85:477–93.

Hougas, R. W. (1956). News and reviews: Foreign potatoes, their introduction and importance *Am. Potato J.* 33:190–9.

Howard, H. W. (1963a). Some potato breeding problems. *Rep. Plant Breed. Inst.*, pp. 5–21.

Howard, H. W. (1963b). The significance of breeding in improving quality and yield. In: Ivins, I. D., Milthorpe, F. L. (eds), *The Growth of the Potato*. Butterworths, London, England, pp. 292–302.

Junior, C. R., Castro, C. M., Medeiros, C. A. B., Pereira, A. da S. and Carvalho, G. C. (2011). Methods for selection for drought tolerance in potatoes. *Acta Hortic.* 889:391–6.

Kerr, R., Li, L., Tier, B., Dutkowski, G. and McRae, T. (2012). Use of the numerator relationship matrix in genetic analysis of autopolyploid species. *Theor. Appl. Genet.* 124:1271–82.

Khan, M. A., Munive, S. and Bonierbale, M. (2015). Early generation in vitro assay to identify potato populations and clones tolerant to heat. *Plant Cell, Tissue Organ Cult.* 121:45–52.

Killick, R. J. (1977). Genetic analysis of several traits in potatoes by means of a diallel cross. *Ann. Appl. Biol.* 86:279–89.

Killick, R. J. and Malcolmson, J. K. (1973). Inheritance in potatoes of field resistance to late blight (*Phytophthora infestans*). *Physiol. Plant Pathol.* 3:121–31.

Kim, H., Cho, K., Cho, J., Park, Y. and Cho, H. (2012). Selection of heat tolerant potato lines via in vitro screening of true potato seeds. *Acta Hortic.* 935:225–30.

Knight, T. A. (1807). On raising of new and early varieties of the potato (*Solanum tuberosum*). *Trans. Hortic. Soc. London* 1: 57–9.

Krantz, F. A. and Hutchins, A. E. (1929). Potato breeding methods II. Selection in inbred lines. *Minnesota Agric. Expt. Sta. Tech. Bull.* 58:3–26.

Landeo, J. A. and Hanneman, Jr. R. E. (1982). Heterosis and combining ability of *Solanum tuberosum* group Andigena haploids. *Potato Res.* 25:227–37.

Levy, D. (1986). Genotypic variation in the response of potatoes (*Solanum tuberosum* L.) to high ambient temperatures and water deficit. *Field Crops Res.* 15:85–96.

Levy, D., Kastenbaum, E. and Itzhak, Y. (1991). Evaluation of parents and selection for heat tolerance in the early generations of a potato (*Solanum tuberosum* L.) breeding program. *Theor. Appl. Genet.* 82:130–6.

Li, L., Tacke, E., Hofferbert, H. R., Lbeck, J., Strahwald, J., Draffehn, A. M., Walkermeier, B. and Gebhardt, C. (2013). Validation of candidate gene markers for marker-assisted selection of potato cultivars with improved tuber quality. *Theor. Appl. Genet.* 126:1039–52.

Lindhout, P., Meijer, D., Schotte, T., Hutten, R. C. B., Visser, R. G. F. and van Eck, H. J. (2011). Towards F1 hybrid seed potato breeding. *Potato Res.* 54:301–12.

Loiselle, F., Tai, G. C. C., Christie, B. R. and Tarn, T. R. (1989). Relationship between inbreeding coefficient and clonal selection in potato cultivar development program. *Am. Potato J.* 66:747–53.

Love, S. L., Werner, B. K. and Pavek, J. J. (1997). Selection for individual traits in the early generations of a potato breeding program dedicated to producing cultivars with tubers having long shape and russet skin. *Am. Potato J.* 74:199–213.

Luthra, S. K., Sharma, P. C., Gopal, J., Khurana, S. M. P. and Pandey, S. K. (2001). Inter-relationships between genetic parameters for tuber yield in potato. *J. Indian Potato Assoc.* 28:15–17.

Luthra, S. K., Gopal, J. and Sharma, P. C. (2005). Genetic divergence and its relationship with heterosis in potato. *Potato J.* 32:37–42.

Maris, B. (1962). Analysis van aardappelpopulaties ten dienste van de veredeling. Versl. Landbouwk. Onderz. 68.8. Wageningen, pp. 208.

Maris, B. (1966). The modifiability of characters important in potato breeding. *Euphytica* 15:18–31.

Maris, B. (1969). Studies on maturity, yield, under water weight and some other characters of potato progenies. *Euphytica* 18:297–319.

Maris, B. (1986). The effect of seed tuber weight on characters in the first and the second clonal generations of potato populations. *Euphytica* 35:465–82.

Maris, B. (1988). Correlations within and between characters between and within generations as a measure for the early generation selection in potato breeding. *Euphytica* 37:205–24.

Maris, B. (1989). Analysis of an incomplete diallel cross among three ssp. *tuberosum* varieties and seven long-day adapted ssp. *andigena* clones of the potato (*Solanum tuberosum* L.). *Euphytica* 41:163–82.

McCann, L. C., Bethke, P. C., Casier, M. D. and Simon, P. W. (2012). Allocation of experimental resources used in potato breeding to minimize the variance of genotype mean chip colour and tuber composition. *Crop Sci.* 52:1475–81.

McHale, N. A. and Lauer, F. I. (1981). Breeding value of 2n pollen from diploid hybrids and *phureja* in 4x-2x crosses in potatoes. *Am. Potato J.* 58:365–74.

Mendoza, H. A. and Haynes, F. L. (1974a). Genetic relationship among potato cultivars grown in the United States. *HortScience.* 9:328–30.

Mendoza, H. A. and Haynes, F. L. (1974b). Genetic basis of heterosis for yield in autotetraploid potato. *Theor. Appl. Genet.* 45:21–5.

Mihovilovich, E., Alarcn, L., Prez, A. L., Alvarado, J., Arellano, C. and Bonierbale, M. (2007). High levels of heritable resistance to potato leafroll virus (PLRV) in *Solanum tuberosum* subsp. *Andigena*. *Crop Sci.* 47:1091–103.

Mok, D. W. S. and Peloquin, S. J. (1975). Breeding value of 2n pollen (dihaploids) in tetraploid x diploid crosses in potatoes. *Theor. Appl. Genet.* 46:307–14.

Morpurgo, R. (1991). Correlation between potato clones grown *in vivo* and *in vitro* under sodium chloride stress conditions. *Plant Breed.* 107:80–2.

Neele, A. E. F. (1990). Study on the inheritance of potato tuber yield by means of harvest index, components and its consequences for choice of parental material. *Euphytica* 48:159–66.

Neele, A. E. F. and Louwes, K. M. (1989). Early selection for chip quality and dry matter content in potato seedling populations in greenhouse or screenhouse. *Potato Res.* 32:293–300.

Neele, A. E. F., Barten, J. H. M. and Louwes, K. M. (1988). Effects of plot size and selection intensity on efficiency of selection in the first Clonal generation of potato. *Euphytica* 39(S3):27–35.

Neele, A. E. F., Nab, H. J., de Jongh, M. J., de Leeuw and Louwes, K. M. (1989). Optimising visual selection in early clonal generations of potato based on genetic and economic considerations. *Theor. Appl. Genet.* 78:665–71.

Neele, A. F. E., Nab, H. J. and Louwes, K. M. (1991). Identification of superior parents in a potato breeding programme. *Theor. Appl. Genet.* 82:264–72.

Ortega, F. and Lopez-Vizcon, C. (2012). Application of molecular marker-assisted selection (MAS) for disease resistance in a practical potato breeding programme. *Potato Res.* 55:1–13.

Ortiz, R. and Golmirzaie, A. M. (2003). Genetic parameters for agronomic characteristics. I. Early and intermediate breeding populations of true potato seed. *Hereditas-Lund* 139:212–16.

Paget, M. F., Apiolaza, L. A., Anderson, J. A., Genet, R. A. and Alspach, P. A. (2015). Appraisal of test location and variety performance for the selection of tuber yield in a potato breeding program. *Crop Sci.* 55:1957–68.

Paget, M. F., Alspach, P. A., Anderson, J. A. D., Genet, R. A., Apiolaza, L. A. and Piepho, H. P. (2015). Trial heterogeneity and variance models in the genetic evaluation of potato tuber yield. *Plant Breed.* 134:203–11.

Pardo, R. L., Barandalla, L., Ritter, E., Galarreta, J. I. R.D. and Wehling, P. (2013). Validation of molecular markers for pathogen resistance in potato. *Plant Breed.* 132:246–51.

Paz, M. M. and Veilleux, R. E. (1997). Genetic diversity based on RAPD polymorphism and its relationship with the performance of diploid potato hybrids. *J. Am. Soc. Hort. Sci.* 122:740–7.

Phillips, M. S. (1981). A method of assessing potato seedling progenies for resistance to the white potato cyst nematodes. *Potato Res.* 24:101–3.

Plaisted, R. L., Sanford, L., Federer, W. T., Kerr, A. E. and Peterson, L. C. (1962). Specific and general combining ability for yield in potatoes. *Am. Potato J.* 39:185–97.

Provan, J., Powell, W. and Waugh, R. (1996). Microsatellite analysis of relationships within cultivated potato (*Solanum tuberosum*) *Theor. Appl. Genet.* 92:1078–84.

Ramakrishnan, A. P., Ritland, C. E., Sevillano, R. H. B. and Riseman, A. (2015). Review of potato molecular markers to enhance trait selection. *Am. J. Potato Res.* 92:455–72.

Rowe, P. R. (1967). Performance and variability of diploid and tetraploid potato families. *Am. Potato J.* 44:263–71.

Schmiediche, P., Hawkes, J. G. and Ochoa, C. M. (1982). The breeding of the cultivated potato species Solanum x juzepczukii and S. x curtilobum. II. The resynthesis of S x juzepczukii and S. x curtilobum. *Euphytica* 31:695–708.

Schroeder, S. H. and Peloquin, S. J. (1983). Parental effects for yield and tuber appearance on 4x families from 4x x 2x crosses. *Am. Potato J.* 60:819.

Seiffert, M. (1957). Die Bedutung der Zuchtung fur die Ertragssteigerung im Kartoflelbau in den letzten Jahrzehnten. Ein Beitrag zur Methodik der Ermittlung des zuchterischen *Fortschritts. Zuchter* 27:1–22.

Simmonds, N. W. (1962). Variability in crop plants, its use and conservation. *Biol. Rev.* 37:422–65.

Slater, A. T., Cogan, N. O. L., Hayes, B. J., Schultz, L., Dale, M. F. B., Bryan, G. J. and Forster, J. W. (2014a). Improving breeding efficiency in potato using molecular and quantitative genetics. *Theor. Appl. Genet.* 127: 2279–92.

Slater, A. T., Wilson, G. M., Cogan, N. O. L., Forster, J. W. and Hayes, B. J. (2014b). Improving the analysis of low heritability complex traits for enhanced genetic gain. *Theor. Appl. Genet.* 127: 809–20.

Stewart, H. E., Taylor, K. and Wastie, R. L. (1983). Resistance to late blight in foliage (*Phytophthora infestans*) of potatoes assessed as true seedlings and as adult plants in the glasshouse. *Potato Res.* 26:363–6.

Sudha, R., Ventakasalam, E. P., Bairwa, A., Bhardwaj, V. and Dalamu, Sharma, R. (2016). Identification of potato cyst nematode resistant genotypes using molecular markers. *Sci. Hortic.* 198:21–6.

Swiezynski, K. M. (1968). Field production of first year potato seedlings in the breeding of early varieties. *Eur. Potato J.* 11:141–9.

Swiezynski, K. M. (1978). Selection of individual tubers in potato breeding. *Theor. Appl. Genet.* 53:71–80.

Swiezynski, K. M. (1984). Early generation selection methods used in polish potato breeding. *Am. Potato J.* 61:385–94.

Tai, G. C. C. (1975). Effectiveness of visual selection for early clonal generation seedlings in potato. *Crop Sci.* 15:15–18.

Tai, G. C. C. (1994). Use of 2n gametes. In: Bradshaw, J. E. and Mackay, G. R. (eds), *Potato Genetics.* CAB International, Wallingford, UK, pp. 109–12.

Tai, G. C. C. and Young, D. A. (1984). Early generation selection for important agronomic characters in a potato breeding population. *Am. Potato J.* 61:419–34.

Tarn, T. R. and Tai, G. C. C. (1983). *Tuberosum* x *Tuberosum* and *Tuberosum* x *Andigena* potato hybrids: comparisons of families and parents, and breeding strategies for *Andigena* potatoes in long-day temperate environments. *Theor. Appl. Genet.* 66:87–91.

Tiwari, J. K., Sidappa, S., Singh, B. P., Kaushik, S. K., Chakrabarti, S. K., Bhardwaj, V., Chandel, P. and Wehling, P. (2013). Molecular markers for late blight resistance breeding of potato: an update. *Plant Breed.* 132:237–45.

Tung, P. X., Rasco, E. T., Vander Zaag, P. and Schmiediche, Jr. P. (1990). Resistance to *Pseudomonas solanacearum* in the potato: 1. Effects of sources of resistance and adaption. *Euphytica* 45:203–10.

Uitdewilligen, J. (2012). Discovery and genotyping of existing and induced DNA sequence variation in potato. PhD thesis. Wageningen University, Wageningen, The Netherlnads, 166p.

Van der Plank, J. E. (1946). Origin of the first European potatoes and their reaction to length of day. *Nature* 157:503–15.

Vielleux, R. E. and Lauer, F. I. (1981a). Variation for 2n pollen production in clones of *Solanum phureja* Juz. and Buk. *Theor. Appl. Genet.* 59:95–100.

Veilleux, R. E. and Lauer, F. I. (1981b). Breeding behaviour of yield components and hollow heart in tetraploid-diploid vs conventionally derived potato hybrids. *Euphytica* 30:547–61.

von Kameke, K. (1975). Untersuchungen zur quantitativen Variabilität in Kreuzungsnachkommen-schaften der Kartoffel. *Hefte fur den Kartoffelbau* 19:58.

von Kameke, K. (1978). Untersuchungen zur Erblichkeit einiger Merkmale bei der Kartoffel. *Der Kartoffelbau* 5:172–3.

Vos, P., Hogers, R., Bleeker, M., Reijans, M., Vandelee, T., Hornes, M., Frijters, A., Pot, J., Peleman, J., Kuiper, M. and Zabeau, M. (1995). AFLP: a new technique for DNA fingerprinting. *Nucleic Acid Res.* 23:4407–14.

Williams, J. G. K., Kubelik, A. R., Livak, K. J., Rafalski, J. A. and Tingey, S. V. (1990). DNA polymorphisms amplified by arbitrary primers are useful as genetic markers. *Nucleic Acid Res.* 18: 6531–5.

Zadina, J. (1971). Selecting highly productive potato crosses according to performance of the seedlings. *Genetika a Slechteni* 7:269–73.

Chapter 5

Hybrid potato breeding for improved varieties

Pim Lindhout, Michiel de Vries, Menno ter Maat, Su Ying, Marcela Viquez-Zamora and Sjaak van Heusden, Solynta, The Netherlands

1 Introduction

The cultivated potato, *Solanum tuberosum*, can be reproduced generatively through seeds and vegetatively through tubers. This may have evolutionary advantages: seeds may provide better survival under extreme conditions, such as frost or drought, and can remain viable in the soil for years. When conditions are mild, tubers survive in a dormant state for a couple of months. When conditions become favourable again, their fast and strong sprouting provides a clear competitive advantage over other plants in the same ecological niche.

In traditional potato breeding, each breeding cycle starts with a cross between two genotypes, usually tetraploid varieties, followed by many years of selection and multiplication. The advantage of this approach is uniformity: the tubers are clones and thus genetically identical. The disadvantage is the low genetic gain in each lengthy breeding cycle, as the genetic composition of the two parental genotypes is just reshuffled, including alleles which negatively affect plant growth and development. As a result, potato yield has not significantly been improved over the past century (Douches

http://dx.doi.org/10.19103/AS.2016.0016.04

et al., 1996; Vos et al., 2015). In addition, the reproduction of seed tubers is less than a factor 10 per season. It takes many years to build up sufficient quantities of seed tubers for commercial production, and the risk of contamination by pathogens increases with each multiplication step.

True potato seed (TPS) has been promoted as an alternative for seed tubers because TPS is easy to store and devoid of most soil-borne pathogens. In South Asia, East Africa and the Andes, TPS is used mainly by subsistent farmers (Almekinders et al., 1996). TPS is produced by crossing parent plants that have been selected to produce a hybrid variety. The parents are propagated vegetatively, similar to seed tuber propagation. As the parents of a TPS variety are heterozygous, all seeds of a TPS cultivar are genetically different. This results in a highly variable crop that is not acceptable in most markets, such as the high-value markets of Europe and North America.

Since the success of hybrid breeding in corn in the 1930s, breeders have adapted a hybrid breeding system for many crops (Fig. 1; Crow, 1998; Troyer, 2006; Hua et al., 2003). Typically, hybrid cultivars produce higher yields and show high crop uniformity (Rijk et al., 2013). In addition, the breeding system is fast and efficient and new traits can rapidly be introduced by marker-assisted introgression.

These advantages are also expected for potato: hybrid potato varieties will be higher yielding, will need less crop protection chemicals due to disease resistance and will have better quality for processors and consumers (FAO et al., 2015). A hybrid breeding system for potato offers two additional advantages: fast multiplication of hybrid seeds and easier logistics, as clean true seeds can easily be produced, transported and stored (Duvick et al., 2005).

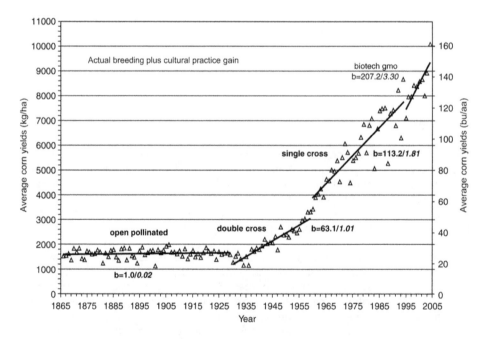

Figure 1 Fivefold increase in corn yields since the introduction of hybrids (Troyer, 2006).

However, self-incompatibility and inbreeding depression have for long hindered progress in hybrid potato breeding (De Jong and Rowe, 1971; Charlesworth and Willis, 2009; Jansky et al., 2016). These two limitations have recently been overcome by introducing a self-compatibility restorer gene and by large-scale and consistent breeding (Lindhout et al., 2011a).

This chapter describes the scientific principles and applied aspects of hybrid potato breeding. The successful introduction of the principle of hybrid potato breeding was described in 2011 (Lindhout et al., 2011a). Since then, we have focused on further developing the potato hybrid breeding system, especially genetic studies to establish a genetics-driven hybrid breeding system. In a recent paper, Jansky et al. (2016) have confirmed the possibilities of such approach.

This is the first publication on the state of the art of a hybrid potato breeding programme. The authors, all working at Solynta, want to emphasize that scientific papers on this topic are not available yet, and hence we have to rely entirely on the results of the Solynta breeding and research programme. Still, in presenting these results, we hope to contribute to a better understanding of the principles and applied aspects of hybrid potato breeding.

2 The scientific basis for hybrid potato breeding

2.1 The principle of hybrid breeding

The basic idea of hybrid breeding is to combine the genes of two parent genotypes, both of which may harbour suboptimal alleles, resulting in weaker performance. If parents have different suboptimal alleles, hybrid offspring can show increased vigour and yield, designated 'heterosis', as the suboptimal alleles in one parent may be compensated by the favourable genes from the other parent (Birchler et al., 2010; Gopal, 2014; Fig. 6). If the parents are completely homozygous, the resulting hybrid offspring will be partially heterozygous and genetically uniform. By testing many hybrid offspring under relevant cultivation conditions, the best combining parents are identified. These are maintained and propagated in separate groups as 'heterotic pools' for further breeding (Brown and Cagliari, 2011).

Thus, hybrid breeding has two distinct processes: development of homozygous parent lines and production and testing of experimental hybrids.

2.2 Diploids are more efficient than tetraploids for hybrid breeding

Homozygous diploids are faster to generate than homozygous tetraploids. For instance, seven generations of selfing are required to obtain 50% homozygous loci starting from a tetragenic tetraploid heterozygote (carrying four different alleles). The same homozygosity level is reached starting from a heterozygous diploid by only one generation of selfing (Haldane, 1930; Fig. 1). For this reason, hybrid potato breeding is more efficient at the diploid level.

2.3 Generation of homozygous diploid potato genotypes via haploidization and via inbreeding

Haploid genotypes can be generated from an egg cell (gynogenesis) or from pollen, often via anther culture (androgenesis). In potato, some haploids have been generated by anther cultures. The resulting haploids were converted into diploids by chromosome doubling. However, the resulting homozygous diploids were very weak and sterile (van Breukelen et al., 1977; Uijtewaal et al., 1987b), hampering their usage in breeding.

Haploidization has been more successful in crossable species like *S. chacoense* (Cappadocia, 1990; Phumichae et al., 2005; Phumichae and Hosaka, 2006) and *S. phureja* (Chani et al., 2000).

A reason for the failure to produce vigorous doubled haploids may be the transition to complete homozygosity in one step. Inbreeding depression may be so severe that homozygous plants are too weak to survive. Repeated selfing, on the other hand, might lead to a more gradual improvement of homozygosity. However, in potato, inbreeding is seriously limited by self-incompatibility, which prevents self-fertilization. Still some rare examples of homozygous diploid plants have been generated by inbreeding but again the homozygous diploid plants always showed a strong inbreeding depression, which limited their usage in breeding (De Jong and Rowe, 1971; Charlesworth and Willis, 2009).

2.4 Large genetic variation in potato causes inbreeding depression

The tetraploid and outcrossing nature of commercial potato is likely responsible for the large genetic variation. In a study on the allelic composition of 800 genes in 83 potato cultivars, an average frequency of 3,2 alleles per locus within a genotype was identified (Uitdewilligen et al., 2013). Among the 83 cultivars, often more than ten alleles per locus

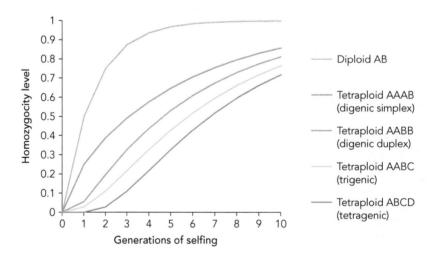

Figure 2 The theoretical increase in homozygosity in diploids and tetraploids through inbreeding, adjusted from Haldane et al. (1930).

were observed. In addition, the frequency of single-nucleotide polymorphisms (SNPs) in potato is 1 in 15–30 base pairs (PGCS, 2011; Visser et al., 2014). This means that the genetic distance between the two sets of chromosomes within one diploid potato genotype is four times larger than the genetic distance between the genomes of man and chimpanzee (CSAC, 2005).

The high frequency of allelic variation has the inevitable consequence that 'weak alleles' that have a negative effect on plant fitness are maintained. Such alleles remain hidden in the large buffer of four genomes, but reveal themselves upon inbreeding when the chance for homozygosity increases. This is even more manifested at the diploid level where the homozygosity level more rapidly increases upon inbreeding (Fig. 2).

This large genetic variation is also helpful for breeding as it forms a genetic reservoir of useful genes. It is a challenge to identify alleles that contribute most to plant performance. As potato has 39 000 genes, the identification and usage of the 'best alleles' per locus, including interactions between them (epistasis), will gradually take place over decades of research and breeding (PGSC, 2011). Corn may serve as a good example, whereby, after a century of dedicated breeding by numerous breeders worldwide, a genetic gain of over 1% per year is still achieved (Troyer, 2006).

2.5 Crossable diploid species and tetraploids increase genetic reservoir for diploid breeding

The potato germplasm available for breeding comprises many species, including diploid species (Jansky and Peloquin, 2006). These have been used as source to introduce resistance genes into cultivated germplasm. Breeders often use diploid potato to rapidly combine favourable traits that can be introduced into the tetraploid germplasm by direct crossings, bridge crossings or via chromosome doubling (De Mainea, 1982; Chauvin et al., 2003). The diploid breeding programme at Wageningen University (Hutten, 1994) has generated donor lines that harbour the most important traits for potato breeding (Table 1). Additional diploid germplasm is available from potato gene banks and public research institutes such as University of Wisconsin-Madison, United States; Potato Germplasm Enhancement Laboratory, Japan; Gene Bank at Gatersleben, Germany; and International Potato Centre, Peru.

Another source of diploid germplasm is tetraploids that can be prickle pollinated to generate diploid offspring, designated dihaploids (Uijtewaal et al., 1987a). A collection of dihaploids obtained from one tetraploid harbours the full set of genes from the tetraploid and can be exploited in a diploid potato breeding programme.

In conclusion, the large genetic variation in potato and in its wild relatives, combined with the technology to switch between ploidy levels, provides a tremendous wealth of germplasm available for diploid hybrid breeding.

2.6 Diploid potato may perform equal to tetraploids

Most important food crops such as rice, corn and soyabean are diploid. Sugar beet cultivars were initially tetraploid, then triploid and since 2000, all new cultivars are diploid. In potato, it has long been assumed that tetraploids outperform diploids (Rowe, 1967; Hutten et al., 1994). Occasional observations have contradicted this assumption: Progeny of diploid potato USW4 with *S. chacoense* M6 produced large tubers and high

Table 1 The *Sli* gene donor, designated DS and 16 diploid potato germplasm, designated D1–D16, used for hybrid breeding at Solynta. The trait abbreviations are according to Hutten (1994)

Abbreviation	Short description
DS	*Sli* gene donor
D1	Early (maturity), long, *Y*, Qcook
D2	Early, *Y*, Qcook
D3	*R3, H1, Gpa2, RXadg, Y* (yellow flesh)
D4	Grp1, early, long (shape), *Ro1 (H1?)*
D5	Early, long, *Y*, Qfry
D6	Long, *Y*, Qfry, *H1*, Qcook
D7	Early, long, *Y, H1*, Qcook, *Zep* (orange flesh)
D8	Early, *y* (white flesh)
D9	Qstarch, *Y*
D10	Wild species hybrid: *phyt avl*
D11	Wild species hybrid: *phyt rch*
D12	Round (shape), Qcook, Qfry
D13	Early, round, *Zep, Y*, Spectacled, Qcook, blue anthocyans
D14	Wild species hybrid: *phyt tar*
D15	Wild species BC1: early, *phyt vnt1*, round, *Y, H1*
D16	Early, round, *y* (white flesh)

yield (Lipman and Zamir, 2007; Jansky et al., 2014). Uijtewaal et al. (1987b) observed that heterozygous diploid potato outperformed all homozygous di- and tetraploid derivatives. These results from potato and from other crops support the expectation that diploid potato will eventually replace tetraploid potato for commercial usage.

3 The state of the art of hybrid potato breeding

In 2008, Solynta started its research by making crosses between diploid potato germplasm, obtained from a pre-breeding programme from Wageningen University (Rutten, 1994), and a homozygous accession of the wild species *S. chacoense*, carrying the dominant self-compatibility controller gene *Sli* (Hosaka and Hanneman, 1998a,b; Phumichai et al., 2005; Lindhout et al., 2011a). The F_1 plants were extremely vigorous and about half of them produced many berries upon self-pollination. This was considered a major breakthrough as these were, to our knowledge, the first vigorous, self-compatible diploid potato plants ever obtained. These F_1 plants were highly heterozygous. The first generation after selfing (designated F_2) should harbour at least 50% homozygous loci. As many of these loci might harbour 'weak alleles', it was expected that F_2 plants would be too weak to survive. We also made BC_1 populations by backcrossing to the *S. chacoense* parent.

As expected, the F_2 and BC_1 showed weak growth and many died in the field. However, a number of plants survived and 10% of the surviving F_2 plants even proved self-compatible (Lindhout et al., 2011a). We generated the second inbred generation, designated F_3 and made crosses between self-compatible individual F_3 plants. These inbred plants were tested with SNP markers to confirm their genetic identity as real inbreds. These results indicated that breeding hybrid potato was now feasible (Lindhout et al., 2011b).

The weak plant performance, the poor tuber quality and the low yield of the inbreds was not only due to inbreeding depression, but was also caused by the wild *S. chacoense*, a species that hardly produces tubers. So, by this approach, we not only started a hybrid potato breeding system, but also the process of domestication of a new 'diploid potato' based on an interspecific cross of diploid *S. tuberosum* and the wild species *S. chacoense*.

We hypothesized that developing vigorous inbred lines is challenging as for each of the 39 000 loci the most favourable alleles should be identified and combined. Unfavourable genes with large effects on plant performance are identified easily and hence selecting increased plant performance is initially easy and fast. Undesired characters from the *S. chacoense* parent such as abundant stolons, small leaves and twisted, small and low-yielding tubers were removed in a few breeding generations.

Sli is a dominant gene (Phumichai et al., 2005; Phumichai and Hosaka, 2006). However, the successful expression of this gene requires a vigorous plant that is fertile and supports self-pollination. Often these criteria are not met. Therefore, the frequency of self-compatible plants is usually lower than expected based on a monogenic trait. Inbreeding depression is exhibited as weaker plant growth upon higher generations of inbreeding (Fig. 3). As a consequence, the self-compatibility level tends to decrease upon further selfings. The first inbred lines obtained by Solynta, containing over 95% homozygous

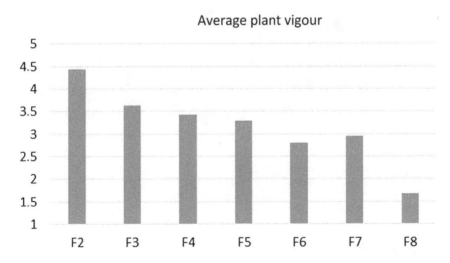

Figure 3 Inbreeding in the diploid Solynta germplasm in winter 2014–15. The Fx indicates x-1 generation of selfing after the last cross was made. Data are averaged over the complete trial consisting of over 5000 plants. The scale of plant vigour ranges from 1 = very weak via 3 = average to 5 = very strong.

Figure 4 Examples of Solynta diploid germplasm. Plants were raised from seedlings and grown in the greenhouse in the summer season of 2014 (left panel) and 2015 (middle and right panel). They were among the best genotypes in a breeding programme comprising over 15 000 plants.

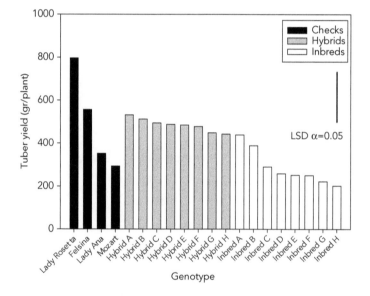

Figure 5 Hybrid performance of the first diploid experimental potato hybrids. Seed tubers were harvested from greenhouse-grown plants, raised in the winter from tubers (checks) and seedlings (hybrids and inbreds). The seed tubers were planted on 8 May 2015 in a trial field on sandy soil in Wageningen and harvested on 17 September 2015.

loci, hardly produced progenies and the seedlings were extremely weak. Therefore, new crosses were made between F_3 and F_5 inbred lines and selfing was started again from these F_1s to continue selecting parent lines that combine beneficial traits. As a result, performance of the inbreds improved over the following generations (Fig. 4).

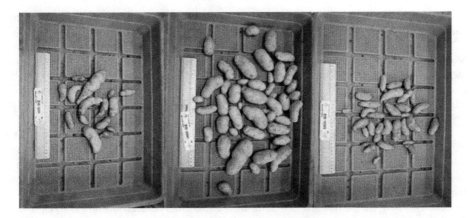

Figure 6 Example of heterosis in diploid hybrid potato. The plants were from the same trial as in Fig. 5. At the left the female F_3 parent and at the right the male F_5 parent, while the hybrid is in the middle.

In mature hybrid breeding systems, parents of the hybrids are selected based on general combining abilities (GCA), whereby molecular markers can be used to better predict the breeding value of the parents (Tobias et al., 2009). As no historic data were available on combining abilities of diploid potato parents, the selection of the first parents was based on the performance of the parents themselves.

Following the predictions of good performing inbred lines, crosses between these lines were made resulting in 45 potato hybrids which were tested in the field together with 20 inbred lines in two replicates of two plants per plot. The yields varied from 83 to 580 g/plant (Fig. 5). Thirteen hybrids scored higher than any of the inbreds, also outperforming the check variety Lady Anna, while some also showed a strong heterosis for yield (Fig. 6). The trials are being repeated in 2016 by using seed tubers raised from the field.

In addition, a new series of 216 experimental hybrids was generated in the winter season of 2014–15, mainly based on F_6 parent lines. The hybrid seeds were sown immediately after harvest and ten seedlings per hybrid were transplanted into the field in June, 2 months later than the usual seed tuber plantings. Still the yield of some hybrids was higher than 500 g/plant and the tubers were similar in size and shape to commercial seed tubers.

The results of the first experimental hybrids illustrate the potential of diploid hybrid potato varieties. As the first hybrids were randomly made without any prior information about the combining abilities of the parents, it is expected that the next series of hybrids based on the results of these field trials will perform better and may show a stronger overlap with commercial controls.

4 Production and commercialization of hybrid seed cultivars

The production of hybrid potato seeds is mainly done by hand pollinations. The seed yield per plant varies from hundreds to many thousands of seeds. Each successful hand pollination generates a berry with 50–150 seeds and each plant produces 5–50 berries. This is similar to other vegetable *Solanum* crops, like tomato and pepper, where commercial seed is produced by hand pollination.

The emasculation of the flowers is a time-consuming part of the hand pollinations. Male sterility systems have been reported that make emasculations redundant and hence reduce seed production costs (Li, 2008). Alternatively, functional male sterility may be used, whereby pollen is prevented to land on the stigma of the same flower as the exerted stigma may be manually pollinated by pollen of the male parent before the own pollen may reach the stigma (Lössl et al., 2000; Abrol et al., 2012).

The transport of potato seeds over the world is very restricted. This is because the dominant breeding systems are based on the production of seed tubers and hardly any rules are in place for potato seeds. As a consequence, most countries consider potato seeds as belonging to the highest risk classes. This is remarkable as seed tubers may harbour any of over 200 species of pathogens that attack potato (Delleman et al., 2004). In contrast, only six pathogens are seed-borne (Solomon-Blackburn and Barker, 2001). These are five viruses and a viroid, which are absent in major potato-growing regions like the Netherlands. So, potato seeds are very safe and regulations will need to be adjusted accordingly.

The registration process for breeders' rights poses a similar situation: in countries which are members of the International Union for the Protection of new Varieties of Plants (UPOV), the path to commercialization of a hybrid starts with the registration process to obtain breeders' rights, which is based on seed tubers. So, protection by breeders' rights of hybrid potato seed cultivars is not yet possible. The European Union (EU) is adjusting the legislation process, but it may still take several years before this is established. Other non-UPOV countries will likely follow later.

5 Inbred lines for genetic research

Inbred lines allow the generation of mapping populations such as F_2, BC_1 and BC_2. As the parents have limited allelic variation, the signal-to-noise ratio is much higher than in studies with heterozygous tetraploid populations. Moreover, putative quantitative trait loci (QTL) can effectively be confirmed in dedicated populations that are selected to segregate for the loci under investigation, and are fixed for other regions on the genome (Wang et al., 2008; Schmalenbach and Pillen, 2009; Fu et al., 2010). In addition, new genetic populations can be generated, which are very powerful for quantitative genetic studies, like nearly isogenic lines (NILs), recombinant inbred lines (RILs) and libraries of introgression lines (Young et al., 1988; Paran et al., 1995; Jeuken and Lindhout, 2004; Finkers et al., 2007; Zhang et al., 2005; Chen et al., 2010; Viquez et al., 2014).

Genetic studies in potato have been done at the tetraploid level and at the diploid level. Tetraploids may support simple genetics like the mapping of resistance genes (Solomon-Blackburn and Barker, 2001), but quantitative studies are less reliable as the genetic noise of the numerous highly heterozygous loci is high (unexplained error). Genome-wide association studies (GWAS) at the tetraploid level will always generate hundreds of potential leads, but only a few hits may be meaningful (Li et al., 2010; D'Hoop et al., 2014). More accurate and reliable quantitative studies were done at the diploid level, initially by crossing heterozygous parents (Prasher et al., 2014) and, more recently, in a diploid F_2 population (Endelman and Jansky, 2016).

The possibility of using self-compatible, homozygous inbred lines for genetic studies offers three powerful advantages:

1 Only one allele per locus is present.
2 Only two alleles per heterozygous locus are present.
3 Backcrosses and selfings are feasible for confirmation studies.

For more than five decades, mutant studies have uncovered new alleles of important genes and have helped to confirm or determine the function of genes. Although advanced technologies like the CRISPR/Cas system (Belhaj et al., 2013) are likely to replace the methods by which mutants are made, mutants will remain powerful tools to discover unknown phenotypic traits or to study induced alleles that also have the advantage to be free of deregulation rules.

5.1 The first completely homozygous self-compatible diploid inbred line in potato

Most *Solanum* species that are crossable with cultivated potato are self-incompatible. An exception is *S. chacoense* (Hosaka and Hanneman, 1998a; Hawkes, 1990; Jansky et al., 2014) and introducing the *Sli* gene from *S. chacoense* into cultivated diploid potato resulted in fertile self-compatible offspring (Lindhout et al., 2011a). After several rounds of inbreeding, highly homozygous self-compatible inbreds were generated. The level of homozygosity was assessed by using SNP markers to investigate the effects of inbreeding (Fig. 7). A strong correlation was observed between the overall level of homozygosity and reduced self-compatibility. By new series of crosses, selections and selfings, the agronomic

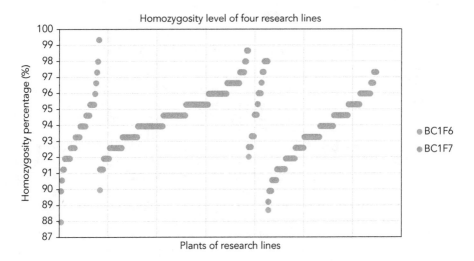

Figure 7 The increase of the level of heterozygosity in the offspring (BC_1F_7) of four BC_1F_6 plants obtained by selfing of the $BC_1(D1 \times F_1[DS \times D1])$ (Table 1). The width of the bar indicates the number of plants in the progeny within the class with plants of the same homozygosity level. The calculations were based on a total of 150 markers.

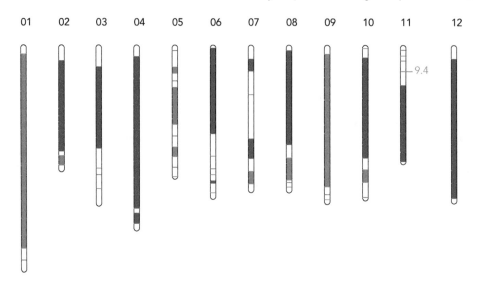

Figure 8 The first essentially homozygous self-compatible diploid potato. The position of the SNP markers is based on the published sequence (PSGC, 2011). The 12 vertical bars indicate the 12 chromosomes. Red bars indicate homozygous D1, blue bars indicate homozygous DS (Table 1). The green line on chromosome 11 indicates a heterozygous scored marker.

performance of the inbred lines continuously improved. After six generations of selfing, a homozygous self-compatible line was generated with only 1 out of 150 SNP markers still heterozygous (Fig. 8). Genotyping by sequencing techniques make it now possible to check the level of homozygosity in more detail.

5.2 Genetic studies in segregating diploid F_2 populations

Recently, Endelman and Jansky (2016) published the first results of a mapping study in an F_2 population of diploid potato. This was based on a cross between the doubled monoploid potato DM1-3 and M6, which is an S_7 inbred line derived from the self-compatible wild relative *S. chacoense*. A single F_1 plant was then self-pollinated and an F_2 population of 109 genotypes was grown, genotyped (>2200 SNPs) and phenotyped. Solynta analysed an F_2 population (108 markers) based on a cross between two clones, namely DS (a homozygous diploid *S. chacoense* clone containing the self-incompatibility overcoming *Sli* gene) and D2 (a partly heterozygous diploid *S. tuberosum* clone; see also Table 1). The results of both studies are comparable: tuber shape is associated with a region on chromosome 10, flesh colour with a region on chromosome 3 and tuber and pigment colour on chromosomes 2, 10 and 11. These QTLs were identified at the same loci as described in literature (van Eck et al., 1993, 1994). In both studies, additional QTLs were identified. Furthermore, there is an overwhelming reservoir of potential useful QTLs in the potato germplasm (Bradshaw et al., 2007) and thus also in the dihaploids that can be made. Such QTLs can now be more reliably studied at the diploid level and this will ultimately lead to the identification of the underlying genes. A limited subset of the diploid germplasm may already harbour many important traits for potato breeding (Table 1). Relevant genes for these traits can be introgressed into vigorous and fertile diploid genotypes to develop progenies with maximum genetic fixation to minimize genetic noise. Such genotypes are unique materials for further detailed genetic studies.

5.3 Marker-assisted backcrossing

Another application for inbred lines in potato is introgression breeding. This requires knowledge of diagnostic markers for the gene of interest, preferably inside the gene, markers for the recurrent parent genome and a self-compatible homozygous backcrossing parent. There are dozens of well-studied resistance genes in potato that can be used for introgression breeding. This paves the way for a marker-assisted backcrossing (MAB) programme in potato (Frisch and Melchinger, 2005).

To this end, homozygous inbred lines are crossed with a diploid donor carrying a specific gene of interest. In two backcrosses, NILs can be generated by selection with diagnostic markers for the gene of interest and against markers in its flanking regions, combined with selection for markers well distributed over the potato genome (whole background selection). Such NILs can harbour over 98% of the recurrent genome in combination with the specific gene. Both parents of a hybrid may have an introgressed gene, resulting in a double stack hybrid. To introgress specific genes in a homozygous parental line will take 2–3 years. These MAB programmes are routinely used in other crops and are also feasible in potato (Mallick et al., 2015; Jeong et al., 2015).

6 Cropping systems based on true seeds

The production of commercial tubers in most parts of the world starts with seed tubers. These have a large reservoir of nutrients for the growing shoots, allow a rapid initial plant growth and fast leaf coverage of the soil, which is one of the most critical factors for potato yield. In contrast, potato seeds are extremely tiny, about 2500 seeds per gram. As a result, during the first period after germination the young seedlings are very vulnerable for abiotic stresses like drought, frost and heat. Field emergence has been reported between 50% and 80% with acceptable tuber yields under different experimental conditions (El-Bedewy et al., 1994; Renia, 1995). However, without a protective environment, the risk is very high that an emerging seedling will not survive, even when it is pelleted or primed.

This chapter describes alternative strategies to circumvent the exposure of week seedlings to harsh conditions.

6.1 Production of seedling tubers in greenhouse

Seedling tubers can be produced under greenhouse conditions by sowing in a medium with sufficient water supply and at optimum germination temperature of 15–20°C (Struik and Wiersema, 2012). As soon as seedlings reach 5–10 cm in length, they are transplanted in pots. The desired tuber size, the available space in the greenhouse and the length of the growing period will determine the pot size, nutrient supply, day/night temperature and light regime (for a detailed protocol see Struik and Wiersema, 2012). Densities of 80–170 plants/m^2 are common in greenhouses (Lommen, 1995; Tierno et al., 2014). There is a trade-off between number of tubers, size of tubers, planting density and time to harvest.

Depending on the exact conditions, end-product requirements and production costs, the optimal combination is chosen. Alternatively, hydroponic techniques are available, whereby the roots are grown in a dark, humid and soilless environment in two layers, one for nutrient uptake and the other layer for producing mini-tubers (Lommen, 2007). The seedling tubers are picked at regular intervals and the total yield per plant may reach dozens of tubers. The soilless culture assures clean seed tubers. The conditions and picking regime are set to have optimal numbers and tuber sizes.

6.2 Production of seedlings for commercial crop

Greenhouse-grown seedlings can also be used to start the cultivation of a commercial crop. This system is equivalent to the one used for lettuce, leek, cabbage and onion (Leskovar et al., 2014). Technically, commercial potato production from seedlings is feasible and maximizes the benefits of true seeds. When potato transplants are grown as a ware potato crop, a whole new cultivation system must be developed. Important elements are transplanting systems, plant spacing including ridging or bedding, the use of soil coverage, weed control, irrigation and harvesting methods. Further mechanization and dedicated cultivation systems will be optimized for cropping systems that start with potato seedlings (Roy et al., 2015). In Kenya, tuber yields of 30 tonnes/ha were obtained, whereby seedling transplants were used as starting material for a commercial cultivation (Muthoni et al., 2014). This already represented 50% higher yield than average in Kenya (Wang'om and van Dijk, 2013).

6.3 Seedling tubers as starting materials for a commercial crop

Seedling tubers are equivalent to mini-tubers that are produced from tissue culture, which is routinely done to start a new multiplication round with clean basic seeds (Amin et al., 2014). They are certified as G1 material. The great advantage of seedlings are the reduced costs, compared with *in vitro* grown plants, and the flexibility to start the production whenever and wherever needed, as seeds can easily be stored and transported. As the cost to produce mini-tubers from *in vitro* grown plants is very high, in the present potato system at least three rounds of field multiplications are needed to dilute these high costs over many seed tubers. For a hybrid seed system, the cost of producing seedling tubers is much lower and hence fewer propagation rounds are needed. Therefore, seedling tubers should be multiplied only 1 year and then released to commercial farmers. Such a system also fits better to the fast introduction of new cultivars, which is typical for a hybrid breeding system.

6.4 Production of seed tubers from seedlings in field

At present, the production of seedling tubers is mostly done under tropical conditions. Seeds are sown in a simple greenhouse or in the field under plastic cover with plant densities of 80–100 plants/m^2 (Kumar, 2014; Struik and Wiersema, 2012; Fig. 9). When the seedlings have reached 5–10 cm in length, they are transplanted to the field, in ridges, at a defined plant density. Additional hilling will increase the number of seed tubers per plant (Wiersema, 1986). To decrease the risk of root damage, the complete substrate is transplanted with the seedling. Plantlets need some time to adapt to outdoor conditions before transplanting directly to the soil (Gopal, 2004). In South Asia, transplants are

Figure 9 The first seedlings of the first diploid hybrid potato hybrids in Democratic Republic of Congo.

planted on the sides of the ridges to benefit from the shade and higher soil humidity. The planting distance may be adjusted to compensate for a shorter growing season compared with seed tubers, if appropriate. The cultivation conditions are similar to traditional potato cultivation systems. In Egypt, seed tuber yields of 40–60 tonnes/ha were obtained in such a system, based on tetraploid TPS populations (El-Yazied et al., 2004).

Seedlings are more sensitive to frost and drought than seed tubers. Thus, transplanting is done in a frost-free season and with irrigation. Compared with the traditional systems of producing seed tubers, whereby the tubers are planted far before the last night with frost, the length of the growing season of transplants may be 2 months shorter. In addition, the plant development may be further delayed due to a transplanting shock and weak initial growth. Hence, tuber numbers are lower and tuber sizes are smaller compared to seed tuber grown plants, causing severe yield reductions. Plant density may be increased to compensate for these reduced yields per plants.

7 Case studies

7.1 Combatting *Phytophthora infestans*

Late blight, caused by the oomycete *Phytophthora infestans*, was responsible for the Irish famine of 1845–47 (Fry, 2008). All potato cultivars were susceptible to the disease and suffered severe yield losses that led to food shortage. Since then, breeders have selected cultivars with field resistance and from the early twentieth century onwards, have introduced specific *R*-genes, often sourced from wild relatives. However, cultivar Pentland Dell, which carries three *R*-genes, already turned out susceptible to a new virulent race of the fungus even before its widespread use and just 4 years after its introduction (Malcomsum, 1969).

At present, the optimal way to control *P. infestans* is a combined approach of clean seed tubers, clean soils, early and preventive chemical protection and destruction of crop debris after harvest. The global annual cost of cultivation measures and yield loss is estimated at US$5 billion (Duncan, 1999).

P. infestans reproduces mainly clonally. With hundreds of billions of spores produced per hectare in an infected crop (Skelsey et al., 2010) and a mutation rate of $1:10^9$, mutations in any given gene of *P. infestans* are likely to occur in a disease-infected field. Sexual recombination, combined with the redundancy of several effectors that are recognized by *R*-genes, explains why *P. infestans* easily mutates effector genes and develops virulence (Jiang and Tyler, 2012).

As a result of the high genetic variation in potato, dozens of resistance sources have been identified and are available for breeding (Park et al., 2009). Whereas single genes are easily defeated by virulent races to *P. infestans*, combinations of *R*-genes are more effective, although the Pentland Dell case indicates that a more dynamic approach may be needed (Niks et al., 2011).

Ideally, isogenic cultivars are developed that only differ in the combination of *R*-genes. This would allow to deploy the most suitable cultivar, dependent on the epidemiology of *P. infestans*. However, the introduction of one gene from a wild related species into a tetraploid cultivar by traditional breeding already takes several decades, and to combine different *R*-genes in a breeding programme is simply too complicated.

Since 1990, many *R*-genes to *P. infestans* have been mapped and cloned (Ballvora et al., 2002; Huang et al., 2005; Park et al., 2009). These all belong to the so-called class of 'NBS/LRR genes' and have a cytoplasmic interaction with effector genes of *P. infestans*, resulting in defence responses that block the growth of the pathogen (Jones and Dangl, 2006). A genetic modification (GM) approach to develop a series of isogenic cultivars with different *R*-genes from crossable species is being pursued (Haverkort et al., 2016, Jacobsen and Schouten, 2008). These so-called cisgenic plants might fall under the highly costly and complex GM legislation, which would hamper their commercial opportunities.

The hybrid breeding system offers a clear path towards resilient resistance to *P. infestans*: *R*-genes can be stacked in a potato hybrid via marker-assisted introgression (Park et al., 2009). Two genes can be combined via the two parents in 2–3 years, and additional *R*-genes can be added within a year to generate multi-stack resistance hybrids. In this way, series of *R*-gene isogenic hybrids can be generated as a dynamic resource to select the best combination of *R*-genes to protect the crop against the prevailing races of *P. infestans*.

7.2 Hybrid potato breeding for East Africa

Hybrid potato cultivars will bring great benefits, not only to modern commercial farmers in the developed world, but they may even have a greater social impact in tropical regions where the population rely on potato as a major source of energy and nutrition (FAOstat, 2016). It is very challenging to start an initiative to develop hybrid potato for these regions (Thomas-Sharma et al., 2016; Kumara et al., 2015). When legal and physical protection of the breeding germplasm is not secured in these regions, the development of inbred lines and the hybrid crosses are done elsewhere. The implementation of hybrid potato cultivars in these regions requires considerable investments and strong cooperation of committed

Table 2 Potato area and production in six East-African countries in 2014 (FAOstat, 2016)

Country	Area (000 ha)	Production (000 tonnes)	Yield (tonnes/ha)
Burundi	24.4	181.2	7.4
Kenya	115.6	1626.0	14.1
Rwanda	166.4	2225.1	13.4
Tanzania	211.5	1761.0	8.3
Uganda	39.0	188.0	4.8
Ethiopia	67.4	921.8	13.7

public and private partners. As an example, the implementation of hybrid potato cultivars for East Africa is briefly described below.

The highlands of East Africa are traditionally important production areas for potato because the crop is an important component of the local diet (Table 2). However, yields remain very low (Table 2). A range of traditional varieties is used from local sources (Kaguongo et al., 2008) as well as improved material from the International Centre for Potato (CIP). Seed tubers are produced by farmers and storage conditions are far from optimal (Kaguongo et al., 2008; Gildemacher et al., 2009). Janssens et al. (2013) concluded that bacterial wilt, lack of clean seed tubers and poor storage are the most prominent production constraints. Gildemacher et al. (2009) showed that only 3% of the seed tubers sold were free of viruses.

True hybrid potato seeds are devoid of contaminating pathogens and therefore offer an excellent opportunity to potato production improvement in East Africa. Such hybrids should be attuned to the needs of the farmers, who grow their crop at the typical local conditions like a short growing cycle of 90–100 days, high temperatures and tuber development under short days. The prerequisites for establishing dedicated hybrid potato breeding system for East Africa are as follows:

- A (private) organization executing a breeding programme tailored to the needs of the region.
- Secured supply of hybrid seeds for the region.
- Regulations supporting imports and exports of seeds, seed tubers and commercial tubers.
- Formal registration system for breeders' rights protection.
- Local organizations testing new experimental hybrids.

Such breeding programme can only become sustainable if the complete downstream part of potato food chain is also well organized. This includes the following:

- Production systems of disease-free seed tubers from seedlings.
- An efficient supply system for farmers to obtain hybrid cultivars.
- Efficient farmers' cropping systems to produce high-quality potato tubers for the target markets.
- Efficient logistics to transport farm produce to consumers and processors.

- A well-developed consumer market.
- A business plan over the whole potato chain, whereby all stakeholders benefit.

Solynta is already testing the first experimental hybrids in the highlands of the Albertine Rift in Ituri Province, Democratic Republic of Congo. These are experimental hybrids derived from the European hybrid breeding programme. At a local farm, established by the Lake Albert Foundation, seeds have been sown and seedlings have been transplanted into the field (Fig. 9). These hybrids are evaluated in good cooperation with local farmers and the results are shared with the Solynta breeding programme. These data are used to adjust the selection of inbred lines to the needs in East Africa and to continuously generate new experimental hybrids, which will be tested at the Congo farm again. This iterative and interactive process may already select the first dedicated hybrids for East Africa in 2–3 years.

This breeding programme is accompanied by research on cropping systems for the region. In addition, training programmes for research institutions, agronomists, local staff and interested farmers in the regions will make the farmers' communities and relevant institutions familiar with the new concepts of hybrid potato cultivars. Also market studies are needed to identify and secure stable and sustainable markets for the farmers' potato products. This market may comprise other countries in the Great Lakes Region (South Sudan, Uganda, Burundi and Rwanda).

The support for this programme by national authorities and development agencies – also in the neighbouring countries like Uganda – is required, but it will take time before the concept of hybrid potato cultivars is fully understood and accepted.

8 Conclusion

Since the first crosses in 2008, the Solynta research efforts have been focused on the development of a hybrid potato breeding system. This research has now reached the stage where hybrid potato breeding will become reality. This has recently been supported by two leading potato breeding companies in EU, KWS and HZPC, who have also expressed their conviction that hybrid potato breeding will be the future main breeding system (KWS, 2016). In addition, 21 leaders in the industry and potato science in the United States have expressed their opinion on 'Reinventing potato as a diploid inbred line-based crop' with a scientific base for diploid hybrid potato breeding (Jansky et al., 2016).

We have made great advances in the development of useful homozygous inbred lines and the first field evaluations of experimental hybrids have shown the potential of hybrid cultivars to harvest heterosis.

The technologies to develop new cropping systems adapted to various climate zones and agronomic practices, which allow the production of commercial seed tubers from seedlings, are already available.

We envision that future hybrid potato varieties, similar to modern tomato hybrids, will harbour up to 15 resistance genes. In addition, the lifetime of new potato hybrid cultivars will be reduced to less than 5 years as is the experience in sugar beet, where the lifetime of new cultivars is only 2–3 years since the first diploid hybrids have been introduced into the market.

Our inbred lines will also be of great value for research purposes as they allow the development of sophisticated populations that are very helpful in genetic and genomic studies as has been shown in other crops. This will give a boost in the exploration and exploitation of the genetic variation in the potato germplasm.

The self-compatible potato inbred lines will greatly stimulate quantitative research on the genetic control of complex traits such as interaction with mycorrhiza, biotic stresses, nutrient uptake, nutrition value and processing quality.

The application of hybrid potato breeding technology will not be restricted to the most advanced research groups in the public or private institutions. New players in the scientific and applied breeding field of the potato business will arise and new cooperations will be established to take full advantage of the hybrid breeding technology in science and product development.

Finally, hybrid potato breeding will require the skills of the breeder as well as the knowledge and tools of scientists. As a result, a new generation of potato breeding teams will be established combining the skills of the breeder with the in-depth knowledge of plant scientists.

9 Where to look for further information

This chapter describes the first implementation of a hybrid potato breeding strategy and the progress that is made since 2008. We direct the interested reader to the papers of Almekinders (1996, 2009) to learn about 'conventional TPS'. As a textbook on plant breeding, we suggest Brown and Cagliari (2011), while more advanced potato genetics and genomics is found in Bradeen and Kole (2011). The history and mechanism of *Phytophthora* attack is well described by Fry (2008). Finally, FAO (FAOstat, 2016) statistics provide numerous data on potato cultivation.

Leading centres of research include Wageningen University in the Netherlands, James Hutton Institute in the United Kingdom, the International Potato Centre in Peru, Wisconsin University and Michigan State University both in the United States, whereby the most recent paper of Jansky et al. (2016) can be considered a must for anybody interested in hybrid potato breeding.

10 Acknowledgements

We are grateful to Jan Leemans and Herman Fleer for critically reading and reviewing this manuscript.

11 References

Abrol, D. P. (2012). Pollination for hybrid seed production. In *Pollination Biology. Biodiversity Conservation and Agricultural Production*. Publisher Springer Netherlands, pp. 397–411.

Almekinders, C. J. M., Chilver, A. S. and Renia, H. M. (1996). Current status of the TPS technology in the world. *Pot. Res.* 39, 289–303.

Almekinders, C. J. M., Chujoy, E. and Thiele, G. (2009). The use of true potato seed as pro-poor technology: The efforts of an international agricultural research institute to innovating potato production. *Pot. Res.* 52, 275–93.

Amin, N., Amin, A. R., Roy, T. S., Ali, M. A., Rashid, M. M., Hossain, M. M. and Hasan, N. (2014). Bulking behavior of seedling tubers derived from true potato seed as affected by its size and harvesting time. *App. Sci. Rep.* 8, 1–8.

Ballvora, A., Ercolano, M. R., Weiss, J., Meksem, K., Bormann, C. A., Oberhagemann, P., Salamini, F. and Gebhardt, C. (2002). The *R1* gene for potato resistance to late blight (*Phytophthora infestans*) belongs to the leucine zipper/NBS/LRR class of plant resistance genes. *Plant J.* 30, 361–71.

Belhaj, K., Chaparro-Garcia, A., Kamoun, S. and Nekrasov, V. (2013). Plant genome editing made easy: targeted mutagenesis in model and crop plants using the CRISPR/Cas system. *Plant Meth.* 9, 39.

Birchler, J. A., Yao, H., Chudalayandi, S., Vaiman, D. and Veitia, R. A. (2010). Heterosis. *Plant Cell* 22, 2105–12.

Bradeen, J. M. and Kole, C. (2011). *Potato Genetics: Genetics, Genomics and Breeding of Potato.* CRC Press, USA

Bradshaw, J. E., Hackett, C. A., Pande, B., Waugh, R. and Bryan, G. J. (2007). QTL mapping of yield, agronomic and quality traits in tetraploid potato (*Solanum tuberosum* subsp. *tuberosum*). *Theor. Appl. Genet.* 116, 193–211.

Brown, J. and Caligari, P. (2011). *An Introduction to Plant Breeding.* Wiley-Blackwell, USA, p. 224.

Cappadocia, M. (1990). Wild Potato (*Solanum chacoense* Bitt.). In Y. P. S. Bajaj (ed.), *Vitro Production of Haploids. Biotechnology in Agriculture and Forestry, Vol. 12 Haploids in Crop Improvement I,* pp. 514–29. Springer-Verlag, Berlin, Heidelberg.

Chani, E., Veilleux, R. E. and Boluarte-Medina, T. (2000). Improved androgenesis of interspecific potato and efficiency of SSR markers to identify homozygous regenerants. *Plant Cell Tissue Organ Cult.* 60, 101–12.

Charlesworth, D. and Willis, J. H. (2009). The genetics of inbreeding depression. *Genetics* 10, 783–96.

Chauvin, J. E., Souchet, C., Dantec, J. P. and Ellisseche, D. (2003). Chromosome doubling of 2x *Solanum* species by oryzalin: method development and comparison with spontaneous chromosome doubling *in vitro*. *Plant Cell Tissue Organ Cult.* 73, 65–73.

Chen, X., Niks, R. E., Hedley, P. E., Morris, J., Druka, A., Marcel, T. C., Vels, A. and Wauh, R. (2010). Differential gene expression in nearly isogenic lines with QTL for partial resistance to *Puccinia hordei* in barley. *Genomics* 11, 629.

Crow, J. F. (1998). 90 Years ago: The beginning of hybrid maize. *Genetics* 148, 923–8.

CSAC (2005). Initial sequence of the chimpanzee genome and comparison with the human genome. *Nature* 437, 69–87.

D'hoop, B. B., Keizer, P. L. C., João Paulo, M., Visser, R. G. F., Van Eeuwijk, F. A. and Van Eck, H. J. (2014). Identification of agronomically important QTL in tetraploid potato cultivars using a marker-trait association analysis. *Theor. Appl. Genet.* 127, 731–48.

De Jong, H. and Rowe, P. R. (1971). Inbreeding in cultivated diploid potatoes. *Pot. Res.* 14, 74–83

De Mainea, M. J. (1982). An evaluation of the use of dihaploids and unreduced gametes in breeding for quantitative resistance to potato pathogens. *J. Agric. Sci.* 99, 79–83.

Delleman, J., Mulder, A. and Turkensteen, L. J. (2004). *Potato Diseases: Diseases, Pests and Defects.* Potatoworld and NIVAP, The Hague, the Netherlands.

Douches, D. S., Maas, D. J., Astrzebski, K. and Chase, R. W. (1996). Assessment of potato breeding progress in the USA over the last century. *Crop Sci.* 36, 1544–52.

Duncan, J. M. (1999). *Phytophthora*-an abiding threat to our crops. *Microbiol. Today* 26, 114–16.

Duvick, D. N. (2005). The contribution of breeding to yield advances in maize (Zea mays L.). *Adv. Agron.* 86, 83–145.

El-Bedewy, R., Crissman, C. and Cortbaoui, R. (1994). Progress report. Egypt's seed system based on true potato seed. *CIP Circular* 20, 5–8.

El-Yazied, A., Elminiawy, S. E., Hamoud, N. K. and El-Kheima, S. (2004). Seed tuber production of some hybrids using true potato seed. Mansoura University. *J. Agric. Sci.* 32, 1329–41.

Endelman, J. B. and Jansky, S. H. (2016). Genetic mapping with an inbred line-derived F_2 population in potato. *Theor. Appl. Genet.* 1–9

FAO, IFAD and WFP. (2015). The State of Food Insecurity in the World 2015. Meeting the 2015 international hunger targets: taking stock of uneven progress. Rome, FAO.

FAOstat. (2016). http://faostat3.fao.org/home/E

Finkers, R., Van Heusden, A. W., Meijer-Dekens, F., Van Kan, J. A. L. and Lindhout, P. (2007). The construction of a Solanum habrochaites LYC4 introgression population and the identification of QTLs for resistance to Botrytis cinerea. Theor. Appl. Genet. 114, 1071–80.

Frisch, M. and Melchinger, A. E. (2005). Selection theory for marker-assisted backcrossing. Genetics 170, 909–17.

Fry, W. (2008). Plant diseases that changed the world Phytophthora infestans: the plant (and R gene) destroyer. Molec. Pl. Path. 9, 385–402.

Gildemacher, P. R., Demo, P., Barker, I., Kaguongo, W., Woldegiorgis, GT., Wagoire, W. W., Wakahiu, M., Leeuwis, C. and Struik, P. C. (2009). A Description of seed potato systems in Kenya, Uganda and Ethiopia. Am. J. Pot. Res. 86, 373–82.

Gopal, J. (2004). True potato seed: Breeding for hardiness. In A. Haneafi (ed.), Sixth Triennial Congress of the African Potato Association. Proc. APA Congr., 5–10 April, Agadir, Morocco, pp. 39–57.

Gopal, J. (2014). Heterosis breeding in potato. Agric. Res. 3, 204–17.

Haldane, J. (1930). Theoretical genetics of autopolyploids. J. Genet. 22, 359–72.

Haverkort, A. J., Boonekamp, P. M., Hutten, R., Jacobsen, E., Lotz, L. A. P., Kessel, G. J. T., Vossen, J. H. and Visser, R. G. F. (2016). Durable late blight resistance in potato through dynamic varieties obtained by cisgenesis: Scientific and societal advances in the DuRPh project. Pot. Res., 59, 35–66.

Hawkes, J. G. (1990). The Potato: Evolution, Biodiversity, and Genetic Resources. Belhaven Press, London.

Hosaka, K. and Hanneman, R. E. (1998a). Genetics of self-compatibility in a self-incompatible wild diploid potato species Solanum chacoense. 1. Detection of an S locus inhibitor (Sli) gene. Euphytica 99, 191–7.

Hosaka, K. and Hanneman, R. E. (1998b). Genetics of self-compatibility in a self-incompatible wild diploid potato species Solanum chacoense. 2. Localization of an S locus inhibitor (Sli) gene on the potato genome using DNA markers. Euphytica 103, 265–71.

Hua, J., Xing, Y., Wu, W., Xu, C., Sun, X., Yu, S. and Zhang, Q. (2003). Single-locus heterotic effects and dominance by dominance interactions can adequately explain the genetic basis of heterosis in an elite rice hybrid. Proc. Natl. Acad. Sci. USA 100, 2574–9.

Huang, S., Van der Vossen, E. A., Kuang, H., Vleeshouwers, V. G., Zhang, N., Borm, T. J., Van Eck, H. J., Baker, B., Jacobsen, E. and Visser, R. G. (2005). Comparative genomics enabled the isolation of the R3a late blight resistance gene in potato. Plant J. 42, 251–61.

Hutten, R. C. B. (1994). Basic Aspects of Potato Breeding Via the Diploid Level. PhD Thesis, Wageningen University, p. 93.

Hutten, R. C. B., Schippers, M. G. M., Hermsen, J. G. Th. and Jacobsen, E. (1994). Comparative performance of diploid and tetraploid progenies from 2x.2x crosses in potato. Euphytica 81, 187–92.

Jacobsen, E. and Schouten, H. J. (2008). Cisgenesis, a new tool for traditional plant breeding, should be exempted from the regulation on genetically modified organisms in a step by step approach. Pot. Res. 51, 75.

Jansky, S. H., Charkowski, A. O., Douches, D. S., Gusmini, G., Richael, C., Bethke, P. C., Spooner, D. M., Novy, R. G., De Jong, H., De Jong, W. S., Bamberg, J. B., Thompson, A. L., Bizimungu, B., Holm, D. G., Brown, C. R., Haynes, K. G., Sathuvalli, V. R., Veilleux, R. E., Miller Jr., J. C., Bradeen, J. M. and Jiang, J. M. (2016). Reinventing potato as a diploid inbred line-based crop. Crop Sci. 56, 1–11.

Jansky, S. H., Chung, Y. S. and Kittipadakul, P. (2014). M6: A diploid potato inbred line for use in breeding and genetics research. J. Plant Registr. 8, 195–9.

Jansky, S. H. and Peloquin, S. J. (2006). Advantages of wild diploid Solanum species over cultivated diploid relatives in potato breeding programs. Genet. Res. Crop Evol. 53, 669–74.

Janssens, S. R. M., Wiersema, S. G., Goos, H. and Wiersema, W. (2013). The value chain for seed and ware potatoes in Kenya; Opportunities for development LEI. Memorandum 13-080, p. 57.

Jeong, H-S., Jang, S., Han, K., Kwon, J-K. and Kang, B-C. (2015). Marker-assisted backcross breeding for development of pepper varieties (Capsicum annuum) containing capsinoids. Molec. Breeding 35, 226.

Jeuken, M. J. W. and Lindhout, P. (2004). The development of lettuce backcross inbred lines (BILs) for exploitation of the Lactuca saligna (wild lettuce) germplasm. Theor. Appl. Genet. 109, 394–401.

Jiang, R. H. Y. and Tyler, B. M. (2012). Mechanisms and evolution of virulence in oomycetes. Ann. Rev. Phytop. 50, 295–318.

Jones, J. D. G and Dangl, J. L. (2006). The plant immune system. Nature 444, 323–9.

Kaguongo, W., Gildemacher, P., Demo, P., Wagoire, W., Kinyae, P., Andrade, J., Forbes, G., Fuglie, K., and Thiele, G. (2008). Farmer practices and adoption of improved potato varieties in Kenya and Uganda. International Potato Center (CIP), Lima, Peru. Social Sciences Working Paper 2008-5. 85 p.

Kumar, V. (2014). True potato seed technology – Prospects and problems. In N. K. Pandey, D. K. Singh and R. Kumar (eds), Current Trends in Quality Potato Production, Processing and Marketing, pp. 175–82. Central Potato Research Institute, India.

Kumara, N. S., Govindakrishnan, P. M., Swarooparani, D. N., Nitin, Ch. Surabhi, J. and Aggarwal, P. K. (2015). Assessment of impact of climate change on potato and potential adaptation gains in the Indo-Gangetic Plains of India. Intern. J. Pl. Prod. 9 (1), 151–70.

KWS press release (2016). http://www.kws.com/aw/KWS/company-info/Products/Potatoes/News-Articles/~hjqx/KWS-to-fully-focus-on-hybrid-potato-bree/

Leskovar, I. D., Crosby, M. K., Palma, A. M. and Edelstein, M. (2014). Vegetable crops: Linking production, breeding and marketing. In R. G. Dixon and E. D. Aldous (eds), Horticulture: Plants for people and places, Volume 1: Production Horticulture, pp. 75–96. Springer Netherlands, Dordrecht.

Li, L., Paulo, M.-J., Van Eeuwijk, F., and Gebhardt, C. (2010). Statistical epistasis between candidate gene alleles for complex tuber traits in an association mapping population of tetraploid potato. Theor. Appl. Genet. 121, 1303–10.

Li, X.-Q. (2008). Male sterility systems for hybrid seed production in Brassica crops. CAB Reviews: Perspectives in Agriculture, Veterinary Science, Nutrition and Natural Resources 3, 1–14.

Lindhout, P., Meijer, D., Schotte, T., Hutten, R. C. B., Visser, R. G. F. and Van Eck, H. J. (2011a). Towards F1 hybrid seed potato breeding. Pot. Res. 54, 301–12.

Lindhout, W. H., Schotte, T. P., Visser, R. G. F., Van Eck, H. J. and Hutten, R. C. B. (2011b). Hybrid seed potato breeding. European Patent Office nr EP 2514303 A1

Lippman, Z. B. and Zamir, D. (2007). Heterosis: revisiting the magic. Trends Genet. 23, 60–6.

Lommen, W. J. M. (1995). Basic studies on the production and performance of potato minitubers. PhD Thesis, Wageningen Agricultural University, Wageningen, The Netherlands, p. 181.

Lommen, W. J. M. (2007). The canon of potato science: Hydroponics. Pot. Res. 50, 315–18.

Lössl, A., Götz, M., Braun, A. and Wenzel, G. (2000). Molecular markers for cytoplasm in potato: Male sterility and contribution of different plastid-mitochondrial configurations to starch production. Euphytica 116, 221–30.

Malcolmson, J. F. (1969). Races of Phytophthora infestans occurring in Great Britain. Trans. Br. Mycol. Soc. 53, 417–23.

Mallick, N., Vinod, Sharma, J. B., Tomar, R. S., Sivasamy, M. and Prabhu, K. V. (2015). Marker-assisted backcross breeding to combine multiple rust resistance in wheat. Plant Breed. 134, 172–7.

Muthoni, J., Shimelis, H., Melis, R. and Kinyua, Z. M. (2014). Response of potato genotypes to bacterial wilt caused by Ralstonia solanacearum (Smith) (Yabuuchi et al.) in the tropical highlands. Am. J. Pot. Res. 91, 215–32.

Niks, R. E., Parlevliet, J. E., Lindhout, P. and Bai, Y. (2011). Breeding Crops with Resistance to Diseases and Pests. Wageningen Academic Publishers, p. 200.

Paran, I., Goldman, I., Tanksley, S. D. and Zamir, D. (1995). Recombinant inbred lines for genetic mapping in tomato. *Theor. Appl. Genet.* 90, 542–8.

Park, T. H., Vleeshouwers, V. G. A. A., Jacobsen, E., Van der Vossen, E. and Visser, R. G. F. (2009). Molecular breeding for resistance to *Phytophthora infestans* (Mont.) de Bary in potato (*Solanum tuberosum* L.): a perspective of cisgenesis. *Plant Breed.* 128, 109–17.

PGSC (2011). Genome sequence and analysis of the tuber crop potato. *Nature* 475, 189–94.

Phumichai, C. and Hosaka, K. (2006). Cryptic improvement for fertility by continuous selfing of diploid potatoes using *Sli* gene. *Euphytica* 149, 251–8.

Phumichai, C., Mori, M., Kobayashi, A., Kamijima, O. and Hosaka, K. (2005). Toward the development of highly homozygous diploid potato lines using the self-compatibility controlling *Sli* gene. *Genome* 48, 977–84.

Prashar, A., Hornyik, C., Young, V., McLean. K., Kumar Sharma, S., Dale, M. F. B. and Bryan, G. J. (2014). Construction of a dense SNP map of a highly heterozygous diploid potato population and QTL analysis of tuber shape and eye depth. *Theor. Appl. Genet.* 127, 2159–71.

Renia, H. (1995). True seed is a commercial reality in USA. *Pot. Rev.* 5, 48–51.

Rijk, B., van Ittersum, M. and Withagen, J. (2013). Genetic progress in Dutch crop yields. *Field Crops Res.* 149, 262–8.

Rowe, P. R. (1967). Performance and variability of diploid and tetraploid potato families. *Am. Pot. J.* 44, 263–71.

Roy, T. S., Baque, M. A., Chakraborty, R., Haque, M. N. and Suter, P. (2015). Yield and economic return of seedling tuber derived from True Potato Seed as influenced by tuber size and plant spacing. *Univ. J. Agric. Res.* 3, 23–30.

Schmalenbach, I. and Pillen, K. (2009). Detection and verification of malting quality QTLs using wild barley introgression lines. *Theor. Appl. Genet.* 118, 1411–27.

Skelsey, P., Rossing, W. A. H., Kessel, G. J. T. and Van der Werf, W. (2010). Invasion of *Phytophthora infestans* at the landscape level: How do spatial scale and weather modulate the consequences of spatial heterogeneity in host resistance? *Phytopath.* 100, 1146–61.

Solomon-Blackburn, R. M. and Barker, H. (2001). A review of host major-gene resistance to potato viruses X, Y, A and V in potato: genes, genetics and mapped locations. *Heredity* 86, 8–16.

Struik, P. C. and Wiersema, S. (2012). *Seed Potato Technology*. Wageningen Academic Publishers, p. 383.

Su, C. F., Lu, W. G., Zhao, T. J. and Gai, J. Y. (2009). Verification and fine-mapping of QTLs conferring days to flowering in soybean using residual heterozygous lines. *Chin. Sci. Bull.* 6, 499–508.

Thomas-Sharma, S., Abdurahman, A., Ali, S., Andrade-Piedra, J. L., Bao, S., Charkowski, A. O., Crook, D., Kadian, M., Kromann, P., Struik, P. C., Torrance, L., Garrett, K. A. and Forbes, G. A. (2016). Seed degeneration in potato: the need for an integrated seed health strategy to mitigate the problem in developing countries. *Plant Path.* 65, 3–16.

Tierno, R., Carrasco, A., Ritter, E. and Ruiz de Galarreta, J. I. (2014). Differential growth response and minituber production of three potato cultivars under aeroponics and greenhouse bed culture. *Amer. J. Pot. Res.* 91, 346–53.

Tobias, A., Schrag, T. A., Möhring, J., Melchinger, A. E., Kusterer, B., Dhillon, B. S., Piepho, H-P. and Frisch, M. (2009). Prediction of hybrid performance in maize using molecular markers and joint analyses of hybrids and parental inbreds. *Theor. Appl. Genet.* 120, 451–61.

Troyer, A. F. (2006). Adaptedness and heterosis in corn and mule hybrids. *Crop Sci.* 46, 528–43.

Uijtewaal, B. A., Huigen, D. J. and Hermsen, J. G. Th. (1987a). Production of potato monohaploids (2n = x = 12) through prickle pollination. *Theor. Appl. Genet.* 73, 751–8.

Uijtewaal, B. A., Jacobsen, E. and Hermsen, J. G. Th (1987b). Morphology and vigour of monohaploid potato clones, their corresponding homozygous diploids and tetraploids and their heterozygous diploid parent. *Euphytica* 36, 745–53.

Uitdewilligen, J. G. A. M. L., Wolters, A. M. A, D'hoop, B. B., Borm, T. J. A., Visser, R. G. F. and Van Eck, H. J. (2013). A next-generation sequencing method for genotyping-by-sequencing of highly heterozygous autotetraploid potato. *PLoS ONE* 8(5), e62355.

Van Breukelen, E. W. M., Ramanna, M. S. and Hermsen, J. G. Th. (1977). Pathenogenetic monohaploids (2n = 2x = 12 from *Solanum tuberosum* L. and *S. verrucosum* Schlechtd. and the production of homozygous potato diploids. *Euphytica* 26, 263–71.

Van Eck, H. J., Jacobs, J. M. E., van den Berg, P. M. M. M., Stiekema, W. J. and Jacobsen, E. (1994), The inheritance of anthocyanin pigmentation in potato (*Solanum tuberosum* L.) and mapping of tuber skin colour loci using RFLPs. *Heredity* 73, 410–21.

Van Eck, H. J., Jacobs, J. M. E., van Dijk, J., Stiekema, W. J. and Jacobsen, E. (1993) Identification and mapping of three flower colour loci of potato (*S. tuberosum* L.) by RFLP analysis. *Theor. Appl. Genet.* 86, 295–300.

Víquez-Zamora, M., Caro, M., Finkers, R., Tikunov, Y., Bovy, A., Visser, R. G. F., Bai, Y. and Van Heusden, S. (2014). Mapping in the era of sequencing: high density genotyping and its application for mapping TYLCV resistance in *Solanum pimpinellifolium*. *BMC Genomics* 15, 1152.

Visser, R. G. F., Bachem, C. W. B., Borm, T., de Boer, J., Van Eck, H. J., Finkers, R., Van der Linden, G., Maliepaard, C. A., Uitdewilligen, J. G. A. M. L., Voorrips, R., Vos, P. and Wolters, A. M. A. (2014). Possibilities and challenges of the potato genome sequence. *Pot. Res.* 57, 327–30.

Vos, P. G., Uitdewilligen, J. G. A. M. L, Voorrips, R. E., Visser, R. G. F. and Van Eck, H. J. (2015). Development and analysis of a 20K SNP array for potato (*Solanum tuberosum*): an insight into the breeding history. *Theor. Appl. Genet.* 128, 2387–401.

Wang, C. M., Lo, L. C, Feng, F., Zhu, Z. Y. and Yue, G. H. (2008). Identification and verification of QTL associated with growth traits in two genetic backgrounds of Barramundi (*Lates calcarifer*). *Animal Genet. Vol.* 39, 34–9.

Wang'om, W. G. and van Dijk, M. P. (2013). Low potato yields in Kenya: do conventional input innovations account for the yields disparity? *Agric. Food Sec.* 2, 14.

Wiersema, S. G. (1986). A method of producing seed tubers from true potato seed. *Pot. Res.* 29, 225–37.

Young, N. D., Zamir, D., Ganal, M. W. and Tanksley, S. D. (1988). Use of isogenic lines and simultaneous probing to identify DNA markers tightly linked to the *tm-2a* gene in tomato. *Genetics* 120, 579–85.

Zhang, Y-M., Mao, Y., Xie, C., Smith, H., Luo, L. and Xu, S (2005). Mapping Quantitative Trait Loci using naturally occurring genetic variance among commercial inbred lines of maize (*Zea mays* L.). *Genetics* 169, 2267–75.

Part 2

Improving particular traits

Chapter 6

Advances in development of potato varieties resistant to abiotic stress

Ankush Prashar and Filipe de Jesus Colwell, Newcastle University, UK; and Csaba Hornyik and Glenn J. Bryan, The James Hutton Institute, UK

1 Introduction

The major challenge in modern agriculture is to sustain crop productivity in the face of ongoing environmental changes (Mahmud et al., 2016). There is also pressure to increase crop productivity to feed the ever-growing global human population. Abiotic stresses namely drought, salinity, high or low temperature, submergence, nutrient deficiency and so forth have an impact on crop yields. These suboptimal conditions restrict crop performance so that the plants do not reach their full genetic potential. Modern agriculture aims to maximise yield and stabilise production by optimising growing conditions and by making it adaptive to soil and climate conditions and resource application in order to prevent stress. The crop domestication traits and selection process has been mainly involved with improvements in yield, so even though the focus of modern agriculture is on maximising the genetic potential of crops, modern crop varieties lack the ability to adapt to less than favourable conditions, crop stresses and low-resource applications (Fess et al., 2011; Shomura et al., 2008). So, one of the challenges for agricultural intensification is breeding for variable environmental conditions and different stress patterns by maintaining crop genetic potential. This may also be referred to as 'sustainable intensification'.

Potato, the most important non-cereal crop, is a plant of temperate climate and cultivated worldwide. Potato cultivation has been adapted to different conditions around the world as the genetic variation of the progenitors at the region of origin (South America) is high, and potatoes can be grown in varied environments, ranging from cool highland to tropical conditions (Hawkes, 1994). Potato breeding programmes started in the nineteenth century, whereby many potato varieties were created using the traditional method of selecting individuals with the desired phenotypic trait and repeated crossings until a stable and improved individual is obtained. This method allowed a substantial increase in productivity and the creation of varieties suitable to a multitude of environmental conditions, most

http://dx.doi.org/10.19103/AS.2017.0016.06

notably a shift from short- to long-day environments. However, the pressure from market forces, human growth and climate change has increased the need for high-yielding varieties which are also resistant to/tolerant of the increased environmental pressures encountered. Traditional methods alone can no longer fulfil this need as breeding programmes can take more than ten years to be successful (Bradshaw, 2009) and the selection process is time-consuming, with the visual assessment of traits often limited in accuracy. The identification of the underlying markers and genes responsible for the desired traits allows for a more accurate and faster selection process, effectively reducing the time taken in a breeding programme.

In the last decade, biotechnological and genomic advances have provided tools that enable the study of many genes, proteins, signalling and metabolic pathways simultaneously. This has opened up new possibilities for elucidating pathways involved in diverse stress responses. These molecular and genetics techniques now allow us not only to better understand the mechanisms behind the desired traits, but also to facilitate propagation of these traits more accurately into new varieties. This chapter aims to look at different abiotic stress improvement targets in the potato and also different tools and techniques being developed and used for crop improvement for abiotic stresses.

2 Abiotic stress improvement targets for potatoes

2.1 Drought stress

Improvement in drought resistance is a key objective worldwide, and it is the most limiting factor of crop production (Jones and Turner, 1980), with predictions that the frequency and severity of drought are likely to increase in future as a result of climate change (Bonierbale et al., 2017). Drought is regarded as one of the major abiotic factors affecting potato production and is mainly attributed to its shallow root system along with its limited ability to recover substantially after periods of water stress (Iwama and Yamaguchi, 2006; Onder et al., 2005). Thus, one of the targets focussed by breeders is the improvement of root characteristics (e.g. density and depth) in potatoes, aiming to increase water uptake and improve yield. Potato crop sensitivity to water stress varies with the developmental stage of the crop. How drought affects the crop growth, development and final yield and different indicators of stress during different growth stages has been published in various papers [e.g. (Obidiegwu et al., 2015)].

The potato plant's response to drought is also regulated by the intensity and duration of the stress. Depending upon the nature of stress, whether it is intermittent or prolonged and its severity, plants perceive and respond to drought stress at the molecular level through shifts in regulatory responses which affect different biochemical pathways, physiological processes and plant development (Mane et al., 2008; Vasquez-Robinet et al., 2008). As an example, photosynthesis reduction is mediated through both stomatal and non-stomatal effects, with decreases in CO_2 availability in the mesophyll during stomatal closure. The other factors which contribute to reduced photosynthesis include changes in electron transport and biochemical pathways whereby there is a reduction in ribulose bisphosphate synthesis and thus a decrease in Rubisco activity and carboxylation (Angelopoulos et al., 1996; Cornic et al., 1983; Gimenez et al., 1992). In order to maintain normal homeostasis and escape permanent wilting, these functional and regulatory proteins also help the plants in rapid and efficient recovery from water stress and maximise their development during drought stress. The varied stress and adaptive response influence physiological response of stomatal characteristics under mild to moderate stress conditions, and photochemical efficiency and biochemical and physiological metabolism under severe stress levels (Liu et al., 2010; Xu et al., 2010) affecting photosynthetic capacity, which then limits crop growth (Haverkort et al., 1991). As mentioned above, drought induces biochemical and molecular changes (Shinozaki and Yamaguchi-Shinozaki, 2007) and drought stress has been shown to severely affect tuberisation, yield and quality. Drought causes osmotic stress and in some cases influences the synthesis of compatible solutes, such as proline, soluble sugars and glycine betaine, which plays a crucial role in osmotic adjustments. Studies have found an interaction between key genes involved in endogenous hormone biosynthesis and signal transduction pathways (Gong et al., 2015).

Plants' response to stress initially appears at the cellular level, followed by physiological and developmental symptoms. In the case of drought, osmotic stress is imposed, which leads to a loss of turgor, loss of membrane integrity, protein denaturation and oxidative damage (Munns and Tester, 2008). Figure 1 highlights some key traits for adaptation for drought stress identified in various research studies. As the genotypic response in potato varies for physiological characteristics, the availability of wide germplasm resources in potato with different physiological and morphological characteristics and photosynthetic assimilation is crucial in improving plants' ability to resist drought and improve yield (Obidiegwu et al., 2015; Wang and Clarke, 1993).

While researching the effects of drought, studies have also highlighted the importance of aspects such as water use efficiency (WUE), traits influencing survival and recovery (e.g.

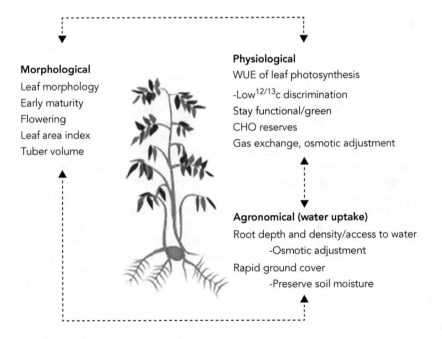

Morphological
Leaf morphology
Early maturity
Flowering
Leaf area index
Tuber volume

Physiological
WUE of leaf photosynthesis
-Low$^{12/13}$c discrimination
Stay functional/green
CHO reserves
Gas exchange, osmotic adjustment

Agronomical (water uptake)
Root depth and density/access to water
-Osmotic adjustment
Rapid ground cover
-Preserve soil moisture

Figure 1 Traits partitioned into different physiological, morphological and agronomical traits associated with main drivers of yield under drought.

dehydration avoidance and tolerance) and early maturity which should become a part of the breeding programmes for drought stress resistance and tolerance (Chaves et al., 2002; Obidiegwu et al., 2015; Xu et al., 2010). During drought conditions, water use efficiency tends to increases as stomata close more frequently (Vos and Groenwold, 1989); so improving drought resistance using WUE traits could have benefits like reduction in the amount of water agriculture uses, but this might have adverse effects on photosynthetic efficiency.

A research study evaluating and characterising new potato genotypes from CIP (International Potato Center) for tolerance to drought for Ugandan conditions found that of all the traits evaluated, yield and number of days to 50% flowering contributed most to drought tolerance (Kesiime et al., 2016). The evaluation of different traits in this study characterised CIP potato clones 391533.1 and 394034.7 as high-yielding clones under stress that were drought tolerant (thus potential drought-tolerant varieties), while clones 395017.242 and 393077.159 were moderately tolerant. Studies have also developed research activities aimed to identify the genes responsible for the traits indirectly linked to drought tolerance. Anithakumari et al. (2011) identified three QTLs linked to shoot-to-root ratio in an *in vitro* study and in a similar study on diploid population identified 28 drought-specific QTLs and another 17 potential QTLs related to post-stress recovery (Anithakumari et al., 2012). Another work on drought stress response resulted in the identification of 45 genomic regions of which 26 are potentially associated with drought stress (Khan et al., 2015). This shows that enough knowledge and some known genomic regions exist for drought stress response, which needs to be replicated and studied further for combatting drought stress.

2.2 Heat stress

Unlike drought tolerance, heat tolerance in potato was not a key improvement target because optimum production occurs mainly in temperate zones with a temperature range of 14–22°C (Figueiredo et al., 2015). Since the introduction of potato cultivation to warmer regions, heat stress is heightened in tropical and subtropical areas where potato crop potential is restricted, having originated from regions with cooler temperatures in the Andes (Hancock et al., 2014); thus, breeding varieties which are heat tolerant is one of the top priorities. Heat stress affects plant productivity because of a reduction in photosynthesis and increased respiration and corresponding limited photosynthate for growth and development. High-temperature stress alters hormonal balances, causing imbalances in the source–sink relation, and disturbs photosynthate partitioning among organs, causing decreases in productivity through delay in tuberisation and tuber bulking, affecting marketable yield (Minhas, 2012). Some studies have shown that the gibberellin content increases, due to heat stress which promotes plant elongation (Mares et al., 1981). Higher levels of potato eukaryotic elongation factor (eEF1A) have been shown to be positively correlated with productivity under both growth culture and field trials when studied in commercial varieties of potato including Agria, Carrera, Desiree, Jelly, Laura and Marabel (Momčilović et al., 2016). The earlier the occurrence of heat stress, the higher is the impact on growth and productivity. Hence, the reduction in potato yield is mainly attributed to the inhibition of physiological and genetic responses during tuberisation and reduction in photoassimilate partitioning to tubers (Lafta and Lorenzen, 1995). Thus one of the main targets for breeding is developing tuber crops with improved thermotolerance. Studies have been carried out in segregating mapping populations to understand the genetics behind thermotolerance of potato, and nine QTLs for internal heat necrosis in tubers were detected (McCord et al., 2011b). Higher temperatures are shown to inhibit tuberisation signals which affect tuber development, thus leading to yield reduction. Recent developments on tuberisation simulation studies have identified Flowering locus T orthologue termed 'StSP6A' as the tuberisation stimulus and studies predict that StSP6A transcript level is sensitive to elevated temperature, with reduced level corresponding to reduced tuber yield (Ewing, 1981; Navarro et al., 2011). Different physiological and biochemical responses have been shown to form a network of complex array of heat stress responses and to affect tuber development and yield (Hancock et al., 2014; Levy and Veilleux, 2007).

Studies in potato species have reported genetic variability for heat tolerance, and some of these have been exploited in breeding programmes (Tai et al., 1994). Heat tolerance is regarded as being under multigenic control with different species, for example, *Solanum chacoense*, *Solanum berthaultii* and *Solanum microdontum* have been explored for having heat tolerance characteristics. In order to combine characteristics from different species, Veilleux and colleagues produced complex tetraploid hybrids by first constructing interspecific diploid hybrids between *Solanum phureja* and *S. chacoense*, *S. berthaultii* and *S. microdontum* (Veilleux et al., 1997) and then sexual polyploidisation between *Solanum tuberosum* cv Atlantic and diploid hybrids. The vigorous cultivar 'Zahov' developed using a superior clone from above crosses was heat tolerant (Levy et al., 2001; Levy and Veilleux, 2007), also suggesting that the potato germplasm is quite diverse and adaptive to different conditions (Hawkes, 1992).

2.3 Salinity stress

Salinity is regarded as one of the major abiotic stresses which reduce not only cultivation area but also crop productivity and quality, which is expected to intensify under anticipated future climatic scenarios (Yamaguchi and Blumwald, 2005). Potato is considered a moderately salt-sensitive crop, and salt stress in potato reduces biomass production, induction of senescence and earlier plant death. This happens due to changes in photosystem apparatus and Calvin cycle (Legay et al., 2009), affecting photosynthesis rate (Chaves et al., 2009), and a decrease in CO_2 assimilation, leading to generation of reactive oxygen species. This leads to increase in NADP+ and hence blocks the repair of photosystem II (Takahashi and Murata, 2008). This has been shown to be linked to repression of genes encoding photosystems I and II and chlorophyll biosynthesis (Legay et al., 2009). Salinity damage affects cellular components in two different ways, as ionic stress and osmotic stress. One of them is similar to drought stress (osmotic regulation and osmoprotectants maintain function of proteins) and the other is related to the removal of Na+ from the cytoplasm. Under salt stress, leaf water and osmotic potentials decline in potato, and there is accumulation of proline, an osmoprotectant, to permit it to cope with the resulting high osmotic potential due to salt stress as proline helps in intracellular osmotic adjustments between the cytoplasm and vacuole (Heuer and Nadler, 1998; Martinez et al., 1996). Similarly, there is an increase in content of soluble sugars which as osmoprotectants are considered to counteract the ionic toxic effects in shoots under salt stress (Zhang et al., 2005). The potato germplasm base is very wide, and greater salinity stress tolerance exists in wild types with examples including *Solanum chacoense*, *Solanum juzepczuckii*, *Solanum kurzianum* and *Solanum curtilobum* (Sabbah and Tal, 1995; Shaterian et al., 2005). Similarly, some varieties which have been reported to be salt-tolerant based on field trial studies using saline water include Cara, Norland, Norchip, Serrana, Alpha, Atica, Patrones and so forth, and those that have been reported to be moderately tolerant include Desiree, Hermes, Maris Peer and Almera (Bilski et al., 1988; Levy, 1992; Levy and Veilleux, 2007; Sabbah and Tal, 1995). Challenges in breeding for salt tolerance arise due to its polygenic control and the variability in salt composition and distribution under field conditions, and this has led to the adoption of a variety of different traits for selection for salinity tolerance. Root growth traits were the first to be discovered, where *in vitro* studies and field performance showed good correlations for potato cultivars like Serrana, Red Pontiac and Norchip, and thus *in vitro* screening of parental material was suggested as a preliminary technique for selection of cultivars tolerant of salinity (Morpurgo, 1991; Naik and Widholm, 1993). Other traits explored using *in vitro* studies for assessing salinity affects and interaction with yield include (1) single-node cutting bioassays; (2) proline accumulation and synthesis; (3) altered activity of antioxidant enzymes and vacuolar sequestration and (4) accumulation of reactive oxygen species (Devi Prasad and Potluri, 1996; Zhang and Donnelly, 1997; Zhang et al., 2005). Studies understanding variable gene response for salt stress highlighted that under salt stress, photosynthesis- and protein-synthesis-related proteins were down-regulated, whereas osmotine-like proteins and heat-shock proteins were up-regulated, suggesting that defence-associated proteins may confer relative salt tolerance to potato plants (Aghaei et al., 2008; Rensink et al., 2005). Thus, there is a considerable variation in germplasm and available varieties of potato for salt stress, and a better understanding of the differences and mechanisms of salt tolerance is needed.

3 Technological advances to develop abiotic stress resistant/tolerant varieties

Abiotic stresses are major contributors in reducing agricultural productivity, and understanding stress-related traits is becoming more crucial under currently predicted climate change scenarios. Hence, the need to produce varieties that are resistant and tolerant to abiotic stress is evident (Yamaguchi and Blumwald, 2005). There are two main approaches which are being used to improve productivity and tolerance for the above-mentioned stresses. The first approach involves exploitation of natural genetic variations mainly through genetic mapping approaches, phenotypic selection or application of marker-assisted selection (MAS), while the second one involves use of genetic engineering approaches to introduce novel genes or alter expression levels (Fig. 2). In the next section, we discuss how these have been explored in potatoes and the opportunities available through recently developed functional tools for further development.

3.1 Genomic tools and techniques

The wide base of potato germplasm has the potential to improve commercial potato cultivars for abiotic stress tolerance (Ramakrishnan et al., 2015). This is becoming possible with the availability and generation of molecular tools and techniques. These techniques allow the time taken for conventional plant breeding process to be vastly reduced and thus genetically elite plants may be recognised much earlier than when using the

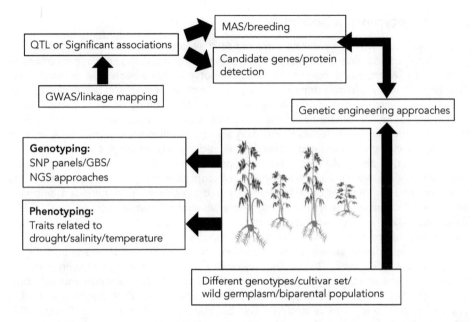

Figure 2 Approaches being used to understand abiotic stress tolerance traits and improve productivity and variety development.

phenotypic selection methods (Barrell et al., 2013). Most of the abiotic stress traits are regarded as quantitative traits, and the use of natural variation available in potato to identify genomic regions associated with drought tolerance would enable breeders to develop improved cultivars with increased drought tolerance using MAS (Gorji et al., 2012). Identification of new and novel genetic markers is possible due to advancement in molecular techniques and technologies. Techniques like genotyping by sequencing, next-generation sequencing, RNA-seq and CHIP-seq techniques allow us to understand expression variation and to expand genome-wide association studies (GWAS) for complex traits like abiotic stress. Advancement in availability of genetic models at both diploid and tetraploid levels is further benefitting linkage analysis and linkage disequilibrium studies for mapping genes (McCord et al., 2011a; Sliwka et al., 2010).

The progress in development of genomics and bioinformatic tools in many plant species including potato offers opportunities for understanding and dissecting the genetic basis of abiotic stress traits. In potato, the availability of genome sequence has further complemented the use of genomic tools and resources (PGSC, 2011; Tuberosa, 2012). Extensive breeding programmes are of paramount importance for alleviating abiotic stress using genetic approaches which are faced with challenges of environmental variation and low heritability (Manavalan et al., 2009). Use of marker-assisted breeding approaches by targeting markers linked or associated with the genomic loci responsible for tolerance is an efficient way of accelerating the achievement of abiotic stress tolerance (Xu et al., 2012). Therefore, the availability of genome sequence information and expression profile data is an invaluable resource for understanding genetic response to abiotic stress tolerance traits.

3.2 Phenotyping: a bottleneck?

Recent developments in genomics have radically altered the landscape for conducting genetic analysis and have great prospects for significantly impacting crop improvement. In contrast, equivalent developments for plant phenotyping have been slow to materialise. Breeding efficiency is largely dependent on accurate phenotyping, and acquisition of high quality phenotypic information is still regarded as a major limiting factor. In certain cases, the quantification for different field traits in potatoes is still measured in discrete values (e.g. a 1–9 breeder scale) without providing any quantitative measurements. Also, with the onset of climate change and increasing unpredictability across the world, improved stress adaptation and yield responses of crops requires the development of strategies which include more accurate measurements of crop phenology under different climate scenarios, and varieties with better and more stable yields with an increased emphasis on genetic 'buffering' against increasing unpredictability (Cabrera-Bosquet et al., 2012; Tuberosa, 2012). Thus, for measuring complex and quantitative traits like abiotic stresses, quantitative information is crucial and for this, rapid and effective phenotyping methods are needed for a large genetic study and linking genetic information (Araus and Cairns, 2014; Furbank and Tester, 2011). The availability of different imaging techniques has allowed real-time image analysis of physiological changes in crop canopies and individual plants, and such imaging techniques have great potential for high-throughput screening of plant populations (Chaerle and Van Der Straeten, 2000; Furbank and Tester, 2011; Tuberosa, 2012). These techniques allow the pre-symptomatic monitoring of plant stress non-destructively long before the actual symptoms are displayed or characterised in plant morphological parts. Thus, the phenotyping bottleneck can now be addressed by combining high-throughput imaging, image analysis and high-performance computing.

In the case of potato, the advancements in using high-throughput and precise phenotypic approaches have lagged behind in contrast to the genomic approaches. A couple of field systems have been reported in potato, one using the RGB imagery to provide an assessment for resistance to potato late blight (Sugiura et al., 2016) and the other using an integrated system of RGB and thermal imaging to infer stress in field conditions (Prashar and Jones, 2014, 2016; Prashar et al., 2013). The advancement and development of these tools and techniques in the potato community will help scientists in linking high-throughput phenotyping with high-throughput tools in genotyping. Although in molecular breeding techniques greater emphasis is on selection based on genotypic information, accurate phenotypic data also obviously play a very important role. For example, in genomic selection to train a prediction model, in marker-assisted recurrent selection, the information provided in a single phenotype session is used to identify markers in subsequent generations, and thus advances in phenotyping are necessary for impacts in breeding and genetics (Araus and Cairns, 2014; Lorenz et al., 2011).

3.3 Genetic engineering approaches

Genetic engineering approaches have been used in potato for engineering stress resistance and tolerance. Osmotic adjustment or accumulation of compatible solutes helps plants to cope with environmental stresses mainly due to drought and salinity, and glycine betaine (GB), one of the important osmolytes (Chen and Murata, 2008), is accumulated in many plant species during stress (and is regarded as an osmoprotectant) (Rhodes and Hanson, 1993). As potato is betaine deficient and studies have shown that plants exhibit enhanced tolerance to drought and salinity under GB accumulation, Zhang and colleagues introduced *BADH* (betaine aldehyde dehydrogenase) gene from spinach responsible for GB biosynthesis and found that transgenic plants acquired drought and salinity tolerance ability (Zhang et al., 2011a). As potato is not a natural accumulator of GB under stress conditions, transgenic potato plants (cv Superior) were developed with the ability to synthesise GB in chloroplasts via introduction of choline oxidase (coda) cDNA derived from *Arthrobacter globiformis* under the control of SWPA2 promoter (Ahmad et al., 2008) and results indicate increased tolerance to a variety of environmental stresses (e.g. drought and salinity). In yet another approach to investigate whether potato synthesises or accumulates glucosylglycerol (GG), a compound found in halotolerant cyanobacteria, potato plants were transformed with the *ggpPS* gene of *Azotobacter vinelandii*, coding the enzyme GG-phosphate phosphate/synthase (GgpPS) and the plants showed increased shoot growth under salt and drought stress conditions (Sievers et al., 2013). In another study, transgenic potato (*S. tuberosum* cv Jowon) constitutively overexpressing AtYUC6 showed enhanced drought tolerance (Kim et al., 2013), and similarly transgenic potato lines produced expressing trehalose-6-phosphate synthase (TPS1) (also an osmolyte related) showed delayed wilting compared to non-transformed lines (Stiller et al., 2008). In addition to some of the transformation approaches for osmolytes mentioned above, the accumulation of mannitol is known to increase response to osmotic stress and improve tolerance to abiotic stresses through osmoregulation and stabilisation of macromolecular structures (Stoop et al., 1996; Tang et al., 2005). Transgenic approach was used to introduce bacterial mannitol 1-phosphate dehydrogenase (*mtlD*) gene into *S. tuberosum* by *Agrobacterium-tumefaciens*-mediated transformation, and mtlD expressed lines had higher amount of mannitol that contributed to enhanced tolerance to NaCl stress (Rahnama et al., 2011). Another osmoprotectant proline had high accumulation and offered salinity tolerance to potato plants, when pyrroline-5-carboxylate synthetase (*P5CS*) gene from *Arabidopsis* was introduced into potato driven by the *35S* promoter (Hmida-Sayari et al., 2005).

Transcription factors or regulatory genes also play an important role in conferring abiotic stress tolerance in potato. Examples including C-repeat binding factors or dehydrin-responsive element binding factors (CBF/DREB) have been shown to increase drought and salt tolerance in different plant species (Kasuga et al., 2004), and studies have confirmed their existence in *S. commersonii* (abiotic stress tolerant species). AtDREB1A gene has been shown to act as a transcriptional factor against abiotic stress in potato, and validation of DREB1 by creating overexpression lines in potato presented a higher salt tolerance (Kim et al., 2014; Shimazaki et al., 2016; Watanabe et al., 2011). *S. tuberosum* and *S. commersonii* transformants generated with *SCCBF1* gene from *S. commersonii* driven by 35S promoter showed better plantlet growth and root development with higher levels of proline in stems and leaves (Pino et al., 2013). There are also genetic approaches which are linked to relative oxidative stress and other transcription factors being studied in potato, and these can be explored more in Kikuchi et al. (2015); Obidiegwu et al. (2015). In some areas, over the last few years, the genetic engineering approaches are being reviewed for crop improvement.

3.4 Gene expression regulators

New, emerging regulators of gene expression can provide novel ways to influence key traits in crops. It is essential for scientists to examine and understand the alternative pathways which can lead to silencing or increase of gene expression especially at post-transcriptional level. One group of these modulators is short (21–24 nt), single-stranded, non-coding RNAs called microRNAs (miRNAs). miRNAs are encoded in the genome, and they go through a relatively well-known RNA processing pathway (Borges and Martienssen, 2015; Reis et al., 2015). The primary miRNA transcript is processed by the Dicer-Like 1 (DCL1) protein (RNase III endoribonuclease), and the established hairpin-like structure is further processed by the same enzyme, establishing a double-stranded intermediate RNA. One strand of this RNA (miRNA) is incorporated into the effector complex (RISC–RNA-induced silencing complex) where the core component is one of the Argonaute proteins. The miRNA can direct this complex to the target RNA in a sequence-specific manner, and the target RNA will be cut and degraded or this complex can direct translational inhibition; both actions will result in the silencing of the gene from which the RNA is transcribed.

miRNAs can target multiple genes, and many of them are conserved among plant species (Jones-Rhoades, 2012; Rogers and Chen, 2013). They offer an additional level of regulation, not only altering one gene, but also affecting complex pathways directing fine-tuning of gene expression. They have important roles in development and show expression level changes upon biotic or abiotic stresses (Li et al., 2016). This is what makes miRNAs a promising target for crop improvement. In the past decade, our basic understanding about the role of miRNAs has increased in model species. But, crop species have only been extensively investigated in recent years, and there is an increasing demand to understand better the action of miRNAs in crops and how they can influence key crop traits (Noman et al., 2017). High-throughput sequencing technologies are used to characterise miRNAs genome-wide in different species including potato (Lakhotia et al., 2014; Zhang et al., 2013). This technological advance increased the knowledge of species-specific and conserved miRNAs, offering an opportunity for deeper investigation into the role of miRNAs in potato. The tuberisation of potato plants is influenced by miR156 and miR172; miR156 can hinder tuber formation regulating squamosa promoter binding-like genes (*SPL3,6,9,13*), but miR172 has a positive effect on tuber development influencing *RAP1* (related to *APETALA2 1*) (Bhogale et al., 2014; Martin et al., 2009). Recent studies highlighted the role of miRNAs in the response for biotic stresses. The miR482 family in potato is in response to *Verticillium dahlia* infection via the regulation of target RNAs of NBS-LRR disease-resistance proteins (Yang et al., 2015). Additionally, in other Solanaceae species, miR482, miR6019 and miR6020 can contribute to biotic stress resistance, influencing key *NB-LRR* genes (Li et al., 2012; Shivaprasad et al., 2012; Zhai et al., 2011). *Phytophthora* infection can strongly influence miRNA expression, which can contribute to the fine-tuning of target genes having an effect on resistance genes upon pathogen infection (Luan et al., 2015; Wong et al., 2014).

Abiotic stresses have a pronounced effect on miRNA expression in plants (Zhang, 2015). Drought, cold and salinity stresses in particular have been investigated in potato and stress-related miRNAs identified. miR159 showed decreased expression upon drought stress, and its targets (*GAMyb-like* genes) had elevated levels, which shows that the target genes are escaping the miRNA-directed regulation, expressing in a higher level compared to non-stress conditions (Yang et al., 2014). Examining the *CPB80* gene in potato showed that miR159 and its targets (*MYB33* and *MYB101*) are involved in the ABA-mediated regulation of plant response to drought (Pieczynski et al., 2013). *Arabidopsis* miR166 and miR159 were reported to have a role in gene regulation under salinity stress in potato (Kitazumi et al., 2015), using network models. The miR166–*HD-ZIP-Phabulosa/Phavulota* network appears to be involved in modulating growth, and vegetative dormancy with possible linkage to defence-related pathways and miR159 via its target *Myb101* might be important for regulating vegetative growth and controlling stress-induced premature transition to reproductive phase. Drought stress could be mimicked by polyethylene glycol treatment (PEG, 6000); miR811, miR814, miR835 and miR4398 showed differential expression in potato upon PEG treatment, and MYB transcription factor, hydroxyproline-rich glycoprotein, aquaporin and WRKY transcription factor genes were identified as target mRNAs (respectively) (Zhang et al., 2014). Leaf tissues originating from control and PEG-treated plants were subjected to RNA isolation, library construction and high-throughput sequencing to identify differentially expressed miRNAs. Their target genes were predicted using bioinformatics tools and the expression levels were validated via reverse transcription quantitative polymerase chain reaction (RT-qPCR). miR169 is a conserved miRNA family in plants having eight members in potato [(Zhang et al., 2013), miRBase release 21]. In tomato, it was reported that the accumulation of miR169 was induced by drought stress, and overexpression of miR169 could direct enhanced drought tolerance, regulating its target genes: three Nuclear Factor Y subunit genes (*NF-YA1/2/3*) and one multidrug-resistance-associated protein gene (*MRP1*) (Zhang et al., 2011b). Interestingly, in potato, drought stress reduced miR169 levels in plants (Yang et al., 2016). Subsequently, its target genes (*NF-YA1,3,4*) showed elevated expression levels in drought conditions. This example shows that miRNAs could affect gene expression differentially in even closely related species, and the regulation of stress connected pathways could be fine-tuned via miRNAs in different ways.

The examples above show that miRNA expression could be modified via transgenic approaches overexpressing or even depleting their levels via miRNA mimicry (Yan et al., 2012). These transgenic approaches extended with precise modification of plant genome (CRISPR/Cas9 system) are promising techniques to alter miRNA levels in crops, but the genetically modified plants are not favoured because of public safety concerns (Ramesh, 2013). It is worth noting that breeders select for favourable alleles of genes for their traits of interest, but in parallel they select for regulatory elements as well, such as miRNAs. This might equally influence the chosen trait in crops. In addition to the miRNAs, long non-coding RNAs (lncRNAs) are emerging regulators of gene expression in plants and animals influencing epigenetic modification of DNA/histones at certain places of the genome, bridging different parts of the genome and/or offering a scaffold to recruit key factors to modify gene expression (Bohmdorfer and Wierzbicki, 2015; Chekanova, 2015). Studying their roles and functions is important in the future, which could lead to better improvement of crops.

4 Future trends and conclusion

Rapid development of new varieties that are tolerant to varying conditions and different stresses will be critical for future crop sustainability and the agriculture industry in the era of climate change. A key aspect in the development of crop phenotypes with increased productivity and tolerance to abiotic stress is understanding the physiological mechanisms underlying morphological traits. So, understanding of plant adaptation to changing environmental conditions involves an array of physiological processes and signalling events, some of which will be common to certain biotic and abiotic stresses (Moshou et al., 2005). The critical point here is understanding functions if possible under multi-field and multi-environmental conditions. A barrier to collecting such data is the cost and the sheer difficulty of measuring phenological traits on large numbers of strains/lines, grown in small plots (10–20 plants). We believe that the use of genetic resources linked with high-throughput phenotyping by developing and using imaging and precision techniques to collect large volumes of 'in field' phenotypic data across several environments can potentially increase both the efficiency of breeding programmes and further understanding of genotype × environment interactions in potato and thus can advance future breeding programmes.

The aim of future breeding programmes is to further improve the precision and efficiency of predicting phenotypes from genotypes (genomic selection) which will enable improved cultivars to be developed with more tolerance to abiotic stresses and improve agronomic performance (Varshney et al., 2009). There is increasing pressure from the public for safe biotechnology (Palmgren et al., 2015), and the availability of site-directed mutagenesis (SDM) and next-generation biotechnological advancements may enable this to happen.

5 References

Aghaei, K., Ehsanpour, A. A. and Komatsu, S. 2008. Proteome analysis of potato under salt stress. *Journal of Proteome Research* 7, 4858–68.

Ahmad, R., Kim, M. D., Back, K.-H., Kim, H.-S., Lee, H.-S., Kwon, S.-Y., Murata, N., Chung, W.-I. and Kwak, S.-S. 2008. Stress-induced expression of choline oxidase in potato plant chloroplasts confers enhanced tolerance to oxidative, salt, and drought stresses. *Plant Cell Reports* 27, 687–98.

Angelopoulos, K., Dichio, B. and Xiloyannis, C. 1996. Inhibition of photosynthesis in olive trees (*Olea europaea* L.) during water stress and rewatering. *Journal of Experimental Botany* 47, 1093–100.

Anithakumari, A. M., Dolstra, O., Vosman, B., Visser, R. G. F. and van der Linden, C. G. 2011. *In vitro* screening and QTL analysis for drought tolerance in diploid potato. *Euphytica* 181, 357–69.

Anithakumari, A. M., Nataraja, K. N., Visser, R. G. F. and van der Linden, C. G. 2012. Genetic dissection of drought tolerance and recovery potential by quantitative trait locus mapping of a diploid potato population. *Molecular Breeding* 30, 1413–29.

Araus, J. L. and Cairns, J. E. 2014. Field high-throughput phenotyping: The new crop breeding frontier. *Trends in Plant Science* 19, 52–61.

Barrell, P. J., Meiyalaghan, S., Jacobs, J. M. E. and Conner, A. J. 2013. Applications of biotechnology and genomics in potato improvement. *Plant Biotechnology Journal* 11, 907.

Bhogale, S., Mahajan, A. S., Natarajan, B., Rajabhoj, M., Thulasiram, H. V. and Banerjee, A. K. 2014. MicroRNA156: A potential graft-transmissible microRNA that modulates plant architecture and tuberization in Solanum tuberosum ssp. andigena. *Plant Physiology* 164, 1011–27.

Bilski, J. J., Nelson, D. C. and Conlon, R. L. 1988. The response of four potato cultivars to chloride salinity, sulfate salinity and calcium in pot experiments. *American Potato Journal* 65, 85–90.

Bohmdorfer, G. and Wierzbicki, A. T. 2015. Control of Chromatin Structure by Long Noncoding RNA. *Trends in Cell Biology* 25, 623–32.

Bonierbale, M., Gastelo, M. and Asfaw, A. 2017. Combining ability of highland tropic adapted potato for tuber yield and yield components under drought. *PLoS One* 12, e0181541.

Borges, F. and Martienssen, R. A. 2015. The expanding world of small RNAs in plants. *Nature Reviews Molecular Cell Biology* 16, 727–41.

Bradshaw, J. E. 2009. Potato breeding at the Scottish Plant Breeding Station and the Scottish Crop Research Institute: 1920–2008. *Potato Research* 52, 141–72.

Cabrera-Bosquet, L., Crossa, J., von Zitzewitz, J., Dolors Serret, M. and Luis Araus, J. 2012. High-throughput phenotyping and genomic selection: The frontiers of crop breeding converge. *Journal of Integrative Plant Biology* 54, 312–20.

Chaerle, L. and Van Der Straeten, D. 2000. Imaging techniques and the early detection of plant stress. *Trends in Plant Science* 5, 495–501.

Chaves, M. M., Pereira, J. S., Maroco, J., Rodrigues, M. L., Ricardo, C. P. P., Osorio, M. L., Carvalho, I., Faria, T. and Pinheiro, C. 2002. How plants cope with water stress in the field. Photosynthesis and growth. *Annals of Botany* 89, 907–16.

Chaves, M. M., Flexas, J. and Pinheiro, C. 2009. Photosynthesis under drought and salt stress: Regulation mechanisms from whole plant to cell. *Annals of Botany* 103, 551–60.

Chekanova, J. A. 2015. Long non-coding RNAs and their functions in plants. *Current Opinion in Plant Biology* 27, 207–16.

Chen, T. H. H. and Murata, N. 2008. Glycinebetaine: An effective protectant against abiotic stress in plants. *Trends in Plant Science* 13, 499–505.

Cornic, G., Prioul, J. L. and Louason, G. 1983. Stomatal and non-stomatal contribution in the decline in leaf net co2 uptake during rapid water-stress. *Physiologia Plantarum* 58, 295–301.

Devi Prasad, P. V. and Potluri, S. D. P. 1996. Influence of proline and hydroxyproline on salt-stressed axillary bud cultures of two varieties of potato (*Solanum tuberosum*). *In Vitro – Plant* 32, 47–50.

Ewing, E. E. 1981. Heat stress and the tuberization stimulus. *American Potato Journal* 58, 31.

Fess, T. L., Kotcon, J. B. and Benedito, V. A. 2011. Crop breeding for low input agriculture: A sustainable response to feed a growing world population. *Sustainability* 3, 1742.

Figueiredo I. C. R. D., Pinto, C. A. B. P., Ribeiro, G. H. M. R., Lino L. d. O., Lyra, D. H. and Moreira, C. M. 2015. Efficiency of selection in early generations of potato families with a view toward heat tolerance. *Crop Breeding and Applied Biotechnology* 15, 210–17.

Furbank, R. T. and Tester, M. 2011. Phenomics – technologies to relieve the phenotyping bottleneck. *Trends in Plant Science* 16, 635–44.

Gimenez, C., Mitchell, V. J. and Lawlor, D. W. 1992. Regulation of photosynthetic rate of 2 sunflower hybrids under water-stress. *Plant Physiology* 98, 516–24.

Gong, L., Zhang, H., Gan, X., Zhang, L., Chen, Y., Nie, F., Shi, L., Li, M., Guo, Z., Zhang, G. and Song, Y. 2015. Transcriptome profiling of the potato (*Solanum tuberosum* L.) plant under drought stress and water-stimulus conditions. *PLoS One* 10, e0128041.

Gorji, A., Matyas, K., Dublecz, Z., Decsi, K., Cernak, I., Hoffmann, B., Taller, J. and Polgar, Z. 2012. In vitro osmotic stress tolerance in potato and identification of major QTLs. *American Journal of Potato Research* 89, 453–64.

Hancock, R. D., Morris, W. L., Ducreux, L. J. M., Morris, J. A., Usman, M., Verrall, S. R., Fuller, J., Simpson, C. G., Zhang, R., Hedley, P. E. and Taylor, M. A. 2014. Physiological, biochemical and molecular responses of the potato (*Solanum tuberosum* L.) plant to moderately elevated temperature. *Plant, Cell & Environment* 37, 439–50.

Haverkort, A., Fasan, T. and Waart, M. 1991. The influence of cyst nematodes and drought on potato growth. 2. Effects on plant water relations under semi- controlled conditions. *Netherlands Journal of Plant Pathology* 97, 162–70.

Hawkes, J. G. 1992. Biosystematics of the potato. In: Harris, P. M. (Ed.), *The Potato Crop: The Scientific Basis for Improvement*. Dordrecht, the Netherlands: Springer Netherlands, pp. 13–64.

Hawkes, J. 1994. Origins of cultivated potatoes and species relationships. In: Bradshaw, J. and Mackay, G. (Eds), *Potato Genetics*. Cambridge, UK: Cab International, pp. 3–42.

Heuer, B. and Nadler, A. 1998. Physiological response of potato plants to soil salinity and water deficit. *Plant Science* 137, 43–51.

Hmida-Sayari, A., Gargouri-Bouzid, R., Bidani, A., Jaoua, L., Savouré, A. and Jaoua, S. 2005. Overexpression of Δ1-pyrroline-5-carboxylate synthetase increases proline production and confers salt tolerance in transgenic potato plants. *Plant Science* 169, 746–52.

Iwama, K. and Yamaguchi, J. 2006. Abiotic stresses. In: Gopal, J. and Khurana, S. M. (Eds), *Handbook of Potato Production, Improvement and Postharvest Management*. New York, NY: Food Product Press, pp. 231–78.

Jones-Rhoades, M. W. 2012. Conservation and divergence in plant microRNAs. *Plant Molecular Biology* 80, 3–16.

Jones, M. M. and Turner, N. C. 1980. Osmotic adjustment in expanding and fully expanded leaves of sunflower in response to water deficits. *Australian Journal of Plant Physiology* 7, 181.

Kasuga, M., Miura, S., Shinozaki, K. and Yamaguchi-Shinozaki, K. 2004. A combination of the Arabidopsis DREB1A gene and stress-inducible rd29A promoter improved drought-and low-temperature stress tolerance in tobacco by gene transfer. *Plant and Cell Physiology* 45, 346–50.

Kesiime, V. E., Tusiime, G., Kashaija, I. N., Edema, R., Gibson, P., Namugga, P. and Kakuhenzire, R. 2016. Characterization and evaluation of potato genotypes (*Solanum tuberosum* L.) for tolerance to drought in Uganda. *American Journal of Potato Research* 93, 543–51.

Khan, M. A., Saravia, D., Munive, S., Lozano, F., Farfan, E., Eyzaguirre, R. and Bonierbale, M. 2015. Multiple QTLs linked to agro-morphological and physiological traits related to drought tolerance in potato. *Plant Molecular Biology Reporter/Ispmb* 33, 1286–98.

Kikuchi, A., Huynh, H. D., Endo, T. and Watanabe, K. 2015. Review of recent transgenic studies on abiotic stress tolerance and future molecular breeding in potato. *Breeding Science* 65, 85–102.

Kim, J. I., Baek, D., Park, H. C., Chun, H. J., Oh, D. H., Lee, M. K., Cha, J. Y., Kim, W. Y., Kim, M. C., Chung, W. S., Bohnert, H. J., Lee, S. Y., Bressan, R. A., Lee, S. W. and Yun, D. J. 2013. Overexpression of Arabidopsis YUCCA6 in potato results in high-auxin developmental phenotypes and enhanced resistance to water deficit. *Molecular Plant* 6, 337–49.

Kim, C.-K., Lim, H.-M., Na, J.-K., Choi, J.-W., Sohn, S.-H., Park, S.-C., Kim, Y.-H., Kim, Y.-K. and Kim, D.-Y. 2014. A multistep screening method to identify genes using evolutionary transcriptome of plants. *Evolutionary Bioinformatics Online* 10, 69–78.

Kitazumi, A., Kawahara, Y., Onda, T. S., De Koeyer, D. and de los Reyes, B. G. 2015. Implications of miR166 and miR159 induction to the basal response mechanisms of an andigena potato (*Solanum tuberosum* subsp. andigena) to salinity stress, predicted from network models in Arabidopsis. *Genome* 58, 13–24.

Lafta, A. M. and Lorenzen, J. H. 1995. Effect of high temperature on plant growth and carbohydrate metabolism in potato. *Plant Physiology* 109, 637–43.

Lakhotia, N., Joshi, G., Bhardwaj, A. R., Katiyar-Agarwal, S., Agarwal, M., Jagannath, A., Goel, S. and Kumar, A. 2014. Identification and characterization of miRNAome in root, stem, leaf and tuber developmental stages of potato (*Solanum tuberosum* L.) by high-throughput sequencing. *BMC Plant Biology* 14, 6.

Legay, S., Lamoureux, D., Hausman, J.-F., Hoffmann, L. and Evers, D. 2009. Monitoring gene expression of potato under salinity using cDNA microarrays. *Plant Cell Reports* 28, 1799–816.

Levy, D. 1992. The response of potatoes (*Solanum tuberosum* L.) to salinity: Plant growth and tuber yields in the arid desert of Israel. *Annals of Applied Biology* 120, 547–55.

Levy, D. and Veilleux, R. E. 2007. Adaptation of potato to high temperatures and salinity-a review. *American Journal of Potato Research* 84, 487–506.

Levy, D., Itzhak, Y., Fogelman, E., Margalit, E. and Veilleux, R. 2001. Ori, Idit, Zohar and Zahov: Tablestock and chipstock cultivars bred for adaptation to Israel. *American Journal of Potato Research* 78, 167–73.

Li, F., Pignatta, D., Bendix, C., Brunkard, J. O., Cohn, M. M., Tung, J., Sun, H., Kumar, P. and Baker, B. 2012. MicroRNA regulation of plant innate immune receptors. *Proceedings of the National Academy of Sciences of the United States of America* 109, 1790–5.

Li, S., Castillo-Gonzalez, C., Yu, B. and Zhang, X. 2016. The functions of plant small RNAs in development and in stress responses. *Plant Journal* 90, 654–70.

Liu, C. C., Liu, Y. G., Guo, K., Zheng, Y. R., Li, G. Q., Yu, L. F. and Yang, R. 2010. Influence of drought intensity on the response of six woody karst species subjected to successive cycles of drought and rewatering. *Physiologia Plantarum* 139, 39–54.

Lorenz, A. J., Chao, S., Asoro, F. G., Heffner, E. L., Hayashi, T., Iwata, H., Smith, K. P., Sorrells, M. and Jannink, J.-L. 2011. Genomic selection in plant breeding: Knowledge and prospects. *Advances in Agronomy* 110, 77–123.

Luan, Y., Cui, J., Zhai, J., Li, J., Han, L. and Meng, J. 2015. High-throughput sequencing reveals differential expression of miRNAs in tomato inoculated with *Phytophthora infestans. Planta* 241, 1405–16.

Mahmud, A.-A., Hossain, M., Karim, M., Mian, M., Zakaria, M. and Kadian, M. 2016. Plant water relations and canopy temperature depression for assessing water stress tolerance of potato. *Indian Journal of Plant Physiology* 21, 56–63.

Manavalan, L. P., Guttikonda, S. K., Phan Tran, L.-S. and Nguyen, H. T. 2009. Physiological and molecular approaches to improve drought resistance in soybean. *Plant and Cell Physiology* 50, 1260–76.

Mane, S. P., Robinet, C. V., Ulanov, A., Schafleitner, R., Tincopa, L., Gaudin, A., Nomberto, G., Alvarado, C., Solis, C., Avila Bolivar, L., Blas, R., Ortega, O., Solis, J., Panta, A., Rivera, C., Samolski, I., Carbajulca, D. H., Bonierbale, M., Pati, A., Heath, L. S., Bohnert, H. J. and Grene, R. 2008. Molecular and physiological adaptation to prolonged drought stress in the leaves of two Andean potato genotypes. *Functional Plant Biology* 35, 669–88.

Mares, D. J., Marscfaner, H. and Krauss, A. 1981. Effect of gibberellic acid on growth and carbohydrate metabolism of developing tubers of potato (*Solanum tuberosum*). *Physiologia Plantarum* 52, 267–74.

Martin, A., Adam, H., Diaz-Mendoza, M., Zurczak, M., Gonzalez-Schain, N. D. and Suarez-Lopez, P. 2009. Graft-transmissible induction of potato tuberization by the microRNA miR172. *Development* 136, 2873–81.

Martinez, C. A., Maestri, M. and Lani, E. G. 1996. In vitro salt tolerance and proline accumulation in Andean potato (*Solanum* spp.) differing in frost resistance. *Plant Science* 116, 177–84.

McCord, P. H., Sosinski, B. R., Haynes, K. G., Clough, M. E. and Yencho, G. C. 2011a. Linkage mapping and QTL analysis of agronomic traits in tetraploid potato (subsp.). *Crop Science* 51, 771.

McCord, P. H., Sosinski, B. R., Haynes, K. G., Clough, M. E. and Yencho, G. C. 2011b. QTL mapping of internal heat necrosis in tetraploid potato. *Theoretical and Applied Genetics* 122, 129–42.

Minhas, J. S. 2012. Potato: Production strategies under abiotic stress. In: Tuteja, N., Gill, S. S., Tiburcio, A. F. and Tuteja, R. (Eds), *Improving Crop Resistance to Abiotic Stress*, Volume 1 and 2. Weinheim, Germany: Wiley-VCH Verlag GmbH & Co. KGaA, pp. 1155–67.

Momčilović, I., Pantelić, D., Zdravković-Korać, S., Oljača, J., Rudić, J. and Fu, J. 2016. Heat-induced accumulation of protein synthesis elongation factor 1A implies an important role in heat tolerance in potato. *Planta* 244, 671–9.

Morpurgo, R. 1991. Correlation between potato clones grown in vivo and in vitro under sodium chloride stress conditions. *Plant Breeding* 107, 80–2.

Moshou, D., Bravo, C., Oberti, R., West, J., Bodria, L., McCartney, A. and Ramon, H. 2005. Plant disease detection based on data fusion of hyper-spectral and multi-spectral fluorescence imaging using Kohonen maps. *Real-Time Imaging* 11, 75–83.

Munns, R. and Tester, M. 2008. Mechanisms of salinity tolerance. *Annual Review of Plant Biology* 59, 651–81.

Naik, P. S. and Widholm, J. M. 1993. Comparison of tissue culture and whole plant responses to salinity in potato. *Plant Cell, Tissue and Organ Culture* 33, 273–80.

Navarro, C., Abelenda, J. A., Cruz-Oro, E., Cuellar, C. A., Tamaki, S., Silva, J., Shimamoto, K. and Prat, S. 2011. Control of flowering and storage organ formation in potato by Flowering locus T. *Nature* 478, 119–22.

Noman, A., Fahad, S., Aqeel, M., Ali, U., Amanullah, Anwar, S., Baloch, S. K. and Zainab, M. 2017. miRNAs: Major modulators for crop growth and development under abiotic stresses. *Biotechnology Letters* 39, 685–700.

Obidiegwu, J. E., Bryan, G. J., Jones, H. G. and Prashar, A. 2015. Coping with drought: Stress and adaptive responses in potato and perspectives for improvement. *Frontiers in Plant Science* 6, 542.

Onder, S., Caliskan, M. E., Onder, D. and Caliskan, S. 2005. Different irrigation methods and water stress effects on potato yield and yield components. *Agricultural Water Management* 73, 73–86.

Palmgren, M. G., Edenbrandt, A. K., Vedel, S. E., Andersen, M. M., Landes, X., Østerberg, J. T., Falhof, J., Olsen, L. I., Christensen, S. B., Sandøe, P., Gamborg, C., Kappel, K., Thorsen, B. J. and Pagh, P. 2015. Are we ready for back-to- nature crop breeding? *Trends in Plant Science* 20, 155.

PGSC. 2011. Genome sequence and analysis of the tuber crop potato. *Nature* 475, 189–95.

Pieczynski, M., Marczewski, W., Hennig, J., Dolata, J., Bielewicz, D., Piontek, P., Wyrzykowska, A., Krusiewicz, D., Strzelczyk-Zyta, D., Konopka-Postupolska, D., Krzeslowska, M., Jarmolowski, A. and Szweykowska-Kulinska, Z. 2013. Down-regulation of CBP80 gene expression as a strategy to engineer a drought-tolerant potato. *Plant Biotechnology Journal* 11, 459–69.

Pino, M. T., Ávila, A., Molina, A., Jeknic, Z. and Chen, T. H. H. 2013. Enhanced in vitro drought tolerance of *Solanum tuberosum* and *Solanum commersonii* plants overexpressing the ScCBF1 gene. *Ciencia E Investigacion Agraria* 40 (1): 171–84.

Prashar, A. and Jones, H. G. 2014. Infra-red thermography as a high-throughput tool for field phenotyping. *Agronomy* 4, 397–417.

Prashar, A. and Jones, H. G. 2016. Assessing drought responses using thermal infrared imaging. *Methods in Molecular Biology* 1398, 209–19.

Prashar, A., Yildiz, J., McNicol, J. W., Bryan, G. J. and Jones, H. G. 2013. Infra-red thermography for high throughput field phenotyping in *Solanum tuberosum*. *PLoS One* 8, e65816, 65811–19.

Rahnama, H., Vakilian, H., Fahimi, H. and Ghareyazie, B. 2011. Enhanced salt stress tolerance in transgenic potato plants (*Solanum tuberosum* L.) expressing a bacterial mtlD gene. *Acta Physiologiae Plantarum* 33, 1521–32.

Ramakrishnan, A., Ritland, C., Blas Sevillano, R. and Riseman, A. 2015. Review of potato molecular markers to enhance trait selection. *American Journal of Potato Research* 92, 455–72.

Ramesh, S. V. 2013. Non-coding RNAs in crop genetic modification: Considerations and predictable environmental risk assessments (ERA). *Molecular Biotechnology* 55, 87–100.

Reis, R. S., Eamens, A. L. and Waterhouse, P. M. 2015. Missing Pieces in the Puzzle of Plant MicroRNAs. *Trends in Plant Science* 20, 721–8.

Rensink, W. A., Iobst, S., Hart, A., Stegalkina, S., Liu, J. and Buell, C. R. 2005. Gene expression profiling of potato responses to cold, heat, and salt stress. *Functional & Integrative Genomics* 5, 201–7.

Rhodes, D. and Hanson, A. D. 1993. Quaternary ammonium and tertiary sulfonium compounds in higher plants. *Annual Review of Plant Physiology and Plant Molecular Biology* 44, 357–84.

Rogers, K. and Chen, X. 2013. Biogenesis, turnover, and mode of action of plant microRNAs. *Plant Cell* 25, 2383–99.

Sabbah, S and Tal, M. 1995. Salt tolerance in *Solanum kurzianum* and *S. tuberosum* cvs Alpha and Russet Burbank. *Potato Research* 38, 319–30.

Shaterian, J., Waterer, D., Jong, H. D. and Tanino, K. K. 2005. Differential stress responses to NaCl salt application in early- and late-maturing diploid potato (*Solanum* sp.) clones. *Environmental and Experimental Botany* 54, 202–12.

Shimazaki, T., Endo, T., Kasuga, M., Yamaguchi-Shinozaki, K., Watanabe, K. N. and Kikuchi, A. 2016. Evaluation of the yield of abiotic-stress-tolerant AtDREB1A transgenic potato under saline conditions in advance of field trials. *Breeding Science* 66, 703–10.

Shinozaki, K and Yamaguchi-Shinozaki, K. 2007. Gene networks involved in drought stress response and tolerance. *Journal of Experimental Botany* 58, 221–7.

Shivaprasad, P. V., Chen, H. M., Patel, K, Bond, D. M., Santos, B. A. and Baulcombe, D. C. 2012. A microRNA superfamily regulates nucleotide binding site-leucine-rich repeats and other mRNAs. *Plant Cell* 24, 859–74.

Shomura, A., Izawa, T., Ebana, K., Ebitani, T., Kanegae, H., Konishi, S. and Yano, M. 2008. Deletion in a gene associated with grain size increased yields during rice domestication. *Nature Genetics* 40, 1023–8.

Sievers, N., Muders, K., Henneberg, M., Klähn, S., Effmert, M., Junghans, H. and Hagemann, M. 2013. Establishing glucosylglycerol synthesis in potato (*Solanum tuberosum* L. cv. Albatros) by expression of the ggpPS gene from Azotobacter vinelandii. *Journal of Plant Science and Molecular Breeding* 2, 1.

Sliwka, J., Jakuczun, H., Kaminski, P. and Zimnoch-Guzowska, E. 2010. Marker-assisted selection of diploid and tetraploid potatoes carrying Rpi-phu1, a major gene for resistance to Phytophthora infestans. *Journal of Applied Genetics* 51, 133–40.

Stiller, I., Dulai, S., Kondrak, M., Tarnai, R., Szabo, L., Toldi, O. and Banfalvi, Z. 2008. Effects of drought on water content and photosynthetic parameters in potato plants expressing the trehalose-6-phosphate synthase gene of Saccharomyces cerevisiae. *Planta* 227, 299–308.

Stoop, J. M. H., Williamson, J. D. and Pharr, D. M. 1996. Mannitol metabolism in plants: A method for coping with stress. *Trends in Plant Science* 1, 139–44.

Sugiura, R., Tsuda, S., Tamiya, S., Itoh, A., Nishiwaki, K., Murakami, N., Shibuya, Y., Hirafuji, M. and Nuske, S. 2016. Field phenotyping system for the assessment of potato late blight resistance using RGB imagery from an unmanned aerial vehicle. *Biosystems Engineering* 148, 1–10.

Tai, G. C. C., Levy, D. and Coleman, W. K. 1994. Path analysis of genotype-environment interactions of potatoes exposed to increasing warm-climate constraints. *Euphytica* 75, 49–61.

Takahashi, S and Murata, N. 2008. How do environmental stresses accelerate photoinhibition? *Trends in Plant Science* 13, 178–82.

Tang, W., Peng, X. and Newton, R. J. 2005. Enhanced tolerance to salt stress in transgenic loblolly pine simultaneously expressing two genes encoding mannitol-1-phosphate dehydrogenase and glucitol-6-phosphate dehydrogenase. *Plant Physiology and Biochemistry* 43, 139–46.

Tuberosa, R. 2012. Phenotyping for drought tolerance of crops in the genomics era. *Frontiers in Physiology* 3, 347.

Varshney, R., Nayak, S., May, G. and Jackson, S. 2009. Next-generation sequencing technologies and their implications for crop genetics and breeding. *Trends in Biotechnology* 27, 522–30.

Vasquez-Robinet, C., Mane, S. P., Ulanov, A. V., Watkinson, J. I., Stromberg, V. K., De Koeyer, D, Schafleitner, R, Willmot, D. B., Bonierbale, M, Bohnert, H. J. and Grene, R. 2008. Physiological and molecular adaptations to drought in Andean potato genotypes. *Journal of Experimental Botany* 59, 2109–23.

Veilleux, R. E., Paz, M. M. and Levy, D. 1997. Potato germplasm development for warm climates: Genetic enhancement of tolerance to heat stress. *Euphytica* 98, 83–92.

Vos, J. and Groenwold, J. 1989. Characteristics of photosynthesis and conductance of potato canopies and the effects of cultivar and transient drought. *Field Crops* 20, 237–50.

Wang, H. and Clarke, J. M. 1993. Relationship of excised-leaf water loss and stomatal frequency in wheat. *Canadian Journal of Plant Science* 73, 93–9.

Watanabe, K. N., Kikuchi, A., Shimazaki, T. and Asahina, M. 2011. Salt and drought stress tolerances in transgenic potatoes and wild species. *Potato Research* 54, 319–24.

Wong, J., Gao, L., Yang, Y., Zhai, J., Arikit, S., Yu, Y., Duan, S., Chan, V., Xiong, Q., Yan, J., Li, S., Liu, R., Wang, Y., Tang, G., Meyers, B. C., Chen, X. and Ma, W. 2014. Roles of small RNAs in soybean defense against Phytophthora sojae infection. *Plant Journal* 79, 928–40.

Xu, Z., Zhou, G. and Shimizu, H. 2010. Plant responses to drought and rewatering. *Plant Signaling & Behavior* 5, 649–54.

Xu, Y., Lu, Y., Xie, C., Gao, S., Wan, J. and Prasanna, B. M. 2012. Erratum to: Whole-genome strategies for marker-assisted plant breeding. *Molecular Breeding* 29, 855.

Yamaguchi, T. and Blumwald, E. 2005. Developing salt-tolerant crop plants: Challenges and opportunities. *Trends in Plant Science* 10, 615–20.

Yan, J., Gu, Y., Jia, X., Kang, W., Pan, S., Tang, X., Chen, X. and Tang, G. 2012. Effective small RNA destruction by the expression of a short tandem target mimic in Arabidopsis. *Plant Cell* 24, 415–27.

Yang, J., Zhang, N., Mi, X., Wu, L., Ma, R., Zhu, X., Yao, L., Jin, X., Si, H. and Wang, D. 2014. Identification of miR159s and their target genes and expression analysis under drought stress in potato. *Computational Biology and Chemistry* 53PB, 204–13.

Yang, L., Mu, X., Liu, C., Cai, J., Shi, K., Zhu, W. and Yang, Q. 2015. Overexpression of potato miR482e enhanced plant sensitivity to *Verticillium dahliae* infection. *Journal of Integrative Plant Biology* 57, 1078–88.

Yang, J. W., Zhang, N., Zhou, X. Y., Si, H. J. and Wang, D. 2016. Identification of four novel stu-miR169s and their target genes in Solanum tuberosum and expression profiles response to drought stress. *Plant Systematics and Evolution* 302, 55–66.

Zhai, J., Jeong, D. H., De Paoli, E., Park, S., Rosen, B. D., Li, Y., Gonzalez, A. J., Yan, Z., Kitto, S. L., Grusak, M. A., Jackson, S. A., Stacey, G., Cook, D. R., Green, P. J., Sherrier, D. J. and Meyers, B. C. 2011. MicroRNAs as master regulators of the plant NB-LRR defense gene family via the production of phased, trans-acting siRNAs. *Genes & Development* 25, 2540–53.

Zhang, B. 2015. MicroRNA: A new target for improving plant tolerance to abiotic stress. *Journal of Experimental Botany* 66, 1749–61.

Zhang, Y. and Donnelly, D. J. 1997. In vitro bioassays for salinity tolerance screening of potato. *Potato Research* 40, 285–95.

Zhang, Z., Mao, B., Li, H., Zhou, W., Takeuchi, Y. and Yoneyama, K. 2005. Effect of salinity on physiological characteristics, yield and quality of microtubers in vitro in potato. *Acta Physiologiae Plantarum* 27, 481–9.

Zhang, N., Si, H.-J., Wen, G., Du, H.-H., Liu B.-L., and Wang, D. 2011a. Enhanced drought and salinity tolerance in transgenic potato plants with a BADH gene from spinach. *Plant Biotechnology Reports* 5, 71–7.

Zhang, X., Zou, Z., Gong, P., Zhang, J., Ziaf, K., Li, H., Xiao, F. and Ye, Z. 2011b. Over-expression of microRNA169 confers enhanced drought tolerance to tomato. *Biotechnology Letters* 33, 403–9.

Zhang, R. X., Marshall, D., Bryan, G. J. and Hornyik, C. 2013. Identification and Characterization of miRNA Transcriptome in Potato by High-Throughput Sequencing. *PLoS One* 8, e57233.

Zhang, N., Yang, J., Wang, Z., Wen, Y., Wang, J., He, W., Liu, B., Si, H. and Wang, D. 2014. Identification of novel and conserved microRNAs related to drought stress in potato by deep sequencing. *PLoS One* 9, e95489.

Chapter 7

Developing early-maturing and stress-resistant potato varieties

Prashant G. Kawar, ICAR-Directorate of Floricultural Research, India; Hemant B. Kardile, S. Raja, Som Dutt and Raj Kumar, ICAR-Central Potato Research Institute, India; P. Manivel, ICAR-Directorate of Medicinal & Aromatic Plants Research, India; and Vinay Bhardwaj, B. P. Singh, P. M. Govindakrishnan and S.K. Chakrabarti, ICAR-Central Potato Research Institute, India

1 Introduction

Potato cultivars are classified as early-, mid- or late-maturing, based on their maturity period. Early-maturing varieties complete their life cycle in 70–80 days in subtropical climates and 90–100 days in temperate climates, whereas 95–110 and 120–135 days are required for the mid- and late-maturing varieties, respectively. Despite the large number of cultivars currently available, there is still a need to develop early-maturing cultivars. Earliness gives the flexibility to include a potato crop in a cropping system and also avoids certain biotic and abiotic stresses that occur later in the season. Early-maturing varieties also allow

http://dx.doi.org/10.19103/AS.2017.0016.07

uniform distribution of potato production by staggering harvesting periods over a longer period. Breeding for earliness and stress resistance started at the International Potato Center (CIP) in 1974, with the aim of extending potato cultivation to the developing world and with its warm, humid and arid environments. Breeding for earliness is worthwhile in the current conditions of global climate change, as the potato is basically a crop of cool and moderate temperatures. The dramatic changes in weather conditions caused by climate change have challenged potato cultivation with heat and drought stress. These stresses can affect plant development and lead to dramatic yield losses, especially at critical stages in the life cycle, such as plant emergence and tuberization (Martinez and Moreno, 1992; Obidiegwu et al., 2015). The effect of climate change on potato production is greater than the effect on cereals, and a 10–16% decline in production has been estimated due to the effect of global climate change in India alone (Singh et al., 2009). Potato crops regularly suffer from drought stress in rainfed growing regions, due to erratic rainfall or inadequate supplemental irrigation (Thiele et al., 2010). This can be clearly seen in China, which has experienced severe and prolonged dry periods since the late 1990s (Zou et al., 2005). Potatoes are increasingly being grown in tropical and subtropical areas as a winter crop, and are therefore experiencing drought stress. The severity and duration of water scarcity vary from year to year, making it difficult for plants to adapt (Easterling et al., 2007). Heat stress is another major constraint affecting potato tuberization and yield (Levy and Veilleux, 2007); it affects sprouting and tuber initiation and development, thereby reducing the crop's yield potential (Ahn and Zimmerman, 2006; Levy and Veilleux, 2007; Papademetriou, 2008).

In developing countries, most potato cultivation is in rainy hilly areas of varying altitudes, with cool and mild climates. Global warming is expected to cause a 4–5°C increase in temperature, and this would have a tremendous effect on potato cultivation. It would disrupt tuberization, extend the growing period and increase susceptibility to major pests and diseases. Irregular rainfall is also creating drought-like conditions that affect potato production (Mendoza et al., 1976). All the existing cultivars suffer significant yield loss in these conditions and have to be replaced with cultivars with fewer growing days and better tolerance to abiotic and biotic stresses. Early-maturing varieties could be a solution for food and nutritional security in Asian countries which have cereal-based cropping systems and face a significant food security challenge to feed their 4.3 billion people. Asian economies and farming systems are linked to a small number of cereals, mainly rice and wheat. Over-intensification from monocropping with cereals, extremes of global climate change in terms of frequent droughts and flooding, and urbanization have degraded arable land, ultimately leading to lower yields. Cereal-based diets are vulnerable to low levels of micronutrients, causing high rates of malnutrition in women and children. This has created a serious threat to the food and nutritional security of the region. Early-maturing potato varieties, with resistance to major biotic and abiotic stresses and good processing quality, are a valid alternative to cereal-based monocropping farming systems. Their short growing period enables these varieties to be cultivated between two consecutive wheat crops, creating huge opportunities for potato cultivation in Asian countries.

Plant development depends on temperature, and plants require a specific amount of heat to develop from one point in their life cycle to another, such as from seeding to harvesting. The growth cycle can be roughly divided into five stages in potatoes: sprout development, vegetative growth, tuber initiation, tuber bulking and maturation. The accumulation of growing degree days (GDDs) and photo-thermal units for each developmental stage is relatively constant and independent of sowing date, but crop variety may affect it considerably (Phadnawis and Saini, 1992). Pushkarnath (1976) reported three maturity groups in the potato: these are early- (<90 days), medium- (91–110 days) and late-maturity (111–130 days), based on tuber maturity. Developing early-maturing potato varieties is of utmost importance in the current situation and it mainly depends on selecting breeding material with the early tuber bulking trait, as well as a high dry matter partitioning capacity. Early tuber initiation and subsequent dry matter partitioning to the tubers lead to early-maturing crops. Tolerance to major biotic and abiotic stresses should also be considered, because these varieties are grown in different agroecological zones. Drought and heat stress are the major abiotic stresses encountered by early-maturing potato varieties, depending on the agroecological zones of cultivation. The Asia-Pacific (AP) region of the world plays a significant role in potato production, mainly China in Southeast Asia and India in Southwest Asia. Although rice is the basic staple food in most AP countries, the potato is considered another important staple in this region. Differences between the agroecological zones in this area are the main constraint on productivity. Northern China has long days (LDs) and dry, flat land, while short-day, humid mountain environments prevail in South and Southeast Asia. Potato production systems in Asia have been classified into four types according to the latitude of the region: the single-cropping system (northern latitudes), the double-cropping system (central China), the mixed-cropping system (southwest China) and the winter-cropping system (extreme south of China, the Indo-Gangetic Plain, Bangladesh and northern Vietnam) (Ezeta, 2008). Of these, cultivation of potatoes as a winter crop is the most common in the extreme south of China, the Indo-Gangetic Plain, Bangladesh and northern Vietnam, where potatoes are planted in the winter between two crops of rice. When grown in this way, it is reasonably likely that the potato crop will experience drought stress, so it is advisable to use a variety with a short growing period and good tolerance to drought so that it can be grown using the residual moisture accumulated during the monsoon season (Haverkort et al., 2004). Screening genotypes for drought resistance is based on a comparison of the drought tolerance index and harvest index (HI), which refers to the weight of the tubers compared to the rest of the plant, for each genotype. Many CIP-bred clones do not produce well in the Central Asian summer, but some virus-resistant genotypes, especially Sarnav (CIP 397077.16) and Pskem (CIP 390478.9), show adaptation to temperate and drought conditions. Similarly, CIP clones were screened for drought tolerance in arid areas of India, and CIP clone CIP 397006.18 was promising, based on its overall yield performance, drought tolerance and acceptability of texture and taste (Sharma et al., 2014). These kinds of genotypes, which are resistant to drought and viruses, can be used to breed for early-maturing varieties. The broad genetic base of these genotypes will adapt to hot and subtropical climates and may be resistant to major biotic stresses. Furthermore, late blight as a biotic stress is the most devastating potato disease and its occurrence is largely dependent on favourable climatic conditions, so resistance to late blight should be considered a high priority, along with the other characteristics mentioned earlier, when selecting parents for earliness.

2 Selecting germplasm and traits for breeding early-maturing varieties

Traits to be selected when breeding early-maturing varieties are tuber initiation, per cent tuberization, foliage senescence, bulking rate, average weight per tuber and average tuber yield per plant, which need to be evaluated at 30, 60 and 90 days after planting or transplanting to the field. On the basis of the general combining ability (GCA) and heritability of the parental material, parents must be selected for the breeding programme. Tuber initiation data can be recorded by visual observation of swollen stolons or tubers on representative plants uprooted from the breeding plot. Per cent tuberization is estimated by counting the number of plants with tubers and dividing by the total number of plants. Similarly, other tuber-related traits like average weight per tuber (total weight of tubers from a plant/number of tubers), average tuber yield per plant (total tuber yield/number of plants) and bulking rate can be calculated by recording the total tuber weight over the growing period.

Selecting for earliness and tolerance to major stresses is a complex process and requires the interaction of several key pathways for normal plant development. Screening genotypes on morphological traits alone is insufficient, and other screening methods, such as metabolite profiling, are required. Metabolite profiling genotypes under a combination of stresses will help identify the metabolic signature involved in multi-pathway adaptation to these kinds of stresses (Zingaretti et al., 2013). A similar approach was successfully used by Drapal et al. (2016) to screen for genotypes that are tolerant to water shortage. It involves a combination of metabolite profiling and physiological/agronomical measurements to explore complex system-level responses to non-lethal water restriction. The metabolites identified have revealed a strong link with physiological responses in three different plant tissues (leaf, root and tuber) in five different potato genotypes, varying in susceptibility/ tolerance to drought. The study explored the potential of metabolite profiling as a tool to unravel sectors of metabolism that react to stress conditions and could mirror the changes in plant physiology. The metabolite results depicted different responses by the three plant tissues to the water deficit, resulting either in different levels of the metabolites detected or different metabolites expressed. Furthermore, the results of these studies revealed genotype-specific signatures for water restriction in all three plant tissues, suggesting that genetics can prevail over environmental conditions, which has important implications for future breeding for earliness.

3 Genetic aspects of earliness and breeding strategy

Photoperiod and temperature are the basis for earliness in potatoes. Results show that long-day reaction for tuberization is recessive to short-day reaction (Mendoza and Haynes, 1977). Potato varieties differ in their critical photoperiod (CPP) requirement and each genotype has its own CPP. Tuberization is favoured by photoperiods shorter than the CPP for the genotype. Both *Solanum tuberosum* ssp. *andigena* and wild species, *S. demissum*, cultivated in the Andes and Mexico, respectively, require short days (SDs) for tuberization. A succession of SDs is usually required for tuber induction; so the term 'inductive SDs' is more accurate. It is not day length but the duration of an uninterrupted dark period (night) that is critical for tuberization. In contrast, exposure to non-inductive LDs impedes tuberization, and may actually promote development of above-ground stolons and their growth as normal aerial shoots. Varieties of Group *Andigena*, cultivated at high altitudes, typically have a CPP of only 12–13 h, while Group *Tuberosum* varieties have undergone centuries of selection for a CPP of 15 h or more (Ewing, 1978).

The physiological mechanism for earliness is governed by early tuberization and subsequent early dry matter partitioning. Photoperiod and temperature are two important factors affecting these traits and ultimately the earliness of the crop (Fig. 1). Selection of elite germplasm is a key step in a successful breeding programme. Breeding for early-maturing potatoes requires parent material with early tuberization and high dry matter partitioning efficiency. In addition, germplasm with heritability for per cent tuberization at 30 days of harvest, foliage senescence at 60 and 90 days and speed of bulking at 30, 60 and 90 days should be selected as parent material when breeding for earliness in potatoes. Previous results show that selection of parent material for earliness has permitted a number of clones to be identified that can transmit these traits to their progenies. The general methodology (Fig. 2) for breeding early-maturing potato varieties involves diallel crossing without reciprocals of early-maturing and medium- to late-maturing clones, followed by evaluation of progenies for tuber initiation in a greenhouse. The progenies are evaluated in the field for tuber initiation, and harvested at 30, 60 and 90 days after transplanting to the field. The tubers obtained from the 90-day harvest are then used for the replicated field evaluation of tuber initiation. In replicated field trials, tubers should be harvested at 30, 60 and 90 days. GCA is estimated for tuber initiation, per cent tuberization, foliage senescence, average weight per tuber and average tuber weight per plant for the seedlings and clones. Similarly, the narrow-sense heritability (h^2) should be estimated for per cent tuberization, foliage senescence, average weight per tuber and average tuber weight per plant from the harvests at 30, 60 and 90 days. Apart from earliness, the variety must be tolerant to pests and diseases and any environmental aberrations. The breeding methodology for earliness must therefore select early clones with resistance to biotic and abiotic stresses (Fig. 3). This can be achieved by inter-crossing selected clones to build a base population for earliness. After progress has been made in earliness, heat tolerance, resistance to other abiotic stresses and resistance to X, Y and PLRV viruses, bacterial wilt, early blight and late blight should be introduced using resistant and adapted progenitors.

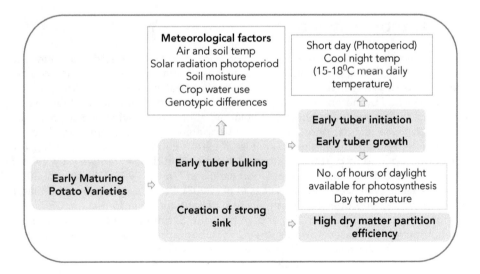

Figure 1 Physiological mechanism governing earliness in potatoes.

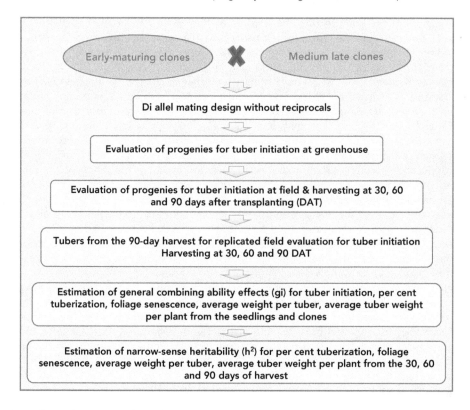

Figure 2 Breeding strategy for early-maturing potato varieties.

4 Early tuber initiation

Understanding the mechanism of tuberization is important when devising a strategy to develop early-maturing potato varieties. Early tuber initiation is a key trait to consider when developing early-maturing varieties. Tuber initiation is a complex developmental process, affected by meteorological, physiological and genetic factors. Meteorological factors include air and soil temperature, solar radiation, photoperiod and soil moisture (Fig. 1). Of these factors, photoperiod and temperature have the greatest effect on tuberization in potatoes (Martinez-Garcia et al., 2001; Lafta and Lorenzen, 1995). The role of photoperiod in potato tuberization has been studied extensively by various authors (Wheeler and Tibbitts, 1986; Dobranszki and Mandi, 1993; Dobranszki, 2001; Martinez-Garcia et al., 2001, 2002; Raices et al., 2003). Potatoes grown at the Equator, require short-day photoperiods (less than 12 h of light) for tuber formation. CPP for *S. tuberosum* subsp. *andigena* is 12 h and 16 h for the *S. tuberosum* subsp. *tuberosum*. As photoperiod increases towards the CPP, tuberization becomes irregular, then retarded and is finally inhibited. CPP increases at high temperatures and low irradiance. Increased photoperiod delays stolon and tuber initiation, reduces photosynthesis, causes small and inconsistent tubers and, finally, reduces the number of tubers.

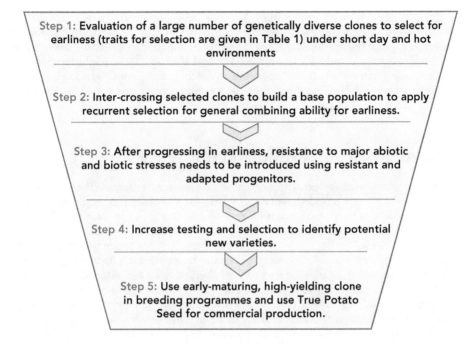

Step 1: Evaluation of a large number of genetically diverse clones to select for earliness (traits for selection are given in Table 1) under short day and hot environments

Step 2: Inter-crossing selected clones to build a base population to apply recurrent selection for general combining ability for earliness.

Step 3: After progressing in earliness, resistance to major abiotic and biotic stresses needs to be introduced using resistant and adapted progenitors.

Step 4: Increase testing and selection to identify potential new varieties.

Step 5: Use early-maturing, high-yielding clone in breeding programmes and use True Potato Seed for commercial production.

Figure 3 Stepwise breeding strategy for early-maturing and stress-resistant potato varieties.

Although the critical night length (dark period) for tuberization and the strength of the photoperiodic response vary with different genotypes, longer days (18 h) and higher temperatures (19°C) retard tuberization and reduce the relative rate of dry matter partitioning to tubers (Snyder and Ewing, 1989). Photoperiod has a significant effect on the earliness or lateness of the potato crop, and the effects of increased photoperiod on traits related to early-maturing potatoes have been demonstrated in previous studies. Day-length response is strong at higher temperatures and less pronounced at lower temperatures (Carrera et al., 2000). Both temperature and photoperiod factors act at the same point, probably by controlling common component(s) of the day length pathway (Rodriguez-Falcon et al., 2006).

The photoperiod-induced timing of tuber initiation determines the earliness or lateness of the potato crop. Earlier tuber initiation, and subsequent early dry matter partitioning to the tubers, leads to earlier-maturing crops; so tuberization induced by short photoperiod affects the earliness of the crop.

The short-day tuberization response (higher tuberization rate or overall rate of development) is affected by temperature and stress (heat and drought stress). The effect of temperature on tuberization has been studied extensively by many authors (Lafta and Lorenzen, 1995; Marinus and Bodlaender, 1975; Kooman et al., 1996; Wheeler and Tibbitts, 1986; Timlin et al., 2006) and high temperatures (over 22°C) inhibit tuberization by affecting photoassimilate partitioning.

All crop growing phases (0–3) and all stages of tuberization, that is, tuber induction, set, bulking and maturation, have been found to be affected by temperature. Plant emergence to tuber initiation (Phase 1) is a critical period and is mainly affected by radiation and temperature. Higher temperatures delay, or even inhibit, tuberization more than lower temperatures (Wheeler and Tibbitts, 1986; Jackson, 1999). Higher temperatures favour foliage development, delay tuber initiation, reduce the rate of tuber bulking and, eventually, reduce the yield (Kumar and Minhas, 2003). Night temperature has a particularly strong effect on tuberization, which is significantly reduced at night temperatures above 20°C. There may not be any tuberization at all at or above 25°C (Muthoni and Kabira, 2015).

5 High dry matter partitioning efficiency

A higher rate of tuber development is indicative of earlier tuberization. The potato crop matures earlier if a larger proportion of daily dry matter is allocated to the tubers, leaving less for the foliage. When all photoassimilates are allocated to the tubers, leaf growth ceases and, after a temperature-dependent period of time, the above-ground portion dies (Kooman et al., 1996). The dry matter produced as a result of photosynthesis is utilized by the plant for its growth and development. For better tuber yield, a substantial part of the assimilated carbon must be translocated to the tubers. It has been discovered that more dry matter is partitioned to the tubers under short-day conditions, and more is partitioned to the foliage under long-day conditions. The dry matter distribution to different parts of the plant also depends on the source/sink interaction. The sink may either be haulms or tubers, and the ability of the sink to attract photoassimilates is known as sink strength. Sink strength also affects the rate of photo assimilation, and the sink strength of the tuber has been found to increase after tuberization, increasing the rate of photosynthesis. Before tuber initiation, most of the photoassimilates are translocated to the haulms for growth and development but, after tuber initiation, a substantial amount of assimilate is partitioned to the tubers. For example, the proportion of dry matter accumulated in potato leaves and stems in one study was found to be 54% at 52 days of growth, and it decreased to about 15% by 80 days, demonstrating greater partitioning of dry matter to the tubers than the haulms at maturity. The growth phase between tuber initiation and the end of leaf growth is important because 90–100% of the daily photoassimilates are partitioned to the tubers during this phase. The duration of this phase depends on the relative tuber growth rate, which determines dry matter partitioning between the tubers and the rest of the plant (Kooman and Haverkort, 1995).

6 Basic factors in breeding for earliness in the potato

6.1 Cultivar maturity or growing degree days (GDDs)

The potato plant requires a specific amount of heat to progress from seeding to the harvesting stage. The variability among cultivars for number of days from planting to physiological maturity has led to the classification of potato cultivars into very early, early, medium, late and very late maturity classes (Ruzukas et al., 2009). Although there is no standard measure for maturity in the potato, because the tubers are produced under ground and monitoring their development is challenging, potato breeders commonly assess potato maturity based on physiological changes in the potato vine. Tuber production is associated with changes in the whole plant, such as a reduction in leaf development, flowering and fruit set (Haga et al., 2012). It is currently thought that judging crop maturity based on GDD is ideal and accurate. GDDs are calculated throughout the season for each planting date. The attainment of the number of GDDs required for each variety is determined by the temperature prevailing at their particular location. The formula GDD = [(minT + maxT)/2–7°C (45°F)] is used to calculate GDD for potato crops, where the base temperature is the temperature below which plant growth is zero (7.2°C or 45°F for potatoes) (Sands et al., 1979). It has been reported that yields may suffer due to both extended cool periods (below 18°C) and extended hot periods (above 20°C). Cooler and hotter temperatures reduce net assimilation to the tubers, and higher temperatures may prevent tuber initiation (Worthington and Hutchinson, 2005). Most potato varieties require more than 800 GDDs to reach maturity, but a minimum of 70–90 days of favourable cool temperatures is required for a profitable potato yield. However, a longer favourable season of 110–120 days, along with enough accumulated heat units, results in substantially higher returns (Khan et al., 2011).

6.2 Heat stress level

Genetic gains in crop yield seem to have remained relatively steady over the last three decades, averaging a little less than 1% per year under favourable conditions (Calderini et al., 1999). In the potato, there has been little yield progress in mid-early and late German (Schuster, 1997) and US potato cultivars (Douches et al., 1996) in that time. However, every newly introduced variety should produce better yields than the existing varieties and perform better agronomically. Selection of genotypes that utilize the available natural resources efficiently and yield better is essential to achieve these improvements (Petrovic et al., 2010). Heat unit accumulations must be converted into biomass accumulations to attain higher yields. The process of photosynthesis in the potato is very sensitive to high temperatures: the optimum temperature is reported to be about 20°C and an increment of 5°C above the optimum is expected to reduce the rate of photosynthesis by 25% (Burton, 1981). Optimum canopy net photosynthesis in potatoes has been reported at 24°C, and maximum biomass accumulation has been reported between 18°C and 20°C (Kooman and Haverkort, 1995; Timlin et al., 2006).

7 Breeding strategies for earliness and stress resistance

Conventional potato breeding involves an initial cross between parents with complementary traits based on their phenotype, followed by selection in subsequent clonal generations (Sleper and Poehlman, 2006). Selection is based on phenotypes, with the aim of identifying clones with as many desirable traits as possible for release as new varieties (Lynch and Walsh, 1998; Bradshaw and Bonierbale, 2010). Since the choice of parents is largely dependent on their performance, it is hard to predict the segregation pattern of the F1 progeny because potatoes are a highly heterozygous crop. Successful breeding combines the desired alleles into a single genotype and tests their stability and adaptation (Acquuah, 2007). Successful earliness should combine early tuber initiation with high dry matter partitioning, or HI, in the shortest possible time.

8 Genetic aspects

8.1 Parental selection

Breeding for earliness in potatoes has also been considered beneficial because early varieties escape some diseases that appear later in the season, such as late blight (Razukas and Jundalas, 2006), and they are also capable of adapting to a variety of environments. The inheritance patterns that condition this trait (El-Bramawy and Shaban, 2007), the gene action involved (additive, nonadditive, dominant or epistatic) and their interaction with the environment have been assessed across a variety of growing environments. Kumar and Kang (2006) evaluated the progeny of seven *S. tuberosum* ssp. *tuberosum* and three *S. tuberosum* ssp. *andigena* accessions comprising 12 *Tuberosum* × *Tuberosum* (T × T) and 9 *Tuberosum* × *Andigena* (T × A) progenies for five important traits for two successive clonal generations under short-day subtropical conditions. The differences in yield, average tuber weight and tuber number between both families in an early (75 days) crop were not significant. Compared to conventional intra-*Tuberosum* families, T × A families had significantly higher % tuber dry matter and specific gravity. In contrast to intra-*Tuberosum* crosses, T × A crosses exhibited positive heterosis for tuber yield. This revealed the usefulness of *Andigena* genotypes as parents when developing early-bulking potato cultivars with a broad genetic base for short-day subtropical plains.

8.2 Combining ability

Combining ability analysis is used to compare the performance of a given parental line in hybrid combinations (Griffing, 1956; Hayder et al., 2009). In potato breeding, both GCA and specific combining ability (SCA) are vital when conditioning traits, and they both are fixed in the F_1 generation because there is no further segregation with clonal breeding stock. GCA seems to be significantly greater than SCA for tuber yield and quality traits in crosses between non-related parents, whereas SCA appears to be more important among related parents (Ortiz and Golmizaie, 2004). With regard to crop maturity, GCA effects tend to be more important than SCA effects (Johansen et al., 1967). Additive and nonadditive effects also influence maturity and total tuber yields (Buso et al., 2008). Paula (2014) reported that parental line 396038.107 exhibited negative GCA effects for growth parameters and positive GCA effects for yield parameters, which are desirable attributes to obtain progenies that combine earliness with high yields. Nakpot-1 had desirable GCA effects for tuber initiation and yield, and undesirable GCA effects for senescence, so it could be used to improve traits where it has desirable combining ability. Similarly, Kiru et al. (2007) reported that potato cultivars Alyona, Bezhitskii, Bryanskii delicatess, Daryonka, Debryansk, Zhavoronok, Zhukovskiirannii, Lakomka, Lina, Ljubava, Pamyaty Osipovoy, Pogarsky, Russkii souvenir, Snegir, Udacha, Kholmogorsky, Effect (Russia); Yavor (Byelorussia), Aster, Bekas, Irga, Harpun, Lena (Poland); Andra, Bonus, Velox (Germany); and Kobra, Korela, Korneta, Krasa, Tegal (Czech Republic) combine earliness with other valuable traits. Vavilov Plant Institute also found that the most useful cultivars for earliness with a high starch content (15–22.7%) were Mariella, Axilia, Binova, Arina, Kurganskii Rannii (Kurgan Early), Laimdota and Lyubimets (Favourite). These cultivars are valuable to improve potato genotypes for earliness and high fresh tuber yield.

8.3 Progeny selection

Progeny can only be identified for earliness after tuber initiation and at harvest. However, seedlings with closed tops are reported to develop into late forms, while those with open tops generally mature early (Zubeldia, 1963). Engel and Moller (1959) found that 80% of seedlings showing long stolons when ready to transplant from boxes to pots (second transplantation) were either early- or medium- to early-maturing. They used this criterion to detect earliness in crosses of early × late forms that were expected to give a high frequency of lates. The mode of inheritance of earliness in potatoes and the magnitude of genotype (G) by environment (E) interactions for earliness and fresh tuber yield in selected potato clones with horizontal resistance to late blight were studied. Both additive and nonadditive gene effects were found to control earliness, although additive gene effects were more dominant (Paula, 2014). Nine advanced potato clones were evaluated across three sites located at different altitudes, ranging from 1400 to 2450 m above sea level, for two seasons to estimate the magnitude of genotype–environment interaction, and stability for earliness and fresh tuber yield (Paula, 2014). Genotypes 396026.103 and 391046.14 were found to be the most stable for earliness and high fresh tuber yield across all the environments. Golmirzaie and Serquen (1992) found that the highest correlation coefficient for earliness obtained in Huancayo was r = −0.146, explaining only 2.13% of the variation in earliness caused by differences in root length. The other coefficients were close to zero, suggesting an almost complete absence of association among these characteristics. Kumar et al. (2008) studied 114 progenies in a line × tester design (19 × 6), and these were studied for ten characters in seedlings and seedling-tuber crops. SCA variance was predominant than GCA variance in the expression of most traits, except for tuber shape, where variances were almost equal. Genotypes JX 90, CP 3125 and JN 1197 were found to be superior in both crops, based on their GCA for all traits. Kufri Bahar × JN 1197, JF 4841 × CP 3125, JX 90 × N 1197, CP 2161 × JN 1197, JX 90 × CP 3125 and JV 67 × JEX/A 592 were superior crosses for developing a high-yielding and uniform true seed population. Kang and Kumar (1999) found that four early-bulking hybrids (J.96-84, J.96-99, J.96-149 and J.96-238) bred for the northern plains had 11.9–16.8% higher tuber yields than the best control (Kufri Ashoka) at 75 days. Kadian et al. (2012) found that, out of nine advanced clones, CIP 301029.18, CIP 388972.22 and CIP 393708.31 had significantly higher yields of marketable tubers (52.6, 49.3 and 52.6 tonnes/ha, respectively), along with desirable tuber shape, size and colour and more than 20% dry matter, screened at Ladol (Gujarat) for heat stress.

9 Case study: developing an early-maturing, moderately late-blight-resistant Kufri Khyati potato variety for Indian plains

In recent times, potato production has been greater in the developing world than in the developed world (Scott and Suarez, 2012). As per the recent report of National Horticultural Research Development Foundation, India is now the second largest producer of potatoes in the world, producing 45.56 million tonnes from 2.06 million hectares, at an average yield of 22.09 tonnes/ha in 2015–16. This has been made possible by adapting the crop to the different agroecological conditions in India. Initially, potatoes were grown at higher altitudes as the climate is similar to European conditions; however, they have now spread to plains and plateau. The potato-growing environments in India have been categorized as hills, plains and plateau (Table 1; Fig. 4). *Hills:* Categorized as very high hills (>2400 m), high hills (1600–2400 m), mid-hills (800–1600 m) and low hills (300–800 m), based on the altitude. In high hills, the higher temperatures at the early stages and lower temperatures after the onset of rains during the bulking period, with LDs, favour heavy haulm growth. The cloudy conditions during the later stages of crop growth are conducive to late blight, which is very severe and causes heavy losses; so yields are low. Therefore, early-, medium- and late-maturing varieties with high light use efficiency and resistance to late blight are required for high yields in this region. In contrast to the very high and high hills, the growing season in the mid- and low hills is warmer, and late blight is not as serious in this region; so this region demands early- and medium-maturing varieties with abiotic stress resistance. *Plains:* Geographically, this represents areas below 300 m mean sea level and it is further subcategorized into North Western, Central and Eastern Plains; the potato-growing season and environmental conditions are different in these three regions. Two distinct, short growing seasons can be delineated in the North Western plains; so this region is called the short-duration, two-crop zone (Pushkarnath, 1976). The short growing season with very low temperatures and incident radiation during the peak growth stage of the potato is not conducive to high yields. Short-duration varieties with high light use efficiency and resistance to late blight, coupled with suitable agricultural techniques for quick emergence and early attainment of full canopy cover, are required to realize high/remunerative yields in the region. The Eastern plains have a short growing period and are therefore referred to as a short-duration, one-crop zone. Late blight and charcoal rot are the notable diseases in this zone, but late blight is not as regular as in the North Western plains. Early-maturing, heat-tolerant varieties are desirable in this zone. *Plateau:* The plateau ranges from 300 to 1200 m in altitude, and is further classified into low-altitude plateau (300–600 m) and high-altitude plateau (600–1200 m). In the high-altitude plateau, two growing seasons (*kharif* and *rabi*) are feasible, but only the winter season crop is possible in the low-altitude plateau. The *kharif* season is fairly long but severely constrained by abiotic and biotic stresses.

As discussed in the previous section, potato plant development depends on temperature and the plant requires a specific amount of heat to develop. The All India Coordinated Research Project (AICRP) potato centres located across the country indicate huge diversity: the mean maximum temperature during the 90-day growing period ranges from 21.1°C to 32.6°C, and the minimum temperature ranges between 8.0°C and 20.9°C. This diversity can be exploited to evaluate hybrids for their performance under different stress levels, and the AICRP centres are classified according to the mean temperature during the growing season for this reason. The organization estimates that crops could experience low-temperature stress at the Pantnagar, Srinagar and Jalandhar centres, as the thermal time accumulated in 90 days is less than 1600 h. Optimum temperatures of 18–20°C would be prevalent at Kanpur, Shillong, Shimla, Patna, Hissar, Dholi, Chindwada, Kota and Modipuram and these are ideal for evaluating genotypes for their yield potential under optimal temperature conditions. The accumulated GDD at these locations ranged from 1600 to 1800. Mild stress (mean temperatures of 20–22°C) is expected at Gwalior, Kalyani, Jorhat and Raipur, and high stress (mean temperatures of >22°C) is expected at Bhubaneswar, Pune, Deesa and Dharwad, which are target locations for evaluating genotypic performance under heat stress. Mean night temperatures during the growing

Table 1 Potato-growing environments in India

Growing environment	Season	Minimum temperature (°C)	Maximum temperature (°C)	Solar radiation (MJ/m²)	Remarks	Varietal requirement
Hills						
High hills	Summer (Mar/Apr to Aug/Sep)		12.9–24.6	14.6–20.7	Rain-fed	Long-day adapted and resistance to late blight
Mid hills	Feb to July	10.3–18.0 (9.5)	13.8–30.3 (23.0)	13.8–35.6 (24.6)	Suitable for processing varieties	Resistance to early blight, late blight, bacterial wilt and viruses
Low hills	Autumn/winter (mid Sep/Oct)	4.2–15.2 (9.3)	15.7–24.8 (20.5)	10.2–14.7 (11.9)		Resistance to early blight, late blight, bacterial wilt, viruses and tuber rot
	Spring (Jan to Feb)	3.86–19.1 (11.6)	14.5–29.7 (22.8)	10.6–20.6 (16.7)	Fully irrigated	
Plains						
North Western	Autumn (Sep/Oct to Dec)	3.5–18.6	19.0–34.0	13.7–23.8		Short-day adapted, early maturity, resistance to early blight, late blight, black scurf, scab, viruses and tolerance to frost
	Spring (Jan to Apr)	3.3–16.1	18.7–34.0	13.8–24.0	Irrigated, severe environmental constraints	
Central	Rabi (Oct/Nov to Mar/Apr)	6.5–17.4	21.0–33.9	15.5–28.5	Zone of high productivity	Short-day adapted, early to medium maturity, resistance to early blight, late blight, viruses and tolerance to frost
Eastern plains	–	–	–	–	–	Short-day adapted, early to medium maturity, resistance to early blight, late blight, charcoal rot, viruses and red tubers are preferred
Plateau						
Low altitude (300–600 m)	Winter	15.5–19.5 (17.2)	28.7–33.8 (30.2)	17.0	Irrigated	Resistance to early blight, late blight, bacterial wilt, viruses, tuber rots and heat tolerance
High altitude (600–1200 m)	Kharif (Jun to Sep/Oct)	20.82–22.0 (19.3)	28.1–33.0 (29.4)	18.9–19.3		

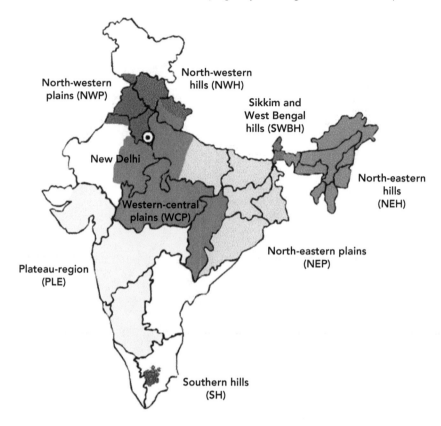

Figure 4 Potato agroecological zones of India. Source: Adapted from Singh et al. (2008).

season also vary widely between the different AICRP centres, and range from 11.3°C (Pantnagar) to 22.9°C (Dharwad). There is similar variability in incident solar radiation; so germplasm is screened for parental lines to develop earliness at these locations. Once the breeding lines are identified, they are tested again at these locations to confirm the yield superiority of these varieties.

The plains now account for more than 80% of the potato acreage in the country, and the rest is accounted for by the hills and plateau, which are together responsible for about 15% of the total acreage under the crop. In the Indian subtropical plains, potatoes are raised under short-day conditions in the winter. Short-duration varieties are required to increase the area under potatoes and enable the crop to compete with intensive agriculture, as well as medium- and long-duration varieties (Shekhawat, 1994). Most of the area covered by the north Indian plains is suitable for growing early potato varieties and there are many advantages of short-duration varieties. Potatoes are usually harvested at about 90–110 days after planting in the Indian plains, but they can be harvested at any time after about 60 days. Potatoes compete with wheat, as both are grown in the *rabi* crop season. Early-bulking varieties can be sandwiched between the crops in different farming systems without limiting the biodiversity of associated crops (Shekhawat et al., 1999). These varieties can enable a rice–potato–wheat crop sequence, and early varieties are profitable for farmers and can escape the vagaries of frost and late blight. Short-duration varieties can enable farmers to harvest the potato crop any time for two months from September to May, resulting in staggered potato production, which can solve the problem of glut/storage to some extent. With this in mind, efforts have been made by ICAR-CPRI, Shimla, since the 1970s to develop early-maturing potato varieties in India.

Based on the above agroecological conditions, India needs early-maturing potato varieties. In the following section, we will discuss in detail the development of the recent potato variety most suited for cultivation in the plains of India, Kufri Khyati (Hybrid J.93-86) as reported by Kumar et al. (2009).

9.1 Parental selection

The cross was first attempted in 1992 at the Central Potato Research Station at Kufri. Six generations of this hybrid's pedigree are shown in Fig. 5. The female parent (MS/82-638) is an Indian advanced hybrid with eight exotic genotypes [Ekishirazu, Katahdin, 134-D, 3069 d(4), 2814 a(1), PI 161695-1, Adina and Sarkov] in its pedigree. The male parent (Kufri Pukhraj) is a released Indian cultivar with *Andigena* accession JEX/B 687 in its parentage. Hybrid J.93-86 was developed by clonal selection from the MS/82-638 × Kufri Pukhraj cross.

9.2 Clonal selection

The hybrid J.93-86 was selected from seedling and clonal generations evaluated at the Central Potato Research Station at Jalandhar in the North Western plains. In trials conducted at Jalandhar before hybridization was started at AICRP on Potato, average total and marketable yields over four crop seasons for Kufri Khyati harvested at 60 days were higher than the control (Kufri Ashoka) by 14.0% and 15.6%, respectively (Table 2). Based on its excellent tuber characters and tuber yield at Jalandhar, it progressed further in multilocation trials.

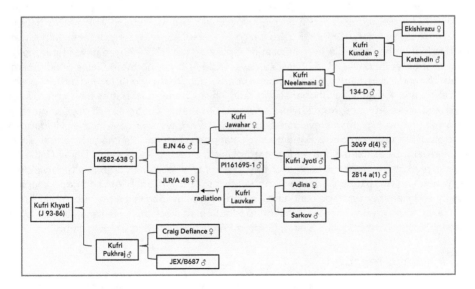

Figure 5 Pedigree of Kufri Khyati. Source: Adapted from Kumar et al. (2009).

Table 2 Yield performance of Kufri Khyati from 1997 to 2001 in replicated yield trials at Jalandhar before initiation in All India Coordinated Research Project on Potato (AICRP on Potato) for multiloca-tion trials

Kufri Khyati/control	Crop season				
	1997–98	1998–99	1999–2000	2000–1	Average
Total yield (tonnes/ha) at 60 days					
Kufri Khyati	27.1	27.6	25.5	20.7	25.2
Kufri Ashoka	24.1	24.8	22.4	17.0	22.1
CD ($p < 0.05$)	3.8	1.2	2.2	1.4	
Marketable yield (tonnes/ha) at 60 days					
Kufri Khyati	26.4	27.0	24.9	19.7	24.5
Kufri Ashoka	22.8	24.2	21.8	16.1	21.2
CD ($p < 0.05$)	3.8	1.1	2.2	1.5	

Source: Adapted from Kumar et al. (2009).

9.3 Multilocation trials

Hybrid J.93-86 was tested in multilocation trials spread over different agroecological regions of the country under the AICRP on Potato from 2002–3 to 2006–7 (Tables 3, 4 and 5).

9.4 Yield performance

The hybrid was evaluated in replicated yield trials during 2002–5 in plots of 7.2 m², then in 1–4 non-replicated yield trials at each location during 2005–7, in plots of 96 m². Yield performance of Kufri Khyati at 60- and 75-day harvests at different AICRP on Potato locations is presented in Tables 3 and 4, respectively. The marketable yields at 75 days in 'On-Farm' trials under AICRP on Potato are given in Table 5. At most locations, Kufri Khyati yielded higher than controls. In multilocation replicated trials, Kufri Khyati gave higher yields than controls at 60 days (Table 3) and 75 days (Table 4) in the northern plains (Modipuram, Jalandhar and Hisar), central plains (Chhindwara, Raipur, Deesa and Kota) and eastern plains (Faizabad and Kalyani). At Bhubaneswar, Kufri Khyati performed better than controls for yield at 60 days. At Patna, Kufri Khyati performed better than controls at 75 days. In on-farm yield trials, Kufri Khyati was evaluated for yield at 75 days and gave higher yields than controls in the northern plains (Modipuram, Jalandhar, Hisar and Pantnagar), central plains (Chhindwara, Deesa and Kota) and eastern plains (Faizabad, Kalyani and Dholi) (Table 5). Overall marketable yield increase over best control was 17.7%. J.93-86 is a high-yielding early-maturing white tuber hybrid with moderate resistance to late blight and good keeping quality under ambient storage conditions, suitable for cultivation in the plains of India. Tuber dormancy is ten weeks. Specific areas for its adoption are Rajasthan, Haryana, Punjab, Uttar Pradesh, Bihar, West Bengal, Gujarat, Madhya Pradesh, Orissa and Chhattisgarh.

9.5 Disease reaction

Kufri Khyati is moderately resistant to late blight. Area under disease progress curve for late blight in Kufri Khyati, under epiphytotic conditions at Kufri, was recorded as 187.5,

Table 3 Total yield of Kufri Khyati at 60 days of harvest at different locations in replicated trials under All India Coordinated Research Project on Potato (AICRP on Potato) (2002–5)

Centre	Total yield (tonnes/ha) during crop season			Average total yield (tonnes/ha)	Per cent yield increase over best control
	2002–3	2003–4	2004–5		
Northern plains					
Modipuram (UP)	24.38 (23.69)	23.96 (21.80)	21.70 (18.92)	23.35 (21.47)	8.8
Jalandhar (Punjab)	22.08 (15.88)	23.84 (20.51)	21.62 (21.64)	22.51 (19.34)	16.4
Hisar (Haryana)	–	12.04 (13.38)	26.48 (18.19)	19.26 (15.78)	22.1
Pantnagar (Uttaranchal)	–	40.32 (47.45)	40.93 (48.33)	40.63 (47.89)	−15.2
Central plains					
Chhindwara (MP)	19.10 (14.40)	17.38 (14.19)	17.73 (15.02)	18.07 (14.55)	24.2
Raipur (Chhattisgarh)	–	22.45 (19.07)	19.82 (20.11)	21.14 (19.59)	7.9
Deesa (Gujarat)	22.66 (15.22)	14.06 (14.78)	26.10 (20.09)	20.94 (16.70)	25.4
Kota (Rajasthan)	–	8.66 (6.90)	8.56 (6.30)	8.61 (6.53)	31.9
Eastern plains					
Faizabad (Eastern UP)	–	20.39 (20.44)	17.90 (14.31)	19.15 (17.37)	10.2
Kalyani (West Bengal)	21.96 (14.58)	21.64 (22.22)	22.68 (21.11)	22.09 (19.30)	14.5
Bhubaneswar (Orissa)	–	18.80 (17.55)	–	18.80 (17.55)	7.1
Patna (Bihar)	17.31 (16.53)	21.11 (23.24)	15.13 (14.32)	17.85 (18.03)	−1.0
Dholi (Bihar)	–	14.81 (17.82)	13.89 (13.29)	14.35 (15.06)	−4.7
Jorhat (Assam)	11.75 (11.95)	11.70 (12.27)	16.12 (17.99)	13.19 (16.10)	−18.1
Average of all	19.89 (16.04)	19.37 (19.17)	20.67 (19.12)	19.99 (18.57)	7.6

Source: Adapted from Kumar et al. (2009).
The yields (tonnes/ha) of best control variety (Kufri Ashoka/Kufri Jawahar) are given in parentheses for comparison. At most of the locations, Kufri Ashoka was the better control than Kufri Jawahar.

compared to 537.5 recorded for the susceptible control cultivar, Kufri Jyoti. Early-maturing varieties are usually highly susceptible to late blight, as late blight resistance is associated with late maturity. Kufri Khyati, however, has moderate resistance to late blight as an additional characteristic. It is also moderately resistant to *Phoma* and early blight.

9.6 Adaptability

Kufri Khyati performed well in the northern, central and eastern plains, and is therefore recommended as a main crop variety for these regions. Its moderate resistance to late blight could be useful in preventing fluctuations in yield over the seasons due to late blight. Kufri Khyati has the potential to fit in with various crop sequences, due to the flexibility of its harvest period any time after 60 days; so it is suitable for the high cropping intensity followed in the Indo-Gangetic plain. In the North Western plains, this variety could fit well into the rice–potato–wheat sequence due to its high yield at early (75 days) and very early (60 days) harvest.

Table 4 Total yield of Kufri Khyati at 75 days of harvest at different locations in replicated trials under All India Coordinated Research Project on Potato (AICPP on Potato; 2002–5)

Centre	Total yield (tonnes/ha) during crop season			Average total yield (tonnes/ha)	Per cent yield increase over best control
	2002–3	2003–4	2004–5		
Northern plains					
Modipuram (UP)	34.78 (32.68)	35.11 (30.12)	30.35 (20.58)	33.41 (27.79)	20.2
Jalandhar (Punjab)	31.99 (26.34)	35.32 (26.06)	36.99 (34.00)	34.77 (28.80)	20.7
Hisar (Haryana)	–	23.24 (21.85)	32.64 (24.77)	27.94 (23.31)	19.9
Pantnagar (Uttaranchal)	–	42.13 (51.76)	46.57 (52.36)	44.35 (50.60)	−12.4
Central plains					
Chhindwara (MP)	23.42 (17.78)	21.90 (17.06)	23.51 (19.44)	22.94 (18.09)	26.8
Raipur (Chhattisgarh)	–	27.78 (22.68)	36.24 (30.29)	32.01 (25.68)	24.6
Deesa (Gujarat)	34.94 (28.77)	47.31 (38.60)	38.29 (31.35)	40.18 (32.96)	21.9
Kota (Rajasthan)	–	20.32 (16.67)	24.45 (17.22)	22.39 (16.95)	32.1
Eastern plains					
Faizabad (Eastern UP)	–	40.70 (32.43)	30.79 (24.54)	35.75 (28.49)	25.5
Kalyani (West Bengal)	23.60 (17.40)	34.03 (30.79)	26.02 (23.66)	27.88 (23.95)	16.4
Bhubaneswar (Orissa)	–	23.61 (21.30)	18.01 (21.34)	20.81 (21.32)	−2.9
Patna (Bihar)	34.82 (27.87)	37.71 (35.56)	25.28 (26.67)	32.60 (30.03)	8.6
Dholi (Bihar)	–	17.82 (20.83)	17.41 (15.56)	17.62 (18.20)	−3.2
Jorhat (Assam)	19.94 (21.44)	17.63 (19.14)	18.02 (19.58)	18.53 (19.12)	−3.1
Average of all	25.77 (21.68)	28.44 (25.63)	27.65 (24.32)	27.74 (24.59)	12.8

Source: Adapted from Kumar et al. (2009).
The yields (tonnes/ha) of best control variety (Kufri Ashoka/Kufri Jawahar) are given in parentheses for comparison. At most of the locations, Kufri Ashoka was the better control than Kufri Jawahar.

9.7 Keeping quality

Keeping quality studies carried out at Jalandhar showed that Kufri Khyati has better keeping quality than the early-maturing varieties used as controls (Table 6). It exhibited lower sprout weight, physiological weight loss and total weight loss than the control varieties (Kufri Ashoka and Kufri Pukhraj) at 90 days after storage at ambient temperature.

9.8 Release

Kufri Khyati was recommended for release as a variety by the 26th Group Meeting of Potato Workers held from 7 to 9 September 2007 at Rajendra Agricultural University, Pusa, Bihar. It was released and named as variety Kufri Khyati (Fig. 6) in 2008 by the Central Sub-Committee on Crop Standards, Notification and Release of Varieties, Ministry of Agriculture, Government of India, for its general cultivation in the plains of India.

Table 5 Marketable yield of Kufri Khyati at 75 days of harvest at different locations in on-farm trials under All India Coordinated Research Project on Potato (AICPP on Potato) (2005–7)

Centre	Marketable yield (tonnes/ha) during crop season		Average marketable yield (tonnes/ha)	% yield increase over best control
	2005–6	2006–7		
Northern plains				
Modipuram (UP)	26.2 (24.3)	31.1 (25.8)	28.7 (25.4)	13.0
Jalandhar (Punjab)	32.3 (29.4)	25.2 (22.4)	28.8 (25.7)	12.1
Hisar (Haryana)	29.6 (22.0)	21.5 (17.0)	25.6 (19.0)	34.7
Pantnagar (Uttaranchal)	30.8 (24.2)	16.3 (17.7)	23.5 (20.0)	17.5
Central plains				
Chhindwara (MP)	24.8 (19.9)	22.5 (19.0)	23.6 (19.4)	21.6
Raipur (Chhattisgarh)	17.6 (21.8)	21.6 (19.9)	19.6 (20.6)	−4.8
Deesa (Gujarat)	34.2 (25.8)	39.9 (36.1)	37.0 (30.9)	19.7
Kota (Rajasthan)	19.3 (10.9)	18.4 (12.0)	18.8 (11.4)	64.9
Eastern plains				
Faizabad (Eastern UP)	25.7 (23.5)	29.0 (26.1)	27.3 (24.1)	13.3
Kalyani (West Bengal)	30.2 (20.1)	18.0 (18.8)	24.1 (19.4)	24.2
Bhubaneswar (Orissa)	16.9 (17.0)	14.1 (24.2)	15.5 (20.3)	−23.6
Patna (Bihar)	32.7 (32.0)	24.9 (23.5)	28.8 (27.5)	4.7
Dholi (Bihar)	23.6 (20.5)	24.7 (20.8)	24.1 (20.6)	17.0
Jorhat (Assam)	10.1 (12.2)	9.3 (9.3)	9.7 (10.7)	−9.3
Average of all	25.3 (20.7)	22.6 (20.2)	23.9 (20.3)	17.7

Source: Adapted from Kumar et al. (2009).
The yields (tonnes/ha) of best control variety (Kufri Ashoka/Kufri Pukhraj) are given in parentheses for comparison. Kufri Pukhraj was the better control at eight locations, while at other six locations, Kufri Ashoka was the better control.

Table 6 Storage behaviour (keeping quality) of Kufri Khyati at ambient temperature at Jalandhar (pooled data of 2006 and 2007 after 90 days of storage)

Hybrid/variety	Per cent sprouting	Per cent rottage by number	Per cent rottage by weight	Sprout wt (g/kg)	Per cent physiological weight loss	Per cent total weight loss
Kufri Khyati	100	0.6	0.5	8.7	10.6	13.4
K. Ashoka	100	2.4	2.3	10.1	14.2	17.2
K. Pukhraj	100	3.5	3.3	15.3	16.5	21.2
CD ($p < 0.05$)	NS	0.29	0.19	0.57	0.57	1.00

Source: Adapted from Kumar et al. (2009).

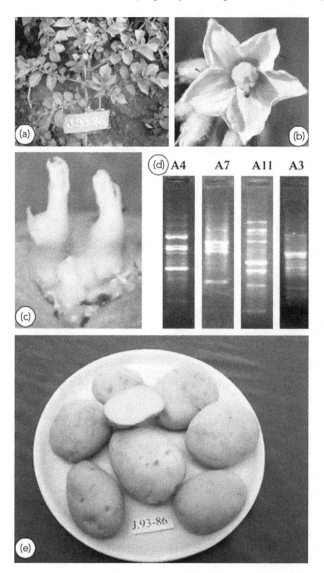

Figure 6 Morphological characteristics of Kufri Khyati. (a) plant; (b) flower; (c) sprouts; (d) DNA finger-prints and (e) tubers. Source: Adapted from Kumar et al. (2009).

AICRP Potato trials were conducted during 2015–16 and 2016–17 to identify the three most promising varieties suitable for the north, central and eastern plains and plateau regions of India. In total, 48 trials were conducted and 15 varieties were evaluated during the winter crop season, through front-line demonstration at 18 AICRP centres (Bhubaneswar, Chhindwara, Deesa, Dholi, Faizabad, Gwalior, Hisar, Jalandhar, Jorhat, Kalyani, Kanpur, Kota, Modipuram, Patna, Pantnagar, Pasighat, Pune and Raipur) and harvested at 60, 75 and 90 days. Kufri Khyati was among the top three high-yielding varieties in the northern and eastern plains and plateau region when harvested at 60, 70 and 90 days. It was among the four highest-yielding varieties in the central plains when harvested at 75 and 90 days.

10 Future trends and conclusion

Earlier attempts to develop potato varieties in India were aimed at high tuber yields with resistance to biotic stresses, such as late blight. Out of necessity, breeding strategies were then extended to develop processing varieties resistant to other diseases when they became a menace to the potato crop, and develop location-specific varieties to solve local issues, for example nematode-resistant varieties for the south Indian hills and so forth. As mentioned earlier, some early-maturing varieties were also released in the process. Now, under changing climatic conditions, many biotic and abiotic stresses are becoming more important and must be dealt with. Some early-maturing parents and sources of resistance to major biotic and abiotic stresses have already been reported among the vast array of genetic resources available in the Indian potato gene bank, and additional sources are to be identified through proper screening. Again, location-specific breeding strategies are to be adopted to develop early-maturing and stress-resistant potato varieties suitable for the particular zones mentioned in this chapter. With the advent of modern biotechnological tools in combination with conventional breeding methods, it would be possible to identify and incorporate the genes responsible for desirable characteristics into new high-yielding early-maturing table and processing potato varieties and develop resistance to biotic and abiotic stress. Despite the challenges, there is ample scope for potato breeders to achieve other success stories; so more early-maturing potato varieties with resistance to biotic and abiotic stress are expected to be developed.

A number of Asian countries are already diversifying their farming systems by making the intensification of prevailing systems more sustainable. Early-maturing potato varieties, particularly a 70- to 90-day potato resistant to heat, viruses, LB and with good processing quality, are a profitable and nutritious complement to low-income cereals. These potato varieties provide flexible planting and harvesting times without putting undue pressure on declining land and water resources.

Under changing agroclimatic conditions, new potato varieties will be required for planting early and in the main crop season to fit potato into diversified cropping systems and agroecologies of subtropical lowlands and highlands of South China, North Vietnam, Bangladesh, India, and the plains of Nepal and East Pakistan. New, early-maturing varieties with heat tolerance, high tuber dry matter and good storability are needed to feed the increasing population of the Asian world. Early-maturing potato would be a crucial solution for timely planting of succeeding crops of rice and wheat. Hence the effects need to be intensified towards developing early-maturing potato varies in consideration with area-wise requirement having earliness, high dry matter, heat tolerance, disease tolerance and good storage qualities. The efforts need to be made to introduce potato in new and non-traditional potato growing areas to improve food security and enhance farmers' income.

11 Acknowledgments

The authors appreciatively recognize Dr. KV Prasad, Director, ICAR-Directorate of Floricultural Research, Pune, India for overall support and Dr. S. Anandhan, ICAR-DOGR,

Rajgurunagar, India, Dr. DVS Raju, ICAR-DFR, Pune, India for their precious suggestions and reviewing the draft of the chapter.

12 Where to look for further information

Potato is a key rotation crop in cereal systems. Good-quality seeds of resilient varieties are in short supply in Asia and greatly limit potato production in the region. To overcome this serious bottleneck, CIP is already working with local, regional and national partners to develop elite tropically adapted bred populations and candidate potato varieties with short growing seasons of 70–80 days in subtropical climates. CIP and its partners are already doing sincere efforts to develop early-maturing varieties with traits for resistance to biotic and abiotic stress, including those required by the market and processing industry, as well as those preferred for home consumption. Collaborative research on the early-maturing potato will explore sustainable cultivation practices and the environmental impact of introducing the potato on cereal-based cropping systems of Asia. This program is being coordinated closely with the CGIAR Research Programs (CRP) in which CIP participates, particularly with Roots, Tubers and Bananas (RTB).

There are a number of current accelerated breeding research projects designed to breed elite tropically adapted early-maturing potato varieties which suit and fit well in extensive cereal-based systems of south Asia by CIP, CPRI and Wageningen University & Research.

https://www.cgiar.org/research/center/cip/
https://cipotato.org/programs/agile-potato-for-asia/
https://research.cip.cgiar.org/confluence/display/SET/Protocol+tuber+bulking+maturity
https://cipotato.org/crops/potato/
https://cipotato.org/annualreport2016/
https://cgspace.cgiar.org/handle/10568/88229/discover
https://www.wur.nl/en/project/Climate-Smart-potato-growing-in-South-Africa-1.htm
https://cgspace.cgiar.org/bitstream/handle/10947/4658/RTB%20Leaflet.pdf?sequence=10
https://cpri.icar.gov.in/?page_id=263
http://gtr.ukri.org/projects?ref=BB%2FP022553%2F1

13 References

Acquaah G. (2007). *Principles of Plant Genetics and Breeding*. Blackwell publishing, Maden, MA.

Ahn Y. J. and Zimmerman J. L. (2006). Introduction of the carrot hsp17.7 into potato (*Solanum tuberosum* l.) enhances cellular membrane stability and tuberization *in vitro*. *Plant Cell Environ.* 29:95–104.

Bradshaw J. E. and Bonierbale M. (2010). Potatoes. In Bradshaw, J. E. (Ed.), *Root and Tuber Crops. Handbook of Plant Breeding*, Vol. 7. Springer Science + Business media, New York, pp. 1–52.

Burton W. G. (1981). Challenges for stress physiology in potato. *Am. Potato J.* 58:3–14.

Buso G. S. C., Paiva M. R., Torres A. C., Resende F. V., Ferreira M. A., Buso J. A. and Dusi A. N. (2008). Genetic diversity studies of Brazilian garlic cultivars and quality control of garlic clover production. *J. Gen. Mol. Res.* 7:534–41.

Calderini D. F., Reynolds M. P. and Slafer G. A. (1999). Genetic gains in wheat yield and main physiological changes associated with them during the 20th century. In: Satorre, E. H. and Slafer,

G. A. (Eds), *Wheat: Ecology and Physiology of Yield Determination*. Food Product Press, New York, pp. 351–77.

Carrera E., Bou J., Garcia-Martinez J. L. and Prat, S. (2000). Changes in GA 20-oxidase gene expression strongly affect stem length, tuber induction and tuber yield of potato plants. *Plant J.* 22:247–56.

Dobranszki J. (2001). Effects of light on *in vitro* tuberization of the potato cultivar Desiree and its relatives. *Acta. Biol. Hung.* 52:137–47.

Dobranszki J. and Mandi M. (1993). Induction of *in vitro* tuberization by short day period and dark treatment of potato shoots grown on hormone-free medium. *Acta. Biol. Hung.* 44:411–20.

Douches D. S., Maas D., Jastrzebski K. and Chase R. W. (1996). Assessment of potato breeding progress in the USA over the last century. *Crop Sci.* 36:1544–52.

Drapal M., Farfan-Vignolo E. R., Gutierrez O. R., Bonierbale M., Mihovilovich E. and Fraser P. D. (2016). Identification of metabolites associated with water stress responses in *Solanum tuberosum* L. clones. *Phytochemistry* 135:1–10.

Easterling W. E., Aggarwal P. K., Batima P., Brander K. M., Erda L., Howden S. M., Kirilenko A., Morton J., Soussana J.-F., Schmidhuber J. and Tubiello F. N. (2007). Food, fibre and forest products. In: Parry, M. L., Canziani, O. F., Palutikof, J. P., van der Linden, P. J. and Hanson, C. E. (Eds), *Climate Change 2007: Impacts, Adaptation and Vulnerability*. Cambridge University Press, Cambridge, UK, pp. 273–313.

El-Bramawy M. A. S. and Shaban W. I. (2007). Nature of gene action for yield, yield components and major disease resistance in Sesame (*Sesamum indicum* L.). *Res. J. Agri. Bio. Sci.* 3:821–6.

Engel K. H. and Moller K. H. (1959). Frühdiagnose auf Reifezeit an Kartoffel Sämlingen. *Züchter* 29:218–20.

Ewing E. E. (1978). Shoot, stolon, and tuber formation on potato (*Solanum tuberosum* L.) cuttings in response to photoperiod. *Plant Physiol.* 61:348–53.

Ezeta F. N. (2008). *Proceedings of the Workshop to Commemorate the International Year of the Potato*. Food and Agricultural Organization of the United Nations, Regional Office for Asia and the Pacific, Bangkok, Thailand.

Golmirzaie A. and Serquen F. (1992). Correlation between early and late growth characters in an improved true potato seed population. *Hortscience* 27(4):350–2.

Griffing B. (1956). Concept of general and specific combining ability in relation to diallel crossing systems. *Aust. J. Biol. Sci.* 9:463–93.

Haga E., Weber B. and Jansky S. (2012). Examination of potential measures of vine maturity in potato *Am. J. Plant Sci.* 3:495–505.

Haverkort A. J., Verhagen A., Grashoff C. and Uithol P. W. J. (2004). Potato-zoning: A decision support system on expanding the potato industry through agroecological zoning using the LINTUL simulation approach. In: MacKerron D. K. L. and Haverkort A. J. (Eds), *Decision Support Systems in Potato Production*. Wageningen Academic Publishers, Wageningen, pp. 29–44.

Haydar A., Alam M. K., Khokan E. H., Ara T. and Khalequzzaman K. M. (2009). Combining ability and genetic variability studies in potato. *J. Soil Nat.* 3(2):1–3.

Jackson S. D. (1999). Multiple signaling pathways control tuber induction in potato. *Plant Physiol.* 119:1–8.

Johansen R. H., Miller J. C., Newsom D. W. and Fontenot J. F. (1967). The influence of environment on the specific gravity, plant maturity and vigour of potato progenies. *Am. Potato J.* 14:107–22.

Kadian M. S., Luthra S. K., Patel N. H., Bonierbale M., Singh S. V., Sharma N., Kumar V., Jai Gopal and Singh B. P. (2012). Identification of short cycle, heat-tolerant potato (*Solanum tuberosum*) clones for the semi-arid agro-ecology. *Indian J. Agr. Sci.* 82(9):814–8.

Kang G. S. and Kumar R. (1999). Breeding early bulking potato varieties for north-western plains of India. Potato, global research & development. *Proceedings of the Global Conference on Potato*, New Delhi, India, 6–11 December 1999, Volume 1, pp. 143–5.

Khan A. A., Jilani M. S., Khan M. Q. and Zubair M. (2011). Effect of seasonal variation on tuber bulking rate of potato. *J. Anim. Plant Sci.* 21(1):31–7. ISSN: 1018–7081.

Kiru S. D., Gavrilenko T. A., Kostina L. I., Rogozina E. V., Antonova O. Y., Truskinov E. V., Shvachko N. A., Krylova E. A. and Smirnova A. B. (2007). Conservation, evaluation and use in breeding of potato genetic diversity at the N.I. Vavilov Institute of Plant Industry (vir). In: Havercort A. and Anisimov B. (Eds), *Potato Production and Innovative Technologies*. Wageningen Academic Publishers, Netherlands, pp. 353–63.

Kooman P. L. and Haverkort A. J. (1995). Modelling development and growth of the potato crop influenced by temperature and daylength: LINTUL-POTATO. Haverkort A. J. and MacKerron D. K. L. (Eds), *Ecology and Modelling of Crops under Conditions Limiting Growth*. Kluwer Academic Publishers, Dordrecht, the Netherlands, pp. 41–60.

Kooman P. L., Fahem M., Tegera P. and Haverkort A. J. (1996). Effects of climate on different potato genotypes 2. Dry matter allocation and duration of the growth cycle. *Eur. J. Agron.* 5:207–17.

Kumar R. and Kang G. S. (2006). Usefulness of Andigena (*Solanum tuberosum* ssp. andigena) genotypes as parents in breeding early bulking potato cultivars. *Euphytica* 150:107–15. doi:10.1007/s10681-006-9098-3.

Kumar D. and Minhas J. S. (2003). Performance of heat tolerant genotypes stored at room temperature. *J. Indian Potato Assoc.* 30:165–66.

Kumar R., Kumar V., Jai Gopal, Luthra S. K. and Pandey S. K. (2008). Inventory of potato germplasm (Group Andigena) Collection. CPRI Tech Bulletin No. 86, Central Potato Research Institute, Shimla, Himachal Pradesh, India, 100p.

Kumar R., Kang G. S., Pandey S. K. and Gopal J. (2009). Kufri Khyati: A new early maturing potato variety for Indian plains. *Potato J.* 36(1–2):14–19.

Lafta A. M. and Lorenzen J. H. (1995). Effect of high temperature on plant growth and carbohydrate metabolism in potato. *Plant Physiol.* 109:637–43.

Levy D. and Veilleux R. E. (2007). Adaptation of potato to high temperatures and salinity-a review. *Am. J. Potato Res.* 84:487–506.

Lynch M. and Walsh B. (1998). *Genetics and Analysis of Quantitative Traits*. Sinauer Associates Incorporated, Sunderland MA, 874p.

Marinus J. and Bodlaender K. B. A. (1975). Response of some potato varieties to temperature. *Potato Res.* 18:189–204.

Martinez C. and Moreno U. (1992). Expresiones fisiologicas de resistencia a la sequia en dos variedades de papa sometidas a estres hídrico en condiciones de campo. *Rev. Bras. Fisiol. Veg.* 4:33–8.

Martinez-Garcia J. F., Garcia-Martinez J. L., Bou J. and Prat S. (2001). The interaction of gibberellins and photoperiod in the control of potato tuberization. *J. Plant Growth Regul.* 20:377–86.

Martinez-Garcia J. F., Virgos-Soler A. and Prat S. (2002). Control of photoperiod-regulated tuberization in potato by the Arabidopsis flowering-time gene CONSTANS. *Proc. Natl. Acad. Sci.* 99:15211–16.

Mendoza H. A. and Haynes F. L. (1976). Variability for photoperiodic reaction in diploid and tetraploid potato clones from three taxonomic groups. *Am. Potato J.* 53:319–32.

Mendoza H. A. and Haynes F. L. (1977). Inheritance of tuber initiation in the tuber bearing Solanum as influenced by photoperiod. *Am. Potato J.* 54:243–52.

Muthoni J. and Kabira J. N. (2015). Potato production in the hot tropical areas of Africa: Progress made in breeding for heat tolerance. *J. Agr. Sci.* 7(9):220–7.

Obidiegwu J. E., Bryan G. J., Jones H. G. and Prashar A. (2015). Coping with drought: Stress and adaptive responses in potato and perspectives for improvement. *Front. Plant Sci.* 6, 542.

Ortiz R. and Golmirzaie A. M. (2004). Combining ability analysis and correlation between breeding values in true potato seed. *Plant Breed. J.* 123:564–7.

Papademetriou M. K. (2008). *Proceedings of the Workshop to Commemorate the International Year of the Potato*. Food and Agricultural Organization of the United Nations, Regional Office for Asia and the Pacific, Bangkok, Thailand.

Paula I. (April 2014). Inheritance and stability of earliness in potato (*Solanum tuberosum* L.). Thesis submitted to Makerere University, Uganda.

Petrovic S., Dimitrijevic M., Belic M., Banjac B., Boskovic J., Zecevic V. and Pejic B. (2010). The variation of yield components in wheat (*Triticum aestivum* L.) in response to stressful growing conditions of alkaline soil. *Genetika* 42(3):545–55.

Phadnawis N. B. and Saini A. D. (1992). Yield models in wheat based on sowing time and phonological development. *Ann. Plant Physiol.*, 6:52–9.

Pushkarnath. (1976). *Potato in Subtropics.* Orient Longman, New Delhi, India.

Raices M., MacIntosh G. C., Ulloa R. M., Gargantini P. R., Vozza N. F. and Tellez-Inon M. T. (2003). Sucrose increases calcium-dependent protein kinase and phosphatase activities in potato plants. *Cell Mol. Biol.* 49:959–64.

Rodriguez-Falcon M., Bou J. and Prat S. (2006). Seasonal control of tuberization in potato: Conserved elements with the flowering response. *Annu. Rev. Plant Biol.* 57:151–80.

Ruzukas A. and Jundulas J. (2006). Potato breeding for nematode and disease resistance. *Žemės ūkio mokslai* 3:26–9.

Ruzukas A., Jankauskiene Z., Jundulas J. and Asakaviciute R. (2009). Research of technical crops (potato and flax) genetic resources in Lithuania. *Agron. Res.* 7:59–72.

Sands P.J., Hackett C. and Nix H. A. (1979). A model of the development and bulking of potatoes (*Solanum tuberosum* L.) I. Deviation from well-managed field crops. *Field Crops Res.* 2:309–31.

Schuster W. H. (1997). How much does plant breeding contribute to yield improvement of crops? (In German, with English abstract.). *Ger. J. Agron.* 1:9–18.

Scott, G. J. and Suarez, V. (2012). The rise of Asia as the centre of global potato production and some Implications for industry. *Potato J.* 39(1):1–22.

Sharma N., Rawal S., Kadian M., Arya S., Bonierbale M. and Singh B. P. (2014). Evaluation of advanced potato clones for drought tolerance in arid zone in Rajasthan, India. *Potato J.* 41:189–93.

Shekhawat G. S. (1994). Future needs of potato crop. *J. Indian Potato Assoc.* 21:1–6.

Shekhawat G. S., Gaur P. C. and Pandey S. K. (1999). Managing biodiversity in potato and associated crops. *J. Indian Potato Assoc.* 26:127–33.

Singh J. P., Dua V. K., Lal S. S. and Pandey S. K. (2008). Agro-economic analysis of potato based cropping systems. *Indian J. Fert.* 4(5):31–3, 35–9.

Singh J. P., Lal S. S. and Pandey S. K. (2009). Effect of climate change on potato production in India. *Central Potato Res. Inst. Shimla Newslett.* 40:17–18.

Sleper D. A. and Poehlman J. M. (2006). *Breeding Field Crops*, 5th edn. Blackwell Publishing Professional, Ames, IA.

Snyder R. G. and Ewing E. E. (1989). Interactive effects of temperature, photoperiod and cultivar on tuberization of potato cuttings. *Hort. Sci.* 24:336–8.

Thiele G., Theisen K., Bonierbale M. and Walker T. (2010). Targeting the poor and hungry with potato science. *Potato J.* 37:75–86.

Timlin D., Lutfor Rahman S. M., Baker J., Reddy V. R., Fleisher D. and Quebedeaux B. (2006). Whole plant photosynthesis development and carbon partitioning in potato as a function of temperature. *Agron. J.* 98:1195–203.

Wheeler R. M. and Tibbitts T. W. (1986). Growth and tuberization of potato (*Solanum tuberosum* L.) under continuous light. *Plant Physiol.* 80:801–4.

Worthington C. M. and Hutchinson C. M. (2005). Accumulated growing degree days as model to determine key developmental stages and evaluate yield and quality of potato in northeast Florida. *Proc. Fla. State Hort. Soc.* 118:98–101.

Zingaretti S. M., Inacio M. C., de Matos Pereira L., Paz T. A. and De Castro França S. (2013). Water stress and agriculture. In: Akinci, D. S. (Ed.), *Responses of Organisms to Water Stress*. In Tech. doi:10.5772/53877. ISBN: 978-953-51-0933-4.

Zou X., Zhai P. and Zhang Q. (2005). Variations in droughts over China: 1951–2003. *Geophys. Res Lett.* 32:1–4.

Zubeldia A. (1963). Selection of young potato seedlings for earliness. *Eur. Potato J.* 6:178–85.

Chapter 8

Developing new sweet potato varieties with improved performance

Peng Zhang, Weijuan Fan, Hongxia Wang, Yinliang Wu and Wenzhi Zhou, Institute of Plant Physiology and Ecology, Chinese Academy of Sciences, China; and Jun Yang, Shanghai Chenshan Plant Science Research Center, Shanghai Chenshan Botanical Garden, China

1 Introduction

Sweet potato, *Ipomoea batatas* (L.) Lam., plays an important role in food security and nutritional improvement as well as serving as a raw material in the processing of feeds, starches and bioethanol in various industries (Antonio et al., 2011; Bovell-Benjamin, 2007; Mukhopadhyay et al., 2011). It is the fifth most important food crop in terms of fresh weight. More than 130 million tons are produced every year, with China accounting for approximately 72% of world's production and more than 40% of the global harvested area (FAO, 2011). Nearly half of the sweet potatoes produced in Asia are used for animal feed, with the remainder primarily used for human consumption. Africa is the second largest sweet-potato-producing region, with almost 17% of the world's production and more than 42% of the world's area, mainly for human consumption. Latin America, the original home

http://dx.doi.org/10.19103/AS.2017.0016.08

of the sweet potato, produces only 1.9 million tons annually. In addition to China, other regions that consider sweet potato a major crop include Cuba and Haiti in the Caribbean region; Java (Indonesia), the island of New Guinea (both in Indonesia and in Papua New Guinea), and Vietnam in Asia; and Africa, particularly in the Lake Victoria area (Burundi, Rwanda, Uganda and the Democratic Republic of Congo), Ghana, Nigeria, Tanzania and Madagascar (Liu et al., 2014a). Sweet potato is enriched with carbohydrates, vitamins and other micronutrients, especially the yellow-orange-fleshed varieties provide particularly high quantities of vitamins A and C (Bovell-Benjamin, 2007). The leaves can also be used as vegetable as they provide additional protein, vitamins and minerals (Islam, 2006).

The unique advantages of sweet potato in subsistence agriculture include short growth cycle, high yield, resistance to stress and adaptation to different environments (Liu et al., 2014a). Traditional breeding has significantly contributed to trait improvement in sweet potato globally (Loebenstein and Thottappilly, 2009). Nevertheless, sweet potato is a hexaploid plant ($2n = 6x = 90$ chromosomes) with a high degree of heterozygosity, high levels of male sterility, and self- and interspecific incompatibility, resulting in strong segregation of hybrid progenies with the loss of valuable traits (Woolfe, 1992). Compared with crops such as potato, the consumer demand for sweet potato falls short in terms of nutrition and starch quality (ß-carotene, anthocyanin, iron and other micronutrients). Sweet potato pests and viral diseases also cause serious damage regionally, restricting further production and industrialization. Novel sweet potato varieties with improved traits are needed, especially in the marginal lands and disease-prone regions. Molecular breeding provides a promising approach for the development of novel varieties with value-added traits, for example, altered starch quality, enhanced nutrition, increased resistance to virus and nematode, and improved salt tolerance. With the development of sweet potato genetic transformation system (Yang et al., 2011), genome sequence (Yang et al., 2016), and genetic engineering tools, for example CRISPR/Cas9 (Woo et al., 2015), transgenic

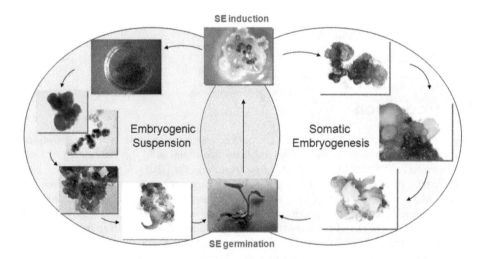

Figure 1 *In vitro* plant regeneration system of sweet potato integrating genetic transformation using embryogenic suspension cells as explants for particle bombardment and *Agrobacterium*-mediated gene delivery (modified from Yang et al., 2011). SE = somatic embryo.

technologies to improve traits provide a promising strategy for molecular breeding of important root crops. This review updates the development and application of genetic transformation and trait improvement in sweet potato (Mohan and Nair, 2012).

2 Genetic transformation of sweet potato from model cultivars to farmer-preferred cultivars

The development of genetically engineered sweet potato using model cultivars (e.g. Jewel, White Star, Kokei 14, Lizixiang) has led to significant improvement in farmer-preferred cultivars using embryogenic callus or suspension (Table 1 and Fig. 1; Mohan and

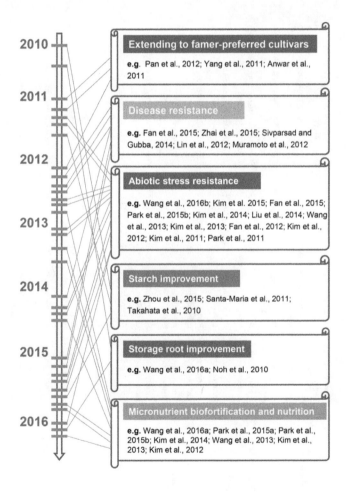

Figure 2 Development of sweet potato genetic engineering for technology adaptation of farmer-preferred cassava cultivars and value-added traits in the last 6 years.

Table 1 Progress and current status of sweet potato genetic transformation and trait improvement since 2010

Cultivar/genotype	Target tissue	Transformation method	Gene	Trait	Molecular analysis	Reference
Taizhong 6	Embryogenic suspension	A. tumefaciens LBA4404	IbVP1, hpt	Enhanced Fe uptake, enlarged root system	SA, RT-PCR, qRT-PCR	Fan et al., 2016b
Sinhwangmi	Embryogenic callus	A. tumefaciens EHA105	IbOr, nptII	Carotenoid accumulation and abiotic stress tolerance	Holdase chaperone activity, BiFC	Park et al., 2016
Ayamurasaki	Embryogenic suspension	A. tumefaciens LBA4404	Lc, hpt	Anthocyanin, storage root development	SA, RT-PCR, qRT-PCR	Wang et al., 2016a
ND98	Embryogenic suspension	A. tumefaciens EHA105	IbNHX2, bar, uidA	Increased salt and drought tolerance	SA, pCR, RT-PCR	Wang et al., 2016b
Nongda 603, Lizixiang	Embryogenic suspension	A. tumefaciens EHA105	IbMIPS1, hpt, uidA	Stem nematode resistance, salt and drought tolerance	SA, qRT-PCR	Zhai et al., 2016
Xushu22	Embryogenic suspension	A. tumefaciens LBA4404	UNC-15-RNAi, hpt	Stem nematode resistance	SA, NA, qRT-PCR	Fan et al., 2015a
Yulmi	Embryogenic callus	Particle bombardment	CuZnSOD, APX, nptII	Sulfur dioxide tolerance	SA, RT-PCR	Kim et al., 2015
Shangshu 19	Embryogenic suspension	A. tumefaciens EHA105	IbSIMT1, PPT	Salt tolerance	PCR	Liu et al., 2015
Xushu22	Embryogenic suspension	A. tumefaciens LBA4404	GBSSI-RNAi, SBE-RNAi, hpt	Waxy and high-amylose starch		Zhou et al., 2015
Sinhwangmi	Embryogenic callus	A. tumefaciens EHA105	IbMYB1, nptII	Increased anthocyanin and carotenoids, antioxidant activity	qRT-PCR	Park et al., 2015a
Sinzami	Embryogenic callus	A. tumefaciens EHA105	IbOr-Ins, hpt	Increased carotenoids	qRT-PCR	Park et al., 2015b
Xushu22	Embryogenic suspension	A. tumefaciens LBA4404	AtNHX1, hpt	Salt and cold stress tolerance	SA, qRT-PCR	Fan et al., 2015b
Yulmi	Embryogenic callus	A. tumefaciens EHA105	IbLCY-β-RNAi, hpt	Increased carotenoid, salt-stress tolerance	RT-PCR	Kim et al., 2014
Kokei No. 14	Embryogenic suspension	A. tumefaciens EHA105	IbP5CR, hpt, uidA	Increased salt-stress tolerance	PCR	Liu et al., 2014b
Lizixiang	Embryogenic suspension	A. tumefaciens EHA105	IbNFU1, hpt, uidA	Salt tolerance	SA, PCR, qRT-PCR	Liu et al., 2014c
Shangshu 19	Embryogenic suspension	A. tumefaciens EHA105	IbMas, bar	Salt tolerance	PCR	Liu et al., 2014d
Blesbok	Apical tip	A. tumefaciens LBA4404	CPs-RNAi	SPFMV, SPCSV, SPVG and SPMMV-resistance	PCR, SA	Sivparsad and Gubba, 2014

Variety	Explant	Strain	Gene/construct	Trait	Analysis	Reference
Ayamurasaki	Embryogenic suspension	A. tumefaciens LBA4404	IbDFR-RNAi, hpt	Reduced anthocyanins and abiotic stress tolerance	SA, qRT-PCR	Wang et al., 2013
Yulmi	Embryogenic callus	A. tumefaciens EHA105	IbOr-Wt, IbOr-Ins, hpt	Increased carotenoids, salt stress tolerance	WA, qRT-PCR	Kim et al., 2013
Yulmi	Embryogenic callus	Particle bombardment	IbEXP1-antisense, nptII	Altered storage root growth	qRT-PCR	Noh et al., 2013
Yulmi	Embryogenic callus	A. tumefaciens EHA105	IbLCY-ε-RNAi, hpt	Increased carotenoids, salt-stress tolerance	qRT-PCR	Kim et al., 2013
Tainung 57	Embryogenic callus	A. tumefaciens LBA4404	miR828 precursor, hpt	Increased lignin and H_2O_2	NA, qRT-PCR	Lin et al., 2012
Sushu2	Embryogenic suspension	A. tumefaciens LBA4404	SoBADH, hpt, uidA	Increased salt, oxidative stress, and low temperature tolerance	SA, qRT-PCR, PAA	Fan et al., 2012
Kokei No. 14	Embryogenic callus	A. tumefaciens EHA101	α-hordothionin, hpt	Increased black rot disease resistance	qRT-PCR	Muramoto et al., 2012
Yulmi	Embryogenic callus	A. tumefaciens EHA105	IbCHY-β RNAi, hpt	Increased β-carotene and total carotenoids, enhanced salt stress tolerance	RT-PCR	Kim et al., 2012
Guangzi No. 1	Stem	A. tumefaciens LBA4404	Resveratrol synthase gene, nptII	Increased resveratrol glucoside	SA, PCR	Pan et al., 2012
Yulmi	Embryogenic callus	A. tumefaciens EHA105	SCOF-1, hpt	Increased low temperature tolerance	RT-PCR	Kim et al., 2011
Ayamurasaki, Sushu2, Sushu9, Sushu11, Wanshu1, Xushu18 and Xushu22	Embryogenic suspension	A. tumefaciens LBA4404	hpt, uidA	None	SA, PCR, qRT-PCR	Yang et al., 2011
Jewel	Leaf with petiole	A. tumefaciens EHA105	amyN26, nptII	Improved hyperthermophilic glycoside hydrolysis	SA, PAA	Santa-Maria et al., 2011
Kokei 14, Shiroyutaka, Kyushu 125, Koutanpakukei 90142-8, and Kyushu 126	Leaf, petiole, stem, root, apical meristem	A. tumefaciens LBA4404	CYP1A1, CYP2B6, CYP2C19, P450 reductase, nptII	Herbicide tolerance.	SA, PCR, RT-PCR	Anwar et al., 2011
White Star	Embryogenic callus	A. tumefaciens EHA105	IbLEA14-Ox, IbLEA14-RNAi, hpt	Increased lignification, osmotic- and salt stress-tolerance	qRT-PCR, WA	Park et al., 2011
White Star	Embryogenic callus	A. tumefaciens EHA101	SSII-RNAi, hpt	Reduced starch pasting temperature	RT-PCR, qRT-PCR	Takahata et al., 2010
Yulmi	Embryogenic callus	A. tumefaciens GV3101	SRD1, nptII	Improved storage root formation	RT-PCR	Noh et al., 2010

SAP = Southern analysis, RT-PCR = reverse transcription polymerase chain reaction, qRT-PCR = quantitative RT-PCR analysis, NA = Northern analysis, PAA = protein activity analysis, WA = Western analysis. For gene abbreviations see text.

Nair, 2012). *Agrobacterium*-mediated transformation of embryogenic cell suspensions is the mainstay of sweet potato transgenic studies. This technology has been transferred to several key laboratories researching sweet potato in China. For instance, our group created a genotype-independent transformation platform suitable for 'Xushu22', 'Sushu2', 'Sushu9', 'Sushu11', 'Aymurasaki', 'Wanshu1' and 'Xushu18' (Yang et al., 2011), and extended several regional farmer-preferred cultivars including 'Taizhong6', 'Yusu303', 'Yuzi7', 'Guangshu87', 'Tainong57' and 'Huachano'.

Routine genetic transformation offers the possibility of functional genomic study and trait improvement in sweet potato (Table 1, Fig. 2). Generally, regulation of target gene expression is achieved using over-expression or RNA interference (RNAi) approaches. Frequently, constitutive promoters or strong tissue-specific promoters are used. The CaMV 35S promoter is frequently utilized in transgenic studies. Several alternatives include improved CaMV 35S promoter (El2Ω promoter, Muramoto et al., 2012), wound-responsive sporamin promoter (Chen et al., 2006), oxidative-stress-inducible SWPA2 promoter (Kim et al., 2011; Park et al., 2015a), potato GBSSI promoter (Xing et al., 2008), storage root-specific sporamin 1 promoter (Park et al., 2015a) and vascular-specific p54/1.0 promoter from cassava (Zhou et al., 2015), which are also components of bioengineering tool kits. Another newly developed but game-changing bioengineering tool includes the CRISPR/Cas9 systems, which are still under development and optimization, especially by plant scientists (Woo et al., 2015). However, nearly all the laboratories with the capacity to generate transgenic sweet potato have developed CRISPR vectors to target different genes or genomic regions of sweet potato. A few transgenic lines of sweet potato with mutated starch biosynthetic genes are already under evaluation in our lab. This technology strongly relies on the availability of full genome sequences of sweet potato. To date, its full genome sequencing is still an extremely challenging task. Nevertheless, significant progress has been made by the release of assembled genomes of *I. trifida* and *I. triloba*, two diploid wild progenitors of sweet potato, using the Illumina and PacBio sequence reads from a Japanese group (http://sweetpotato-garden.kazusa.or.jp) and an American group (https://www.sweetpotatobase.org). More importantly, an assembled genome of sweet potato cultivar 'Taizhong 6' has been released to the public, as the joint effort from our group and Max Planck Institute for Molecular Genetics (http://public-genomes-ngs.molgen.mpg.de/SweetPotato).

3 Production of disease-resistant sweet potato

Sweet potato faces a serious challenge due to viral diseases and pest infestations, for example, stem nematode and sweepoviruses that restrict the production of sweet potato in many parts of the world (Liu et al., 2014a). Genetically engineered sweet potato varieties with enhanced tolerance to multiple diseases are desirable (Fig. 2).

3.1 Stem nematodes

Ditylenchus destructor Thorne, commonly known as stem nematode, is one of the major pathogens of sweet potato regionally (Mohandas and Siji, 2012). Stem necrosis and storage root rot are the main symptoms of stem nematode infection in sweet potato. Reduced or absent yield of sweet potato has been reported in regions of severe infestation (Xu et al.,

2015). Infection of the aerial parts of sweet potato plants generally represents weak disease phenotype, or none at all, during the entire planting period. Until the rotten storage roots are harvested, it is difficult to make early diagnosis and to control the disease (Clark et al., 2013). Therefore, improved resistance to stem nematode infection has become an important goal of sweet potato breeding and field management, especially in East China.

Since movement is essential for stem nematode life cycle, disrupting locomotor function greatly affects nematode infectivity and propagation. Fan et al. (2015a) demonstrated that interference with stem nematode virulence by expressing small interfering RNAs (siRNAs) homologous to the *unc-15* gene in transgenic sweet potato 'Xushu22' plants enhanced their resistance to the pathogen. The *unc-15* encodes paramyosin, which is a major thick filament component, playing an important role in muscle assembly (Benian et al., 2006). Artificial inoculation with stem nematodes in storage roots was used to evaluate the infectivity. The expression of *unc-15* gene in stem nematodes collected from transgenic plant tissues showed great repression when compared with the wild type. The infected areas in the storage roots were significantly smaller than in the wild type. In stem nematode-infested fields, the transgenic plants showed increased yield compared with the wild type. No yield and phenotype differences were observed under normal field conditions, suggesting that interference with stem nematode movement via expression of siRNAs targeting *unc-15* in sweet potato was effective. The results demonstrate that the expression of siRNAs targeting *unc-15* of *D. destructor* is an effective approach to improving stem nematode resistance in sweet potato, in conjunction with global integrated pest management programmes (Fan et al., 2015a).

Another group also reported increased stem nematode resistance in transgenic sweet potato expressing a rice cysteine proteinase inhibitor oryzacystatin-I gene *OCI* (Gao et al., 2011). Proteinase inhibitors were used to determine the role of plants against pathogens and insects. OCI protein inhibits proteinase activity in the insect gut to prevent protein assimilation. Transgenic plants transformed with *OCI* manifest enhanced resistance to plant nematodes such as *Meloidogyne incognita* and *Globodera pallida* (Atkinson et al., 1996; Vain et al., 1998). Using the sweet potato cultivar 'Lizixiang', Gao et al. (2011) evaluated 134 transgenic plants in the field for 2 years and found that only 3 transgenic lines showed resistance to stem nematodes. The resistance was due to the high expression of *OCI* in these lines. Compared with the RNAi approach (Fan et al., 2015a), this approach was much less effective.

Inositol is a precursor of many inositol-containing compounds and participates in various physiological and biochemical processes in plants, such as cell membrane biogenesis, hormonal regulation, growth regulation, programmed cell death, stress signalling, host defence and immunity (Boss and Im, 2012). Myo-inositol-1-phosphate synthase (MIPS) is a key rate-limiting enzyme in myo-inositol biosynthesis. Zhai et al. (2016) reported that the expression of the sweet potato *IbMIPS1* improved tolerance to stem nematodes. After stem nematode infection, storage roots of transgenic plants showed a significant increase in inositol, inositol-1,4,5-trisphosphate (IP3), phosphatidic acid (PA), Ca^{2+}, ABA, callose and lignin content. Significantly enhanced ROS scavenging was also observed. This study indicates that overexpression of the *IbMIPS1* gene improved resistance to biotic stress in plants.

3.2 Viral diseases

Virus diseases reduce yields markedly, often more than 50%. Mixed infection of sweet potato by multiple viruses is a common phenomenon and often results in a synergistic effect on severe disease symptoms (Clark et al., 2012). Currently, more than 15 virus species from different families, including both RNA and DNA viruses, are known to infect this crop, resulting in potential epidemics in certain regions. Sweepoviruses, which are important begomoviruses that infect *Ipomoea* plants worldwide, draw more attention in recent years due to rapid spread by whiteflies and mixed infection with other viruses. Sweet potato is vegetatively propagated from vines or root slips (sprouts), and therefore, virus diseases are transmitted via propagation to the newly planted field, resulting in decreased yield.

Biotechnological approaches for increasing virus resistance have great potential in plant improvement. Most strategies are based on the concept of 'pathogen-derived resistance' (PDR). Although great efforts have been made to develop transgenic sweet potato with virus resistance using the coat protein (*CP*) gene of sweet potato feathery mottle virus (SPFMV; genus Potyvirus, family Potyviridae) and/or sweet potato chlorotic stunt virus (SPCSV; genus Crinivirus, family Closteroviridae) (Okada et al., 2001, 2002; Okada and Saito, 2008), this approach has only limited success under field conditions due to the distribution and genetic diversity of sweet potato viruses. Other strategies to generate transgenic sweet potato resistant to SPVD have also been tested. The *OCI*-mediated resistance to potyviruses has been shown in tobacco to inhibit the viral cysteine proteinase NIa that processes the potyviral polyprotein (Gutierrez-Campos et al., 1999). The *OCI* expression in transgenic sweet potato plants also confers resistance to SPCSV, as closteroviruses also encode cysteine proteinases to modify some of their proteins. However, the resistance was not effective, and typical symptoms of SPVD developed in plants infected with SPCSV and SPFMV (Clark et al., 2012).

RNAi-based virus resistance has also been tested in sweet potato (Kreuze et al., 2008). Transgenic plants transformed with an intron-spliced hairpin construct targeting the replicase-encoding sequences of SPCSV and SPFMV detected mild or no symptoms following SPCSV infection, and accumulation of SPCSV was significantly reduced. None of these plants were resistant to SPCSV, and the high levels of resistance to accumulation of SPCSV failed to prevent the development of synergistic sweet potato virus disease in transgenic plants co-infected with SPFMV. These results showed that sweet potato was protected against the disease caused by SPCSV using RNA silencing. Nevertheless, the authors also found that even low levels of SPCSV, detected in the three transgenic varieties of 'Huachano', were sufficient to disrupt resistance to SPFMV, indicating that immunity to SPCSV may be required to prevent the synergistic diseases caused by SPCSV in plants co-infected with other viruses. Resistance to multiple viruses in transgenic sweet potato was also reported using stacked CP gene segments of SPFMV, SPCSV, SPVG and SPMMV. Six transgenic plants challenged by graft inoculation with SPFMV, SPCSV, SPVG and SPMMV-infected *Ipomoea setosa* Ker displayed delayed and milder symptoms of chlorosis and mottling of lower leaves when compared with the untransformed control plants, despite detection of the virus using nitrocellulose enzyme-linked immunosorbent assay (Sivparsad and Gubba, 2014). These results warrant further investigation of resistance to virus infection under field conditions.

Further molecular understanding of virus interaction with plants is of importance in developing effective strategies to control viral epidemics. Protocols for efficient infection with different viruses still need to be established. The infectivity of cloned sweet potato leaf curl Lanzarote virus (SPLCLaV) in sweet potato plants has been demonstrated using syringe infiltration (Trenado et al., 2011). Sweet potato leaf curl virus-Jiangsu (SPLCV-JS), a sweepovirus cloned from infected sweet potato plants in East China (Bi and Zhang,

2012), has been successfully infected to sweet potato using agroinnoculation (Bi and Zhang, 2014). Studies of host–viral protein interaction are ongoing in several advanced laboratories.

3.3 Black rot disease

The black rot caused by pathogenic fungus *Ceratosistis fimbriata* is one of the most serious diseases of sweet potato resulting in significant losses during storage (Clark et al., 2013). Nevertheless, studies targeting this disease are limited. Muramoto et al. (2012) reported that transgenic sweet potato expressing thionin from barley driven by a strong constitutive promoter of E12Ω or the β-amylase gene promoter in sweet potato provides resistance to black rot disease. The α-hordothionin (αHT) from barley endosperm is an antimicrobial peptide with structural resemblance to purothionin, which displays strong toxicity to *C. fimbriata in vitro*. Leaves of *E12Ω:αHT* plants exhibited reduced yellowing following *C. fimbriata* infection compared with wild-type leaves. Storage roots of both *E12Ω:αHT* and β-*Amy:αHT* plants exhibited reduced lesion areas around the site inoculated with *C. fimbriata* spores compared to wild type. This study demonstrated the advantage of transgenic sweet potato expressing antimicrobial peptide against black rot disease and in reducing the use of agrochemicals (Muramoto et al., 2012).

4 Production of sweet potato resistant to abiotic stresses

The main environmental stresses, such as salinity and low temperature, cause adverse effects on sweet potato growth and development, resulting in yield reduction, especially in marginal lands of salty soils and high-altitude lands. Approaches increasing abiotic stress resistance in sweet potato include regulation of stress-related signal transduction, transcription factors, biosynthesis of soluble protective substances, water channel proteins and antioxidants, and activated-oxygen-scavenging systems (Table 1).

Glycine betaine (GB) plays an important role in protecting plant cells from damage caused by various abiotic stresses (Fan et al., 2016a). The gene encoding betaine aldehyde dehydrogenase (*BADH*) is involved in the biosynthesis of GB in plants. GB accumulation via heterologous overexpression of *BADH* improves abiotic stress tolerance in plants (Chen and Murata, 2008). Recently, Fan et al. (2012) demonstrated that enhanced GB biosynthesis in sweet potato improved its tolerance to multiple abiotic stresses including salt, oxidative stress and low temperature without causing phenotypic defects (Fig. 3a and b). A chloroplastic *BADH* gene from *Spinacia oleracea* was introduced into the sweet potato cultivar Sushu2. The increased BADH activity and GB accumulation in the transgenic plant lines under normal and multiple environmental stresses resulted in increased protection against cell damage via maintenance of cell membrane integrity, stronger photosynthetic activity, reduced reactive oxygen species (ROS) production, and induction or activation of ROS scavenging via increased activity of free radical-scavenging enzymes. The increased proline accumulation and systemic upregulation of many ROS-scavenging genes in stress-treated transgenic plants also indicated that GB accumulation stimulated the ROS-scavenging system and proline biosynthesis via an integrative mechanism (Fan et al., 2012).

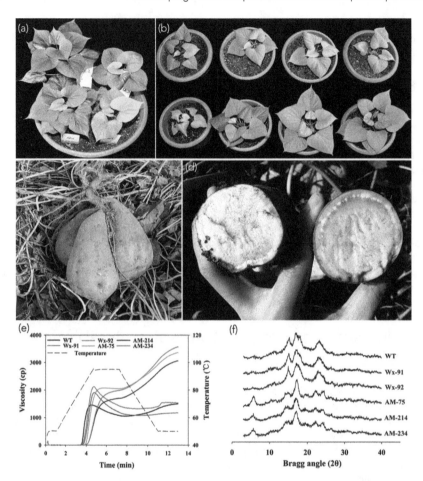

Figure 3 Genetically improved sweet potato. (a) Phenotypic analyses of the *SoBADH* transgenic plants (OE lines) under 150 mM NaCl stress for 16 days; (b) transgenic OE lines showed notable differences from wild-type Sushu-2 in foliage growth after salt stress (a and b adapted from Fan et al., 2012); (c) field-harvested storage roots of transgenic Xushu22 with downregulated expression of GBSSI; (d) iodine staining of storage roots between wild type and waxy transgenic plants; (e) Waxy (Wx-91, WX-92) and high-amylose (AM-75, AM-214, and AM-234) starches from transgenic sweet potato plants showed different Rapid Visco-Analyser pasting profiles from those of the wild-type (WT); (f) X-ray diffraction analysis showing the crystal structure of sweet potato starches transformed from the A-type to the B-type with increased amylose content (c–f adapted from Zhou et al., 2015).

Indeed, under stressful conditions, proline accumulation and upregulated expression of pyrroline-5-carboxylate reductase (P5CR), a critical enzyme in proline biosynthesis, in sweet potato have been observed. Transgenic sweet potato (cv. Kokei No. 14) overexpressing *IbP5CR* showed improved salt tolerance compared with the wild type (Liu et al., 2014b). The systemic upregulation of ROS-scavenging genes was also confirmed in the transgenic plants under salt stress. These findings suggest that overexpression of *IbP5CR* increases proline accumulation, which enhances salt tolerance of the transgenic sweet potato plants by regulating osmotic balance, protecting membrane integrity and photosynthesis and activating ROS-scavenging system. This study indicates that the *IbP5CR* gene has the potential to improve salt tolerance in plants.

Fan et al. (2015b) also demonstrated that overexpression of the *Arabidopsis NHX1* encoding a vacuolar Na^+/H^+ antiporter, in sweet potato enhanced Na^+/H^+ exchange activity of tonoplast membrane in transgenic sweet potato lines compared with wild-type plants. Under salt stress, the transgenic plants accumulated higher Na^+ and K^+ levels in their tissues compared with the wild-type plants, maintaining high K^+/Na^+ ratios. Consequently, the transgenic plants showed enhanced protection against cellular damage due to increased proline accumulation, preserved cell membrane integrity, enhanced ROS scavenging (e.g. increased superoxide dismutase activity), and reduced H_2O_2 and malondialdehyde (MDA) production. Further, the transgenic plants showed improved cold tolerance through multiple mechanisms of action, revealing the first molecular evidence supporting NHX1 function in cold response. The transgenic plants showed better biomass production and root yield under stressful conditions (Fan et al., 2015b). Similarly, a new vacuolar Na^+/H^+ antiporter gene, *IbNHX2*, was cloned from sweet potato and its overexpression in transgenic sweet potato significantly improved tolerance to salt and drought stresses (Wang et al., 2016b). These findings demonstrate that overexpressing plant vacuolar Na^+/H^+ antiporters in sweet potato renders the crop tolerant to various stresses, providing greater capacity for the use of NHX in improving crop performance under combined abiotic stress conditions.

Enhanced ROS-scavenging capacity in transgenic sweet potato improved tolerance to abiotic stresses. Transgenic sweet potato expressing CuZn superoxide dismutase (*CuZnSOD*) and ascorbate peroxidase (*APX*) genes under the stress-inducible SWPA2 promoter showed significant improvement following treatment with 100 mmol/l NaCl (salt stress). The activity of SOD, ascorbate peroxidease (APX), peroxidase (POD) and catalase (CAT) in leaves of transgenic plants was higher than in wild-type plants (Wang et al., 2012). These plants also showed better growth, photosynthetic activity (Fv/Fm) and water status under drought stress, but tuberization was poor (Lu et al., 2010). These transgenic plants also proved to be tolerant to sulphur dioxide (Kim et al., 2015). The results indicate that maintenance of ROS homeostasis is an important remedy against water stress-induced ROS generation via mobilizing enzymatic defence systems.

Abscisic acid (ABA) plays an important role in response to various abiotic stresses. ABA biosynthesis overlaps with β-carotene biosynthesis. Therefore, enhancing carotenoid synthesis via the β-branch-specific pathway (yielding β-carotene and ABA) in sweet potato might affect ABA-dependent stress response. Kim et al. (2013) reported that the downregulation of lycopene ε-cyclase (LCY-ε) expression in sweet potato activated β-branch carotenoids and its related genes. Transgenic calli showed a twofold increase in ABA content compared with control calli and showed significant tolerance to 200 mM NaCl. These results suggest that, in sweet potato, downregulation of the ε-cyclization of lycopene increases carotenoid synthesis via the β-branch-specific pathway and may positively regulate cellular defences against salt-mediated oxidative stress. Similarly, transgenic sweet potato calli with RNAi of gene expression of lycopene β-cyclase (LCY-β), a key enzyme involved in the synthesis of α- and β-branch carotenoids such as α-carotene and β-carotene via cyclization of lycopene, showed increased total carotenoid and ABA content. These transgenic calli exhibited less salt-induced oxidative-stress damage and greater tolerance for polyethylene glycol (PEG)-mediated drought compared with that of wild-type cells (Kim et al., 2014). Indeed, increased β-carotene and total carotenoids in transgenic sweet potato cell cultures via RNAi-mediated downregulation of β-carotene hydroxylase, a key regulatory enzyme in the beta-beta-branch of carotenoid biosynthesis, enhanced their salt stress tolerance (Kim et al., 2012). Similarly, in transgenic sweet potato calli and plants expressing the sweet potato Orange (IbOr) protein, which functions as a molecular chaperone to stabilise phytoene synthase (Park et al., 2015b, 2016), conferred tolerance against salinity and methyl viologen (MV) treatment in transgenic sweet potato along with enhanced carotenoid content (Kim et al., 2013). These studies demonstrate that manipulation of carotenoid biosynthetic pathway in sweet potato enable the development of transgenic sweet potato enriched with nutritional carotenoids and with greater tolerance to abiotic stresses.

In addition, transgenic sweet potatoes expressing other genes, such as α/β-hydrolase gene IbMas (Liu et al., 2014d), LOW OSMOTIC STRESS 5 (LOS5, encodes a molybdenum cofactor sulphurase) (Gao et al., 2011), myo-inositol-1-phosphate synthase gene IbMIPS1 (Zhai et al., 2015), late embryogenesis abundant gene IbLEA14 (Park et al., 2011), iron-sulphur cluster scaffold protein gene IbNFU1 (Liu et al., 2014c) and salt-induced methyltransferase gene IbSIMT1 (Liu et al., 2015), also confirmed improved abiotic stress resistance, especially to drought and salt. Transgenic sweet potato plants expressing type I H$^+$-pyrophosphatase (H$^+$-PPase) gene IbVP1 showed better growth, including enlarged root systems, and improved Fe acquisition under Fe-sufficient or -deficient conditions (Fan et al., 2016b). More importantly, expression of transcription factors including soybean cold-inducible zinc finger protein SCOF-1 and sweet potato R2R3-type MYB1 (IbMYB1) in sweet potato showed enhanced tolerance to low-temperature stress and antioxidant capacity, respectively (Kim et al., 2011; Park et al., 2015a). Therefore, cloning and identification of native transcription factors that regulate the stress response in sweet potato might be a useful approach to improve abiotic stress resistance.

5 Starch modification for industrial applications

Sweet potato starch has been widely used industrially to produce different types of processed foods or biomaterials (Antonio et al., 2011). These applications indicate that sweet potato plays an increasingly important role in food security and bioindustrial development (Bovell-Benjamin, 2007). Genetic engineering of sweet potato for the development of novel starches is in demand. Therefore, a better understanding of starch biosynthetic mechanisms and the key genes involved enable us to breed sweet potato rationally to produce novel starches with improved functionality (Jobling, 2004).

Starch is the major component of sweet potato storage root and accounts for 50% to 80% of their dry matter (Bovell-Benjamin, 2007). Sweet potato starch granules are normally round, oval or polygonal with sizes ranging from 4 μm to 40 μm, with an average size of 19 μm (Hoover, 2001). Sweet potato starch is normally composed of 20–30% linear and slightly branched amylose and 70–80% highly branched amylopectin with a large molar mass. The application of different sweet potato cultivars mainly depends on the starch content and amylose-to-amylopectin ratio. Starches with varying amylose content are of interest because of their structural diversity and physicochemical properties, facilitating different applications (Kitahara et al., 2007). Amylose-free starch, also known as waxy starch, is relatively easy to gelatinize, produces a clear viscous paste with a sticky or tacky surface, and has a lower tendency for retrogradation, and therefore, increased and stable viscosity. These features of waxy starch enable wide application in food industry. High-amylose starch is primarily used in candy manufacture as high-amylose gel to promote the shape and integrity of the candy. High-amylose starch films are widely used in industrial products, such as degradable plastics.

Naturally, the ratio of amylose to amylopectin varies among the different sweet potato varieties and growth environments across a limited range. For example, the amylose content of sweet potato ranges between 20% and 35%. Natural starches with amylose content lower than 20% and higher than 35% are limited. The ratio of amylose to amylopectin is predominantly controlled by several starch biosynthetic enzymes, including granule-bound starch synthase I (GBSSI), soluble starch synthase (SSS), and branching enzyme (BE). Several groups reported the production of transgenic sweet potato plants with altered amylose content via downregulation of *GBSSI* or *BE* expression (Kimura et al., 2001; Kitahara et al., 2007; Noda et al., 2002; Shimada et al., 2006). Nevertheless, no further investigation of the relationship between amylose content and starch physicochemical properties has been studied in these studies due to the very narrow range of amylose content in the sweet potatoes.

Recently, our group generated hundreds of transgenic sweet potato lines with a wide range of amylose content (0–70%) by downregulating the expression of key starch biosynthetic enzymes GBSSI or BEs using RNAi technology (Fig. 3c and d; Zhou et al., 2015). The expression of hairpin siRNA targeting either *GBSSI* or *BEs* was driven by two different promoters, vascular-specific p54/1.0 and constitutive CaMV 35S. The amylose content of these transgenic sweet potatoes ranged from 0% to 85% by colorimetric method. Subsequently, Zhou et al. (2015) further investigated the properties of starch with different amylose content (Fig. 3e and f). The results show that the amylose content affects the starch structure and physicochemical properties significantly. The average granule size is increased in both amylose-free and high-amylose starches. Compared with the wild-type starch, the waxy and high-amylose starches contain fewer short chains and more medium and long chains. The changes in high-amylose starches are more significant than in the waxy starches. X-ray diffraction studies reveal that high-amylose starches manifest the type-B crystal structure with a markedly decreased level of crystallinity, in contrast to the type-A of wild-type and waxy starches (Fig. 3f). The shear stress resistance of starch gel after gelatinization is also enhanced with increasing amylose content, as shown by rapid viscosity analysis and rheological tests. This study provided robust evidence supporting the relationship between amylose content, starch ultrastructure, and physicochemical properties in sweet potato starches, which are used to guide sweet potato breeding and starch applications.

6 Increased understanding of storage root development for better yield

The yield of storage roots directly reflects the productivity of sweet potato. Further understanding of the mechanism related to storage root formation is of importance to improving the yield of root and tuber crops. Typically, the sweet potato root system consists of three different types of roots: fibrous roots, pencil roots and storage roots (Villordon et al., 2014; Wang et al., 2016a). All the three types of roots originate from adventitious roots, which are derived from the nodal region of stem cuttings or cut sprouts or from the cut ends of stem

2–5 days after transplantation (Belehu et al., 2004). Anatomically, the fibrous roots are 1–2 mm thick, containing a central core of xylem with no central pith and four protoxylem elements with alternate phloem tissue within the stele. These fibrous roots have no secondary growth, and function in water and nutrient absorption. Pencil roots are adventitious roots, 2- to 5-cm thick, with pentarch or hexarch, heavily lignified stele, a broad secondary cortex, and a limited amount of secondary xylem and phloem. Storage roots develop as secondary growths with the proliferation of anomalous cambial and parenchymal cells (Villordon et al., 2014; Ravi et al., 2014). Starch accumulation occurs predominantly during storage root development, as evidenced by transcriptomic analysis of sweet potato root during development. Upregulation of starch biosynthesis-related genes and downregulation of lignin biosynthesis-related genes occur in developing storage roots (Firon et al., 2013).

Storage root development is a complex process representing a balance between cambium propagation and lignification, which is affected by genetic factors (transcription factors, lignification and starch biosynthetic genes), endogenous growth regulators (auxin, cytokinins, ABA, jasmonic acid and ethylene) and environmental factors (temperature, soil, illumination, CO_2, and water) (Ravi et al., 2014). For example, the IAA content controls the pentarch or hexarch condition of the adventitious roots, which is a prerequisite for storage root initiation in sweet potato (Nakatani and Komeichi, 1991; Noh et al., 2010); The ABA may be related to the activity of vascular and anomalous cambia and promotes cell differentiation and thickening of storage roots per se or via interaction with cytokinins (Nakatani and Komeichi, 1991; Wang et al., 2006). JA may be involved in cessation of elongation during the early period of storage root development (Kim et al., 2002). Under stressful conditions, for example, drought and poor soil fertility, sweet potato shows retarded storage root development with more pencil root production, suggesting that sufficient supply of photo-assimilates from the source to the sink is important for starch metabolism and storage root development.

Using transcriptomic approach, dozens of transcription factors have been found to be involved in storage root initiation and development or pencil root formation in sweet potato (Firon et al., 2013), which required further functional validation. Using proteomic analysis, the differential expression of 30 proteins in the pencil and storage roots was identified. Among the differentially expressed proteins, several enzymes of the lignin biosynthesis pathway were found to exhibit higher activity in pencil roots (catechol oxidase and peroxidase), implying that lignin biosynthesis was upregulated or uniquely expressed in pencil roots. The results also indicate that the reduction of carbon flow towards phenylpropanoid biosynthesis and its delivery to carbohydrate metabolism is a major event in storage root formation (Lee et al., 2015).

Recently, several key genes regulating storage root formation, such as the Dof zing finger transcriptional factor *SRF1*, MADS-box protein *SRD*, *KNOX1* and *expansins*, have been intensively studied (Noh et al., 2010, 2013). *SRF1* regulates carbohydrate metabolism in the storage roots via negative regulation of a vacuolar invertase gene1. *SRD1* controls the formation of storage roots by inducing the proliferation of cambium and metaxylem cells (Noh et al., 2010), a process that is negatively regulated by an expansin gene (*IbEXP1*) (Noh et al., 2013). *KNOX1* genes may activate cytokinin biosynthesis, as indicated by the high t-ZR content in the developing storage roots (Tanaka et al., 2008). All these findings indicated that the balance between lignification and cambium proliferation/development determines the final storage root number and yield.

Our group recently found that the heterologous expression of the maize leaf colour (*Lc*) gene in sweet potato increased anthocyanin pigmentation in the whole plant. It resulted in reduced size with an increased length/width ratio, low yield and less starch content. The expression analysis revealed dramatic upregulation of the genes involved in the lignin biosynthesis pathway in developing storage roots, leading to greater lignin content in the *Lc* transgenic lines, compared with the wild type. Enhanced lignification of vascular cells in the early storage roots was observed. Increased expression of the β-amylase gene in the leaves and storage roots accelerated starch degradation and increased the sugar use efficiency, providing more energy and carbohydrate sources for lignin biosynthesis in the *Lc* transgenic sweet potato. Consequently, lesser starch accumulation was observed in the developing storage roots at the initiation stage of the *Lc* plants. Collectively, this study, for the first time, provides direct evidence supporting the critical transition from the biosynthesis of lignin to starch in storage root development of sweet potato (Wang et al., 2016a). Our findings indicate that the shift in carbon flux towards lignin biosynthesis is associated with increased starch degradation at both the sink and the source, and retarded starch accumulation in the parenchymal cells of the developing storage roots.

7 Production of purple sweet potato with increased anthocyanin content

Purple sweet potato accumulates anthocyanins in the storage roots. Many anthocyanins have been isolated and identified in purple sweet potato (Montilla et al., 2000; Terahara et al., 2004; Jin et al., 2013; Lee et al., 2013). For example, twenty-six anthocyanins were detected and characterized in the aqueous extract of the purple cell line. These anthocyanins are exclusively cyanidin or peonidin 3-sophoroside-5-glucosides and their acylated derivatives (Tian et al., 2005). Purple sweet potato anthocyanins protect the rat liver from hepatoxin-induced injury (Zhang et al., 2009) and show antioxidative ability to scavenge active oxygen radicals (Zhu et al., 2010). Anthocyanins extracted from sweet potato are used extensively as dietary supplements and additives in food industry. Genetic engineering approaches to increase anthocyanin accumulation offer huge economic value (Zhang et al., 2014).

Anthocyanin biosynthesis has been characterized in several model plants drawing the attention of sweet potato research community (Wang et al., 2013, 2016). In plants, regulation of structural gene expression is tightly organized spatially and temporally during plant development, and is orchestrated by a ternary complex involving transcription factors from the R2R3-MYB, basic helix–loop–helix (bHLH) and WD40 classes. This MYB–bHLH–WD40 (MBW) complex regulates the genes that encode enzymes specifically involved in the late stages of the pathway leading to the biosynthesis of anthocyanins (Petroni and Tonelli, 2011). In sweet potato, an R2R3-MYB transcription factor, IbMBY1, was isolated predominantly from the tuberous roots of purple-fleshed sweet potato (cv Ayamurasaki), showing its organ specificity. Only the purple-fleshed sweet potato cultivars expressed IbMYB1 in the tuberous roots, indicating its role in anthocyanin biosynthesis (Mano et al., 2007).

Structural and functional characterization of *IbMYB1* genes in recent Japanese purple-fleshed sweet potato cultivars also suggests that *IbMYB1* genes were established through multiple gene-duplication events (Tanaka et al., 2012). In sweet potato, the *IbMYB* expression was repressed by miR828 during the wounding response (Lin et al., 2012). Interestingly, novel wounding-induced small RNA (sRNAs), sRNA8105, targeted the first intron of IbMYB1 (MYB domain protein 1) before RNA splicing, and mediated RNA cleavage and DNA methylation of IbMYB1, indicating that wounding causes metabolic flux in the phenylpropanoid pathway of lignin biosynthesis rather than the flavonoid biosynthesis pathway (Lin et al., 2013). The *IbMYB1* gene has been used as a visible marker of sweet potato transformation with intragenic vectors as well as anthocyanin production in other plant species (Kim et al., 2010). Park et al. (2014) reported that the expression of *IbMYB1*, driven by either the storage root-specific sporamin 1 (SPO1) promoter or the oxidative-stress-inducible peroxidase anionic 2 (SWPA2) promoter, in an orange-fleshed cultivar with high carotenoid levels, resulted in the generation of a dual-pigmented transgenic sweet potato with improved antioxidant activity. Additional investigations are required to reveal the regulatory role of the MBW complex in sweet potato, especially in storage roots.

In purple-fleshed sweet potato cv. Ayamurasaki, expression of *IbDFR* was strongly associated with anthocyanin accumulation in leaves, stems and roots. Downregulation of *IbDFR* expression in transgenic sweet potato (DFRi) using an RNAi approach dramatically reduced anthocyanin accumulation in young leaves, stems and storage roots. In contrast, the content of flavonols quercetin-3-O-hexose-hexoside and quercetin-3-O-glucoside in the leaves and roots of DFRi plants was increased significantly. Therefore, the metabolic pathway increased flavonol influx in the DFRi plants following decreased anthocyanin and proanthocyanidin accumulation. These plants also displayed reduced antioxidant capacity compared with wild type. These results directly demonstrated that anthocyanins played a key role in protection against oxidative stress in sweet potato (Wang et al., 2013). Heterologous expression of the maize *Lc* gene in sweet potato dramatically increased anthocyanin accumulation and the structural genes involved in the anthocyanin biosynthetic pathway were systemically upregulated in *Lc* transgenic sweet potato plants. The Lc transcription factor bound to the DNA element G-box (CACGTG) in the promotor of anthocyanin genes such as *ANS* and *DFR* and regulated their expression (Wang et al., 2016a). Furthermore, the biosynthesis of flavonols (quercetin-3-O-hexose-hexoside and quercetin-3-O-hexoside) was also increased in leaf, stem and root of the *Lc* transgenic plants, which is correlated with the upregulation of FLS expression (Wang et al., 2016a). The study increases our understanding of anthocyanin biosynthesis mechanism and provides additional insights into the engineering of this metabolic pathway and novel approaches for anthocyanin biofortification in crops.

8 Conclusion and perspectives

Overall, genetic engineering improved the performance of sweet potato resulting in better yield, improved biotic and abiotic stress resistance and enhanced nutritional value. With the development of sweet potato functional genomics and proteomics, especially, the current initiative on sweet potato genome sequencing, further investigations are needed to determine the key native genes underlying the value-added traits observed in field conditions and in commercial application. However, compared with other major crops or model plants, this topic is heavily ignored due to the methodological challenges. Nevertheless, sweet potato manifests unique features and biological questions related to development and regulation of root system, which cannot be addressed using other model plants. The development of sweet potato transformation system switching from models to farmer-preferred cultivars facilitates studies to address the specific issues.

From a commercialization and bio-safety perspective, the current transformation system needs further improvement to meet the potential concerns of genetically modified crops. Genome-editing technologies, such as TALLEN and CRISPR/Cas9, provide new strategies for genetic manipulation of targeting genes and production of marker-free plants, which have been used in advanced research on sweet potato. These technologies, together with traditional hybridization and marker-assisted breeding, enable the construction of an integrated breeding system to promote the development of sweet potato industrialization and food security.

9 Where to look for further information

Sweet potato genetic transformation protocols are available by several key references:

1 Efficient embryogenic suspension culturing and rapid transformation of a range of elite genotypes of sweet potato (*Ipomoea batatas* [L.] Lam.), http://dx.doi. org/10.1016/j.plantsci.2011.01.005.
2 Transgenic plant production from embryogenic callus of sweet potato (*Ipomoea batatas* (L.) Lam.) using *Agrobacterium tumefaciens*, http://doi.org/10.5511/ plantbiotechnology.15.11.

Sweet potato reviews related to genetic modification are available by several key references:

1 Characterization of genes and promoters, transformation and transgenic development in sweet potato, http://www.globalsciencebooks.info/Online/GSBOnline/images/2012/ FVCSB_6(SI1)/FVCSB_6(SI1)43-56o.pdf.
2 Genetic transformation of sweet potato for improved tolerance to stress: A review, http://repository.embuni.ac.ke/bitstream/handle/123456789/1116/Imbo%20 Budambula%20et%20al%202016.pdf?sequence=1.
3 Exploiting the use of biotechnology in sweet potato for improved nutrition and food security: Progress and future outlook, https://cgspace.cgiar.org/handle/10568/67667.

Sweet potato genome sequences are available by several key public databases:

1 IPOMOEA BATATAS GENOME BROWSER, http://public-genomes-ngs.molgen. mpg.de/SweetPotato/.
2 Sweetpotato GARDEN, http://sweetpotato-garden.kazusa.or.jp/.
3 Sweetpotato Gene Index, https://research.cip.cgiar.org/confluence/display/SPGI/ Home.

10 Acknowledgements

This work was supported by grants from the National Natural Science Foundation of China (31201254, 31501356) and the International Science and Technology Cooperation Program of China (2015DFG32370).

11 References

Antonio, G. C., Takeiti, C. Y., de Oliveira, R. A. and Park, K. J. (2011) Sweet potato: Production, morphological and physicochemical characteristics, and technological process. *Fruit, Vegetable and Cereal Science* 5(2): 1–18.

Anwar, N., Watanabe, K. N. and Watanabe, J. A. (2011) Transgenic sweet potato expressing mammalian cytochrome P450. *Plant Cell, Tissue and Organ Culture* 105(2): 219–31.

Atkinson, H. J., Urwin, P. E., Clarke, M. C. and McPherson, M. J. (1996) Image analysis of the growth of *Globodera pallida* and *Meloidogyne incognita* on transgenic tomato roots expressing cystatins. *Journal of Nematology* 28: 209–15.

Belehu, T., Hammes, P. S. and Robbertse, P. J. (2004) The origin and structure of adventitious roots in sweet potato (*Ipomoea batatas*). *Australian Journal of Botany* 52: 551–8.

Benian, G. M., Mercer, K. B., Miller, R. K., Tinley, T. L., Sheth, S. and Qadota, H. (2006) Caenorhabditis elegans UNC-96 is a new component of M-lines that interacts with UNC-98 and paramyosin and is required in adult muscle for assembly and/or maintenance of thick filaments. *Molecular Biology of the Cell* 17: 3832–47.

Bi, H. and Zhang, P. (2012) Molecular characterization of two sweepoviruses identified from China and infectivity evaluation of cloned SPLCV-JS in *Nicotiana benthamiana*. *Archives of Virology* 157: 441–454.

Bi, H. and Zhang, P. (2014) Agroinfection of sweet potato by vacuum infiltration of an infectious sweepovirus. *Virologica Sinica* 29(3): 148–54.

Boss, W. F. and Im, Y. J. (2012) Phosphoinositide signaling. *Annual Review of Plant Biology* 63: 409–29.

Bovell-Benjamin, A. C. (2007) Sweet potato: a review of its past, present, and future role in human nutrition. *Advances in Food and Nutrition Research* 52: 1–59.

Chen, H. J., Wang, S. J., Chen, C. C. and Yeh, K. W. (2006) New gene construction strategy in T-DNA vector to enhance expression level of sweet potato sporamin and insect resistance in transgenic Brassica oleracea. *Plant Science* 171(3): 367–74.

Chen, T. H. H. and Murata, N. (2008) Glycinebetaine: an effective protectant against abiotic stress in plants. *Trends in Plant Science* 13: 499–505.

Clark, C. A., Davis, J. A., Abad, J. A., Cuellar, W. J., Fuentes, S., Kreuze, J. F., Gibson, R. W., Mukasa, S. B., Tugume, A. K., Tairo, F. D. and Valkonen, J. P. T. (2012) Sweetpotato viruses: 15 years of progress on understanding and managing complex diseases. *Plant Disease* 96(2): 168–85.

Clark, C. A., Ferrin, D. M., Smith, T. P. and Holmes, G. J. (2013) *Compendium of Sweetpotato Diseases, Pests, and Disorders*, Second Ed., APS Press.

Fan, W. J., Wei, Z. R., Zhang, M., Ma, P. Y., Liu, G. L., Zheng, J. L., Guo, X. D. and Zhang, P. (2015a) Resistance to *Ditylenchus destructor* infection in sweet potato by the expression of small interfering RNAs targeting unc-15, a movement-related gene. *Phytopathology* 105: 1458–65.

Fan, W., Deng, G., Wang, H., Zhang, H. and Zhang, P. (2015b) Elevated compartmentalization of Na$^+$ into vacuoles improves salt and cold stress tolerance in sweet potato (*Ipomoea batatas*). *Physiologia Plantarum* 154: 560–571.

Fan, W., Wang, H. and Zhang, P. (2016a) Engineering glycinebetaine metabolism for enhanced drought stress tolerance in plants. In M.A. Hossain et al. (Eds.), *Drought Stress Tolerance in Plants, Volume 2, Molecular and Genetic Perspectives*, chapter 18, 513–30, Springer International Publishing Switzerland.

Fan, W. J., Wang, H. X., Wu, Y. L., Yang, N., Yang, J. and Zhang, P. (2016b) H$^+$-pyrophosphatase IbVP1 promotes efficient iron use in sweet potato [*Ipomoea batatas* (L.) Lam.]. *Plant Biotechnology Journal* Doi: 10.1111/pbi.12667.

Firon, N., LaBonte, D., Villordon, A., Kfir, Y.,Solis, J., Lapis, E., Perlman, T. S., Doron-Faigenboim, A., Hetzroni, A., Althan, L. and Nadi, L. A. (2013) Transcriptional profiling of sweetpotato (*Ipomoea batatas*) roots indicates down-regulation of lignin biosynthesis and up-regulation of starch biosynthesis at an early stage of storage root formation. *BMC Genomics* 14: 460.

Gao, S., Yu, B., Yuan, L., Zhai, H., He, S. Z. and Liu, Q. C. (2011) Production of transgenic sweet potato plants resistant to stem nematodes using Oryzacystatin-I gene. *Scientia Horticulturae* 128: 408–14.

Gutierrez-Campos, R., Torres-Acosta, J. A., Saucedo-Arias, L. J. and Gomez-Lim, M. A. (1999) The use of cysteine proteinase inhibitors to engineer resistance against potyviruses in transgenic tobacco plants. *Nature Biotechnology* 17: 1223–6.

Hoover, R. (2001) Composition, molecular structure, and physicochemical properties of tuber and root starches: a review. *Carbohydate Polymers* 45(3): 253–67.

Islam, S. (2006) Sweetpotato (*Ipomoea batatas* L.) leaf: its potential effect on human health and nutrition. *Journal of Food Science* 71(2): R13–21.

Jobling, S. (2004) Improving starch for food and industrial applications. *Current Opinion in Plant Biology* 7(2): 210–18.

Kim, S. H., Mizuno, K. and Fujimura, T. (2002) Isolation of MADS-box genes from sweet potato (*Ipomoea batatas* (L.) lam.) expressed specifically in vegetative tissues. *Plant and Cell Physiology* 43: 314–22.

Kim, C. Y., Ahn, Y. O., Kim, S. H., Kim Y-H, Lee H-S, Catanach, A. S., Jacobs, J. M. E., Conner, A. J. and Kwak, S.-S. (2010) The sweet potato *IbMYB1* gene as a potential visible marker for sweet potato intragenic vector system. *Physiologia Plantarum* 139: 229–40.

Kim, Y. H., Kim, M. D., Park, S. C., Yang, K. S., Jeong, J. C., Lee, H. S. and Kwak, S. S. (2011) SCOF-1-expressing transgenic sweetpotato plants show enhanced tolerance to low-temperature stress. *Plant Physiology & Biochemistry* 49(12): 1436–41.

Kim, S. H., Ahn, Y. O., Ahn, M. J., Lee, H. S. and Kwak, S. S. (2012) Down-regulation of β-carotene hydroxylase increases β-carotene and total carotenoids enhancing salt stress tolerance in transgenic cultured cells of sweetpotato. *Phytochemistry* 74: 69–78.

Kim, S. H., Ahn, Y. O., Ahn, M. J., Jeong, J. C., Lee, H. S. and Kwak, S. S. (2013) Downregulation of the lycopene ε-cyclase gene increases carotenoid synthesis via the β-branch-specific pathway and enhances salt-stress tolerance in sweetpotato transgenic calli. *Physiologia Plantarum* 147: 432–42.

Kim, S. H., Jeong, J. C., Park, S., Bae, J. Y., Ahn, M. J., Lee, H. S. and Kwak, S. S. (2014) Down-regulation of sweetpotato lycopene β-cyclase gene enhances tolerance to abiotic stress in transgenic calli. *Molecular Biology Reports* 41(12): 8137–48.

Kim, Y. H., Lim, S., Han, S. H., Lee, J. J., Nam, K. J., Jeong, J. C., Lee, H. S. and Kwak, S. S. (2015) Expression of both CuZnSOD and APX in chloroplasts enhances tolerance to sulfur dioxide in transgenic sweet potato plants. *Comptes Rendus Biologies* 338(5): 307–13.

Kimura, T., Otani, M., Noda, T., Ideta, O., Shimada, T. and Saito, A. (2001) Absence of amylose in sweet potato [*Ipomoea batatas* (L.) Lam.] following the introduction of granule-bound starch synthase I cDNA. *Plant Cell Reports* 20(7): 663–6.

Kitahara, K., Hamasuna, K., Nozuma, K., Otani, M., Hamada, T., Shimada, T., Fujita, K. and Suganuma, T. (2007) Physicochemical properties of amylose-free and high-amylose starches from transgenic sweetpotatoes modified by RNA interference. *Carbohydate Polymers* 69(2): 233–40.

Kreuze, J. F., Klein, I. S., Lazaro, M. U., Chuquiyuri, W. J.. C., Morgan, G. L., Mejia, P. G. C., Ghislain, M. and Valkonen, J. P. T. (2008) RNA silencing-mediated resistance to a crinivirus (Closteroviridae) in cultivated sweetpotato (*Ipomoea batatas* L.) and development of sweetpotato virus disease following co-infection with a potyvirus. *Molecular Plant Pathology* 9: 589–98.

Lee, M. J., Park, J. C., Choi, D. S. and Jung, M. Y. (2013) Characterization and quantitation of anthocyanins in purple-fleshed sweet potatoes cultivated in korea by HPLC-DAD and HPLC-ESI-QTOF-MS/MS. *Journal of Agricultural & Food Chemistry* 61(12): 3148–58.

Lee, J. J., Kim, Y. H., Kwak, Y. S., An, J. Y., Kim, P. J., Lee, B. H., Kumar, V., Park, K. W., Chang, E. S., Jeong, J. C., Lee, H. S. and Kwak, S. S. (2015) A comparative study of proteomic differences between pencil and storage roots of sweetpotato (*Ipomoea batatas* (L.) lam.). *Plant Physiology & Biochemistry* 87: 92–101.

Lin, J. S., Lin, C. C., Lin, H. H., Chen, Y. C. and Jeng, S. T. (2012) MicroR828 regulates lignin and H2O2 accumulation in sweet potato on wounding. *New Phytologist* 196(2): 427–40.

Lin, J. S., Lin, C. C., Li, Y. C., Wu, M. T., Tsai, M. H., Hsing, Y. I. C. and Jeng, S. T. (2013) Interaction of small RNA–8105 and the intron of *IbMYB1, R. N.A* regulates *IbMYB1* family genes through secondary siRNAs and DNA methylation after wounding. *Plant Journal* 75: 781–94.

Liu, Q., Liu, J., Zhang, P. and He, S. (2014a) Root and tuber crops. In N. Van Alfen (Ed.), *Encyclopedia of Agriculture and Food Systems,* Vol. 5, 46–61. Elsevier.

Liu, D. G., He, S. Z., Zhai, H., Wang, L. J., Zhao, Y., Wang, B., Li, R. J. and Liu, Q. C. (2014b) Overexpression of IbP5CR enhances salt tolerance in transgenic sweetpotato. *Plant Cell, Tissue and Organ Culture* 117: 1–16.

Liu, D. G., Wang, L. J., Zhai, H., Song, X. J., He, S. Z. and Liu, Q. C. (2014c) A novel α/β-hydrolase gene IbMas enhances salt tolerance in transgenic sweetpotato. *PLoS, O. N.E* 9: e115128.

Liu, D. G., Wang, L. J., Liu, C. L., Song, X. J., He, S. Z., Zhai, H. and Liu, Q. C. (2014d) An *Ipomoea batatas* iron-sulfur cluster scaffold protein gene, *IbNFU1*, is involved in salt tolerance. *PLoS, O. N.E* 9: e93935.

Liu, D. G., He, S. Z., Song, X. J., Zhai, H., Liu, N., Zhang, D. D., Ren, Z. T. and Liu, Q. C. (2015) *IbSIMT1*, a novel salt-induced methyltransferase gene from *Ipomoea batatas*, is involved in salt tolerance. *Plant Cell Tissue and Organ Culture* 120: 701–15.

Loebenstein, G. and Thottappilly, G. (2009) *The Sweetpotato.* Springer Dordrecht, The Netherlands.

Lu, Y. Y., Deng, X. P. and Sangsoo, K. (2010) Over expression of cuzn superoxide dismutase (CuZn, S. O. D) and ascorbate peroxidase (APX) in transgenic sweet potato enhances tolerance and recovery from drought stress. *African Journal of Biotechnology* 9(49): 8378–91.

Mano, H., Ogasawara, F., Sato, K., Higo, H. and Minobe, Y. (2007) Isolation of a regulatory gene of anthocyanin biosynthesis in tuberous roots of purple-fleshed sweet potato. *Plant Physiology* 143: 1252–68.

Mohan, C. and Nair, A. G. (2012) Characterization of genes and promoters, transformation and transgenic development in sweet potato. *Fruit, Vegetable and Cereal Science and Biotechnology* 6(Special Issue 1): 43–56.

Mohandas, C. and Siji, J. V. (2012) Nematode problems in sweet potato and their management. *Fruit, Vegetable and Cereal Science and Biotechnology* 6(Special Issue 1): 139–42.

Montilla, E. C., Hillebrand, S., Butschbach, D., Baldermann, S., Watanabe, N. and Winterhalter, P. (2010) Preparative isolation of anthocyanins from Japanese purple sweet potato (*Ipomoea batatas* L.) varieties by high-speed countercurrent chromatography. *Journal of Agricultural and Food Chemistry* 58: 9899–904.

Mukhopadhyay, S., Chattopadhyay, A., Chakraborty, I. and Bhattacharya, I. (2011) Crops that feed the world 5. Sweetpotato. Sweetpotatoes for income and food security. *Food Security* 3(3): 283–305.

Muramoto, N., Tanaka, T., Shimamura, T., Mitsukawa, N., Hori, E., Koda, K., Otani, M., Hirai, M., Nakamura, K. and Imaeda, T. (2012) Transgenic sweet potato expressing thionin from barley gives resistance to black rot disease caused by *Ceratocystis fimbriata* in leaves and storage roots. *Plant Cell Reports* 31(6): 987–97.

Nakatani, M. and Komeichi, M. (1991) Changes in the endogenous level of zeatin riboside, abscisic acid and indole acetic acid during formation and thickening of tuberous roots in sweet potato. *Japanese Journal of Crop Science* 60: 91–100.

Noda, T., Kimura, T., Otani, M., Ideta, O., Shimada, T., Saito, A. and Suda, I. (2002) Physicochemical properties of amylose-free starch from transgenic sweet potato. *Carbohydate Polymers* 49(3): 253–60.

Noh, S. A., Lee H-S, Huh, E. J., Huh, G. H., Paek K-H, Shin, J. S. and Bae, J. M. (2010) SRD1 is involved in the auxin-mediated initial thickening growth of storage root by enhancing proliferation of metaxylem and cambium cells in sweetpotato (*Ipomoea batatas*). *Journal of Experimental Botany* 61: 1337–49.

Noh, S. A., Lee H-S, Kim Y-S, Paek K-H, Shin, J. S. and Bae, J. M. (2013) Down-regulation of the IbEXP1 gene enhanced storage root development in sweetpotato. *Journal of Experimental Botany* 64: 129–42.

Okada, Y., Saito, A., Nishiguchi, M., Kimura, T., Mori, M., Hanada, K., Sakai, J., Miyazaki, C., Matsuda, Y. and Murata, T. (2001) Virus resistance in transgenic sweetpotato [*Ipomoea batatas* L. (Lam)] expressing the coat protein gene of sweet potato feathery mottle virus. *Theoritical and Applied Genetics* 103:743–51.

Okada, Y., Nishiguchi, M., Saito, A., Kimura, T., Mori, M., Hanada, K., Sakai, J., Matsuda, Y. and Murata, T. (2002) Inheritance and stability of the virus-resistant gene in the progeny of transgenic sweet potato. *Plant Breeding* 121: 249–53.

Okada, Y. and Saito, A. (2008) Evaluation of resistance to complex infection of SPFMV in transgenic sweet potato. *Breeding Science* 58: 243–50.

Pan, L. P., Yu, S. L., Chen, C. J., Li, H., Wu, Y. L. and Li, H. H. (2012) Cloning a peanut resveratrol synthase gene and its expression in purple sweet potato. *Plant Cell Reports* 31(1): 121–31.

Park, S. C., Kim, Y. H., Jeong, J. C., Kim, C. Y., Lee, H. S., Bang, J. W. and Kwak, S. S. (2011) Sweetpotato late embryogenesis abundant 14 (IbLEA14) gene influences lignification and increases osmotic- and salt stress-tolerance of transgenic calli. *Planta* 233(3): 621–34.

Park, S. C., Kim, Y. H., Kim, S. H., Jeong, Y. J., Kim, C. Y., Lee, J. S., Bae, J. Y., Ahn, M. J., Jeong, J. C., Lee, H. S. and Kwak, S. S. (2015a) Overexpression of the IbMYB1 gene in an orange-fleshed sweet potato cultivar produces a dual-pigmented transgenic sweet potato with improved antioxidant activity. *Physiologia Plantarum* 153: 525–37.

Park, S. C., Kim, S. H., Park, S., Lee, H. U., Lee, J. S., Park, W. S., Ahn, M. J., Kim, Y. H., Jeong, J. C., Lee, H. S. and Kwak, S. S. (2015b) Enhanced accumulation of carotenoids in sweetpotato plants overexpressing IbOr-Ins gene in purple-fleshed sweetpotato cultivar. *Plant Physiology and Biochemistry* 86: 82–90.

Park, S., Kim, H. S., Jung, Y. J., Kim, S. H., Ji, C. Y., Wang, Z., Jeong, J. C., Lee, H. S., Lee, S. Y. and Kwak, S. S. (2016) Orange protein has a role in phytoene synthase stabilization in sweetpotato. *Scientific Reports* 6: 33563.

Petroni, K. and Tonelli, C. (2011) Recent advances on the regulation of anthocyanin synthesis in reproductive organs. *Plant Science* 181: 219–29.

Ravi, V., Chakrabarti, S. K., Makeshkumar, T. and Saravanan, R. (2014) Molecular regulation of storage root formation and development in sweet potato.In J. Janick (Ed.), *Horticultural Reviews*, Volume 42, 157–207. John Wiley & Sons, Inc., Hoboken, NJ.

Shimada, T., Otani, M., Hamada, T. and Kim, S. H. (2006) Increase of amylose content of sweetpotato starch by RNA interference of the starch branching enzyme II gene (IbSBEII). *Plant Biotechnology* 23(1): 85–90.

Sivparsad, B. J. and Gubba, A. (2014) Development of transgenic sweet potato with multiple virus resistance in South Africa (SA). *Transgenic Research* 23(2): 377–88.

Takahata, Y., Tanaka, M., Otani, M., Katayama, K., Kitahara, K., Nakayachi, O., Nakayama, H. and Yoshinaga, M. (2010) Inhibition of the expression of the starch synthase II gene leads to lower pasting temperature in sweetpotato starch. *Plant Cell Reports* 29(6): 535–43.

Tanaka, M., Kato, N., Nakayama, H., Nakatani, M. and Takahata, Y. (2008) Expression of class I knotted1-like homeobox genes in the storage roots of sweetpotato (*Ipomoea batatas*). *Journal of Plant Physiology* 165: 1726–35.

Tanaka, M., Takahata, Y., Kurata, R., Nakayama, H. and Yoshinaga, M. (2012). Structural and functional characterization of ibmyb1 genes in recent japanese purple-fleshed sweetpotato cultivars. *Molecular Breeding* 29(3): 565–74.

Terahara, N., Konczak, I., Ono, H., Yoshimoto, M. and Yamakawa, O. (2004) Characterization of acylated anthocyanins in callus induced from storage root of purple-fleshed sweet potato, *Ipomoea batatas* L.. *BioMed Research International* 5: 279–86.

Tian, Q., Konczak, I. and Schwartz, S. J. (2005) Probing anthocyanin profiles in purple sweet potato cell line (*Ipomoea batatas* L. Cv. Ayamurasaki) by high-performance liquid chromatography and electrospray ionization tandem mass spectrometry. *Journal of Agricultural and Food Chemistry* 53: 6503–9.

Trenado, H. P., Orílio, A. F., Belén, M. M., Enrique, M. and Jesús, N. C. (2011) Sweepoviruses cause disease in sweet potato and related *Ipomoea* spp.: fulfilling koch's postulates for a divergent group in the genus begomovirus. *PLoS, O. N.E* 6(11): e27329.

Vain, P., Worland, B., Clarke, M. C., Richard, G., Beavis, M., Liu, H., Kohli, A., Leech, M., Snape, J., Christou, P. and Atkinson, H. (1998) Expression of an engineered cysteine proteinase inhibitor (Oryzacystatin-IΔD86) for nematode resistance in transgenic rice plants. *Theoretical and Applied Genetics* 96: 266–71.

Villordon, A. O., Ginzberg, I. and Firon, N. (2014) Root architecture and root and tuber crop productivity. *Trends in Plant Science* 19: 419–25.

Wang, H., Fan, W., Li, H., Yang, J., Huang, J. and Zhang, P. (2013) Functional characterization of Dihydroflavonol-4-reductase in anthocyanin biosynthesis of purple sweet potato underlies the direct evidence of anthocyanins function against abiotic stresses. *PLoS, O. N.E* 8: e78484.

Wang, H., Yang, J., Zhang, M., Fan, W., Firon, N., Pattanaik, S., Yuan, L. and Zhang, P. (2016a) Altered Phenylpropanoid Metabolism in the Maize Lc-Expressed Sweet Potato (Ipomoea batatas) Affects Storage Root Development. Scientific Reports 6: 18645.

Wang, B., Zhai, H., He, S., Zhang, H., Ren, Z., Zhang, D. and Liu, Q. (2016b) A vacuolar Na+/H+ antiporter gene, IbNHX2, enhances salt and drought tolerance in transgenic sweetpotato. *Scientia Horticulturae* 201: 153–66.

Wang, Q., Zhang, L., Guan, Y. and Wang, Z. (2006) Endogenous hormone concentration in developing tuberous roots of different sweet potato genotypes. *Agricultural Science in China* 5: 919–27.

Woo, J. W., Kim, J., Kwon, S. I., Corvalán, C., Cho, S. W., Kim, H., Kim, S. G., Kim, S. T., Choe, S. and Kim, J. S. (2015) DNA-free genome editing in plants with preassembled CRISPR-Cas9 ribonucleoproteins. *Nature Biotechnology* 33(11): 1162–4.

Woolfe, J. A. (1992) *Sweet Potato: An Untapped Food Resource*. Cambridge University Press, UK.

Xing, Y. J., Ji, Q., Yand, Q., Luo, Y. M., Li, Q. and Wang, X. (2008) Studies on Agrobacterium-mediated genetic transformation of embryogenic suspension cultures of sweet potato. *African Journal of Biotechnology* 7(5): 534–40.

Xu, Z., Zhao, Y. Q., Yang, D. J., Sun, H. J., Zhang, C. L. and Xie, Y. P. (2015) Attractant and repellent effects of sweet potato root exudates on the potato rot nematode, *Ditylenchus destructor*. *Nematology* 17: 117–24.

Yang, J., Bi, H. P., Fan, W. J., Zhang, M., Wang, H. X. and Zhang, P. (2011) Efficient embryogenic suspension culturing and rapid transformation of a range of elite genotypes of sweet potato (*Ipomoea batatas* [L.] Lam.). *Plant Science* 181(6): 701–11.

Yang, J., Moeinzadeh M-H, Kuhl, H., Helmuth, J., Xiao, P., Liu, G., Zheng, J., Sun, Z., Fan, W., Deng, G., Wang, H., Hu, F., Fernie, A. R., Timmermann, B., Zhang, P. and Vingron, M. (2016) The haplotype-resolved genome sequence of hexaploid *Ipomoea batatas* reveals its evolutionary history. *bioRxiv* 064428.

Zhai, H., Wang, F., Si, Z., Huo, J., Xing, L., An, Y., He, S. and Liu, Q. (2016) A myo-inositol-1-phosphate synthase gene, IbMIPS1, enhances salt and drought tolerance and stem nematode resistance in transgenic sweet potato. *Plant Biotechnol Journal* 14: 592–602.

Zhang, Z. F., Fan, S. H., Zheng, Y. L., Lu, J., Wu, D. M., Shan, Q. and Hu, B. (2009) Purple sweet potato color attenuates oxidative stress and inflammatory response induced by D-galactose in mouse liver. *Food and Chemical Toxicology* 47: 496–501.

Zhang, Y., Butelli, E. and Martin, C. (2014) Engineering anthocyanin biosynthesis in plants. *Current Opinion in Plant Biology* 19: 81–90.

Zhou, W., Yang, J., Hong, Y., Liu, G., Zheng, J., Gu, Z. and Zhang, P. (2015) Impact of amylose content on starch physicochemical properties in transgenic sweet potato. *Carbohydrate Polymers* 122: 417–27.

Zhu, F., Cai, Y. Z., Yang, X., Ke, J. and Corke, H. (2010) Anthocyanins, hydroxycinnamic acid derivatives, and antioxidant activity in roots of different Chinese purple fleshed sweetpotato genotypes. *Journal of Agricultural and Food Chemistry* 58: 7588–96.

Chapter 9

Nutritional properties and enhancement/biofortification of potatoes

Duroy A. Navarre, Washington State University and USDA-ARS, USA; and M. Moehninsi, Sen Lin and Hanjo Hellmann, Washington State University, USA

1 Introduction

Potatoes, along with other staple foods, are consumed in much greater quantities than most fruits and vegetables. Consequently, staples like potatoes, rice, wheat and corn are ideally positioned to deliver vitamins, minerals and phytonutrients into our diets. In light of the high consumption, staple foods are logical targets for traditional and precision nutritional breeding efforts because increasing phytonutrients in a food with a per capita consumption of 120 pounds per year will have a bigger impact on public health than increasing phytonutrients in a food eaten a handful of times a year.

In recent years, the relationship between diet and health has been increasingly studied by scientists. Concurrently, public awareness and media coverage of diet and health have increased. Part of this increased attention likely comes from the scope of the obesity problem. Obesity was classified as a disease in 2013 by the American Medical Association and continues to increase in much of the world. In the United States, 155 million adults, roughly 50% of the population, are overweight or obese, with an estimated health cost to the US economy in the hundreds of billions annually. In addition to toll on the health of individuals battling obesity-related chronic diseases, the emotional and financial

http://dx.doi.org/10.19103/AS.2017.0016.09
Published by Burleigh Dodds Science Publishing Limited, 2018.

well-being of the families is affected. One in five deaths worldwide is related to diet according to the Global Burden of Disease Study (GBDS, 2017). Common diseases including cancer, diabetes, arthritis and arteriosclerosis are influenced by diet. A recent study of 600 overweight adults evaluated the effect of low-fat versus low-carbohydrate diets. No significant difference was seen between the two diets; instead diet quality seemed to be a key factor in weight loss (Gardner et al., 2018).

A result of all of this is an increasingly health-conscious public, a growing number of whom are prioritizing the perceived nutritional value of foods more than ever before. To be sure not all people are basing their food choices on health, or there would not be an obesity epidemic in the first place. Nevertheless, for a significant portion of the population, health considerations are influencing foods purchased both in stores and in restaurants, which offers incentives for companies and restaurants to provide healthy options. Even the fast food industry has been impacted by evolving consumer preferences 'as the USA experiences a "sea change" in fast-food consumption' (Horovitz and Waggoner, 2014), and is altering menus to provide healthier options. Negative publicity, whether scientifically based or not, can depress sales, as has happened to commodities including beef, potatoes, dairy, wheat and eggs.

One consequence of this new emphasis on health and diet is that it can spur plant scientists and breeding programmes to place increased effort towards increasing the nutritional value of crops. Such efforts may be especially important for foods that have been subject to negative publicity, such as potatoes which were linked to obesity in one study (Mozaffarian et al., 2011). Criticism from nutritionists can cause the public to perceive potatoes as unhealthy, leading to decreased sales. Per capita potato consumption is about 25 lbs per person less today than 10 years ago.

As discussed in this review, unlike the common misperception that potatoes provide 'empty calories', they are in actuality a nutrient-dense food with strong scientific data supporting their nutritional value. A nutrient-dense food must provide an equal or greater amount of nutrients as they do calories. For example, a 3.5 ounce portion of baked potato provides only about 97 calories or 5% of the daily recommended intake. But this same portion provides 5% or greater amounts of vitamins and minerals including vitamin C, vitamin B6, vitamin B9, potassium, fibre and more (SR28, 2016). In addition, potatoes with yellow flesh provide carotenoids, including some important for eye health, whereas red- and purple-flesh potatoes tend to contain high amounts of antioxidants.

Another reason that potato nutritional composition is important is because potatoes will have a key role in providing global food security in the future, as they have multiple times in the past, including when they fuelled dramatic population growth in Europe (Nunn and Qian, 2011). A combination of factors, including population growth, loss of farmland to urbanization, emerging middle classes in the developing world and climate change, have led to estimates that crop yields must double by 2050 to feed the world (Ray et al., 2013). Potatoes are unusually high-yielding plants; for example, 30–50 ton yields are common in the Pacific Northwest of the United States. Other advantages of potatoes contributing to their food security potential are their storability and affordability. Historically, even in the absence of modern refrigerated cold storage, potatoes were routinely stored in cellars for months after harvest, helping ensure a stable food supply. Affordability also impacts food supply, as low-income populations can struggle to afford healthy diets (Drewnowski and Eichelsdoerfer, 2010). Potatoes and beans provided the most nutrients per dollar in a study of 90 foods (Drewnowski and Rehm, 2013). The affordability, nutritional value, storability and high yields of potatoes have made them important components for future global food

Published by Burleigh Dodds Science Publishing Limited, 2018.

security strategies, including in China, which leads the world in potato production (Devaux et al., 2014; Hairong, 2015).

The anticipated challenge to provide food security over the next few decades, along with the increasing consumer demand for healthy food choices, combines to make the nutritional value of potatoes an important goal for potato breeding programmes. This chapter will review the nutritional composition of potatoes from diverse germplasm including vitamin C, B vitamins, potassium, carotenoids, phenylpropanoids and glycoalkaloids (GAs). We will briefly discuss the biosynthesis and regulation of some phytonutrient metabolic pathways to provide a synopsis of information useful for traditional and precision breeding efforts. In many cases, as described, there is an extensive knowledge of the biochemical pathways involved, which can facilitate development of new cultivars with even higher amounts of vitamins, minerals and phytonutrients.

2 The vitamin B family

Potatoes provide generous amounts of B vitamins, especially B_6. The family of B vitamins belongs to a diverse group of water-soluble chemical compounds that were historically classified based on their common, positive impacts on health and growth, as well as preventing skin diseases such as pellagra in human and animals (Birch et al., 1935). Currently, eight of these compounds are recognized as B vitamins, namely thiamine (vitamin B_1), riboflavin (B_2), niacin or niacin amide (B_3), pantothenic acid (B_5), pyridoxine (B_6), biotin (B_7), folic acid (B_9) and cobalamins (B_{12}) (Fig. 1; Roje, 2007; Hellmann and Mooney, 2010). Humans that experience deficiencies in any of these vitamins will suffer from severe health problems. Some B vitamins, such as niacin, riboflavin and pyridoxine, fulfil a wide range of functions in the cell, and are needed in larger amounts and may not be stored effectively in the body. These are recommended by the 19th World Health Organization (WHO) Model List of Essential Medicines as 'minimum medicine needs for a basic health-care system' (http://www.who.int/medicines/publications/essentialmedicines/en/). Other vitamins recommended in this context by the WHO are ascorbate (vitamin C) and retinol, as a precursor of vitamin A.

This section will focus on the five B vitamins (thiamine, niacin, pantothenic acid, pyridoxine and folic acid; Table 1) for which potato tubers have been demonstrated to be a good source, and briefly discuss their cellular functions, biosynthesis, tuber content and ways to increase their amounts in tubers.

2.1 Thiamin (vitamin B_1)

Thiamin (also known as thiamine) is composed of a thiazole ring and an aminopyrimidine, which are connected through a methylene bridge (Fig. 1). The vitamin is primarily active in its di-phosphorylated form (thiamine diphosphate; TDP) and is needed as an essential cofactor for enzymes active in carbohydrate and branched-chain amino acid metabolism, and nucleic acid, ATP and NADPH biosynthesis (Manzetti et al., 2014). In plants, the aminopyrimidine moiety is synthesized from 5-aminoimidazole ribonucleotide, while the thiazole ring derives from a sulphur donor protein, NAD^+ and glycine (Goyer, 2010). The biosynthesis requires activities of two enzymes, a 4-amino-5-hydroxymethyl-2-methylpyrimidine phosphate synthase (THIC) and a thiamin biosynthesis protein (TH1), which together produce thiamine

Published by Burleigh Dodds Science Publishing Limited, 2018.

monophosphate (TMP). Surprisingly, TMP is first dephosphorylated before a thiamine pyrophosphokinase pyrophosphorylates thiamin to generate TDP (Ajjawi et al., 2007).

The National Institute of Health (NIH) Recommended Dietary Allowances (RDA) for thiamin are 1.2 mg/day for men and 1.1 mg/day for women and are easy to reach with a balanced diet (Table 1). If thiamine deficiency occurs, it can cause cardiovascular and neurological problems commonly referred to as beriberi (Abdou and Hazell, 2015). Such deficiencies may primarily arise in areas where the population mainly thrives on a single staple food such as rice (Juguan et al., 1999). In most countries, thiamine deficiency is not a threat, but may pose a problem in people who consume large amounts of alcohol (Rees and Gowing, 2013). Alcoholics may suffer from Wernicke-Korsakoff syndrome, a genetic disease that causes neurological disorders and brain damage (Donnelly, 2017). The syndrome is based on a recessive mutation in a thiamin-dependent transketolase that affects affinity of the enzyme for the vitamin.

Potato is a good source of thiamin, and 100 g of some varieties contains nearly 20% of the RDA values (Table 1; Goyer and Haynes, 2011; Goyer and Sweek, 2011). However, because thiamin is sensitive to heat (Kandutsch and Baumann, 1953), processed potato products that are exposed to high temperatures, such as French fries and baked potatoes, may contain reduced amounts that are less than 10% of the RDA. Even with milder

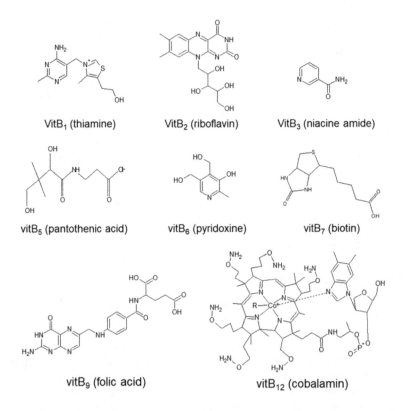

Figure 1 Chemical structures of eight B vitamins. The chapter focuses on vitamins B_1, B_3, B_5, B_6 and B_9 which are the B vitamins most abundant in potato.

Published by Burleigh Dodds Science Publishing Limited, 2018.

Table 1 Summary of recommended daily uptake, biological function and content in potato for B vitamins covered in this chapter

Vitamin	Recommended uptake (adults)	Function	Content in potato (selected examples)
B$_1$ (thiamin)	1.1–1.2 mg/day	Carbohydrate and branched-chain amino acid metabolism, biosynthesis of nucleic acids, ATP and NADPH	~0.2 mg/100 g fresh tuber; ~0.067 mg/100 g baked potato
B$_3$ (niacin)	14–16 mg/day	Carbohydrate and fatty acid metabolism	~1.6 mg/100 g baked or boiled potato
B$_5$ (pantothenic acid)	5 mg/day	Biosynthesis of fatty acids, carbohydrates and amino acids	~0.5 mg/100 g of boiled potato
B$_6$ (pyridoxine)	1.3–1.7 mg/day	Amino acid metabolism, biosynthesis of fatty acids and some vitamins, catabolism of glycogen and starch Antioxidant	~0.39 mg/100 g baked potato; ~1 mg/100 g chips
B$_9$ (folate)	0.4 mg/day	Sulphur–iron cluster metabolism, biosynthesis of thymidylate, purines, methionine, pantothenate and formyl-Met-tRNA	~0.01 mg/100 g boiled potato; ~0.026 mg/100 g baked potato

treatments, such as boiling, only around 10% of the RDA (0.106 mg) remains. Thus, in potatoes and any cooked food, a reduction in thiamin content is unavoidable.

There is a high potential to increase thiamin content through breeding efforts, as well as through bioengineering approaches (Dong et al., 2015; Goyer and Sweek, 2011). Analysis of 33 accessions of primitive cultivars (*Solanum tuberosum* group Andigenum) and three modern potato varieties showed that thiamin content among these varieties differed up to fourfold (Goyer and Haynes, 2011). *Arabidopsis* co-overexpressing the two key enzymes in thiamin biosynthesis, THIC and TH1, had up to six-fold increases in thiamin content in their leaves (Dong et al., 2015). These plants were less susceptible to pathogen infection (*Pseudomonas syringae*), indicating that increasing thiamin content in plants may benefit the crop (Dong et al., 2015).

Overall, the current knowledge on thiamine underscores the importance of this vitamin for human well-being, and suggests that potato tubers can likely be fortified with this vitamin without negatively impacting general plant growth. In fact, elevated levels may even have beneficial impacts on pathogen resistance of the tuber. Thus, increasing thiamin phytonutrient content in potato appears to be a desirable goal for breeders.

2.2 Niacin (vitamin B$_3$)

Niacin is the generic term for nicotinamide and nicotinic acid (Fig. 1), and is converted in the cell to NAD$^+$ or NADP$^+$. It is part of the NAD$^+$ salvage pathway in the cell to replenish NAD$^+$ via nicotinate mononucleotide (Ashihara et al., 2005). NAD$^+$ and NADP$^+$ play critical roles in photosynthesis, carbohydrate and fatty acid metabolism

(Wahlberg et al., 2000), and lack of either one is lethal for the plant (Katoh et al., 2006). In humans, niacin may reduce cholesterol, widen blood vessels and lower blood pressure (Gehring, 2004; Rolfe, 2014; Williams and Ramsden, 2005). Chronic deficiency can lead to pellagra, a disease associated with dementia, diarrhoea and dermatitis (Fu et al., 2014; Kohn, 1938). Pellagra is common in countries where the population is mainly thriving on a corn-based diet. This can be avoided if the corn goes through a process called nixtamalization which releases the vitamin from the seeds and makes it bioavailable (Gwirtz and Garcia-Casal, 2014).

Potatoes contain good amounts of niacin relative to the RDA values which are 14 and 16 mg/day for women and men, respectively. As such, 100 g of baked or boiled potatoes, for example contains around 10% of the RDA (Table 1). Surprisingly, little research has been done on elevating niacin content in plants, and it remains open about the impact on plant growth and performance. The limited available data on potato do not indicate major variation among the few tested genotypes (Wills et al., 1984). However, increasing niacin content in tubers could be of interest to the processing industry because the vitamin may reduce acrylamide formation in processed potato products (Sansano et al., 2015; Zeng et al., 2010). Consequently, it may be important for breeders to gain more insight into the variation among different cultivars and better assess the potential to raise niacin content.

2.3 Pantothenic acid (vitamin B$_5$)

Pantothenic acid (Fig. 1) is synthesized in the cytosol from alanine and pantoic acid, then imported into the chloroplast, phosphorylated to 4-phosphopantothenate, and further metabolized to function as a core component of the coenzyme A (CoA) and the prosthetic group of the acyl carrier protein (ACP) (Webb et al., 2004; Ottenhof et al., 2004). As part of CoA and ACP, it has an integral role in the biosynthesis of fatty acids, carbohydrates and amino acids (Depeint et al., 2006). Pantothenic acid is also a potent plant growth regulator, and has been shown in combination with gibberellic acid to induce and effectively promote shoot and root regeneration in tissue culture, which may be helpful for efficient mass propagation of potato clones (Yasmin et al., 2011). There is evidence it may have a protective function against oxidative stress (Wojtczak and Slyshenkov, 2003; Walczak-Jedrzejowska et al., 2013).

Pantothenic acid can generally be found in all food sources, and though deficiency of vitamin B$_5$ is uncommon in humans, reports from other mammals or birds showed that an insufficient supply can cause a variety of health issues including anaemia, dermatitis and neurological disorders (Tahiliani and Beinlich, 1991; Depeint et al., 2006; Bender, 1999). Current Dietary Reference Intakes (DRIs) that were developed by the Food and Nutrition Board at the Institute of Medicine in the United States propose values of 5 mg/day for adult men and women (Table 1; Tahiliani and Beinlich, 1991).

Like niacin, information about the content of pantothenic acid in tubers and how it varies among different varieties is missing. The vitamin is sensitive to heat treatment, and is likely unstable when exposed to reactive oxygen species and high light (Riaz et al., 2009). Still, 100 g of boiled potatoes contains around 10% of the recommended DRI values, demonstrating that potato is a good nutritional source. Increasing the content of the vitamin through genetic engineering had been demonstrated in oilseed rape (*Brassica napus*), where levels were 1.5–2.5 fold higher compared to wild-type plants (Chakauya et al., 2008). These findings demonstrated the feasibility of this approach, and may be a reasonable approach for potato as well.

2.4 Pyridoxine (vitamin B$_6$)

Vitamin B$_6$ is the generic name for a group of six chemically closely related compounds. They vary in their 4'-position and carry either an aldehyde {pyridoxal [PL], and aminomethyl [pyridoxamine (PM)] or a hydroxyl (pyridoxine (PN) group (Fig. 1). All three derivatives can be phosphorylated, but pyridoxal-phosphate (PLP) is the main active form of vitamin B$_6$ (Mooney and Hellmann, 2010). As such, it participates in more than 140 different biochemical reactions that are mainly connected with amino acid metabolism. PLP is also required for the biosynthesis of fatty acids, some vitamins, compounds needed in photosynthesis and photorespiration, and in the catabolism of the animal and plant storage carbohydrates glycogen and starch (Mooney and Hellmann, 2010). Vitamin B$_6$ is a potent antioxidant and has been demonstrated in plants to be critical for ameliorating high-light stress (Havaux et al., 2009). In addition, several groups reported that the vitamin contributes to pathogen resistance in plants (Denslow et al., 2007; Herrero et al., 2007; Titiz et al., 2006; Zhang et al., 2014, 2015). For example, tomato and *Arabidopsis* mutants with reduced PLP biosynthesis become hyper-susceptible to infection with the fungus *Botrytis cinerea* or the bacterium *P. syringae*, respectively (Zhang et al., 2014). Because these mutant plants showed significantly higher levels of reactive oxygen species when compared to wild type, their lowered resistance may be a consequence of reduced antioxidative capacity (Zhang et al., 2014, 2015). Increasing the level of vitamin B$_6$ in plants by overexpressing the biosynthetic machinery has been accomplished (Raschke et al., 2011), but unfortunately whether such elevated levels provide enhanced disease resistance has not been tested.

Given the breadth of functions vitamin B$_6$ has in the cell, it is not surprising that it is essential for all living organisms. Deficiencies widely impact human health, with symptoms ranging from neurological disorders, such as depression, to anaemia, certain dermatological problems and weakening of the immune system (Hellmann and Mooney, 2010).

The current RDA for vitamin B$_6$ varies, but in general between 1.3 and 1.7 mg/day is recommended for adults (Table 1). Based on these values, potatoes are excellent sources of this vitamin. Because pyridoxine is light-labile but not so much sensitive to heat, processing of potato by baking or frying does not considerably degrade its content. Consequently, baked potatoes or potato chips contain 23–60% of the RDA per 100 g, respectively. The content of vitamin B$_6$ in the tuber varies among different cultivars, and can range from 18 to 27 µg B$_6$/g dry weight in mature tubers, or more than 30% on average (Mooney et al., 2013). This emphasizes the potential to increase the B$_6$ content in potatoes.

Potatoes use starch as a carbohydrate storage compound, and PLP availability likely plays an important role in the starch turnover rate. A key enzyme involved in the breakdown of starch is α-glucan phosphorylase, which requires PLP as a cofactor to generate glucose-1-phosphate from linear glucan chains (Kossmann and Lloyd, 2000). Potatoes are often stored at low temperatures ranging from 4°C to 8°C, and under these conditions, many varieties start to break down the starch and accumulate reducing sugars like fructose and glucose (Chen et al., 2012; Kossmann and Lloyd, 2000). This process is known as 'cold-induced sweetening' and is undesired by the processing industry because exposing the stored tubers to high temperatures during cooking processes can result in the Maillard reaction, leading to brown discoloration of the food and may even result in the formation of acrylamide (Arribas-Lorenzo and Morales, 2009; Lojzova et al., 2009). Interestingly, some potato varieties also accumulate vitamin B$_6$ when stored over longer periods under cold conditions, but it is currently poorly understood to what extent starch degradation and cold-induced sweetening depend on the availability of the vitamin (Mooney et al., 2013). Because pyridoxine has such wide-ranging effects on cell function, including being a potent antioxidant, increasing its content in potato is likely feasible and beneficial for both the consumer and the plant itself.

2.5 Folic acid (vitamin B$_9$)

Folate is the generic name for tetrahydrofolate (THF) (Fig. 1), a compound composed of a pterin ring, an aminobenzoic acid and a glutamic acid tail (Hanson and Gregory, 2011). THF is readily available in the cell and is mainly required for 1-carbon metabolism. As such, it participates in sulphur–iron cluster metabolism, the biosynthesis of thymidylate, purines, methionine, pantothenate and formyl-Met-tRNA (da Silva et al., 2014; Li et al., 2003; Tibbetts and Appling, 2010). In plants, it is a critical component of the photorespiratory pathway, where it is needed in mitochondria for the activity of glycine decarboxylase complex (Douce and Neuburger, 1999).

Folate biosynthesis is highly compartmentalized in the cell, as the mitochondria, chloroplast and cytosol all contribute to its biosynthesis. The pterin ring is synthesized in the cytosol and imported as 6-hydroxymethyl dihydropterin (6-HMDP) into the mitochondria. The aminobenzoic acid is made in the chloroplasts and imported from there into the mitochondria to be condensed with 6-HMDP and glutamate to THF [for an overview, see Gorelova et al. (2017)].

Folate deficiency is a common problem in human populations worldwide. It can be lost during cooking, by either leaching or sensitivity to heat in conjunction with reducing agents, oxygen availability or pH shifts that can lead to oxygenation of certain C- and N- bonds within THF, rendering the vitamin non-functional (de Paiva et al., 2015; Delchier et al., 2014). Hence, the kind of foods consumed, as well as the method of processing, strongly impacts the retention volume of folate (McKillop et al., 2002). Deficiency is connected with a variety of symptoms that include anaemia, cardiovascular diseases and even a higher risk for certain cancers (Blancquaert et al., 2010). Pregnant women must be especially cautious of folate deficiency based on the higher nutritional demands that the embryo imposes on the body. Here, deficiencies over longer periods can cause megaloblastic anaemia or birth defects such as spina bifida in the newborn (Copp and Greene, 2010; Oakley, 2009; Beaudin and Stover, 2009). The RDA values for adults are 400 µg of folate equivalents, while for pregnant women, slightly higher amounts of 600 µg/day are recommended (Table 1). Although these RDA values are significantly lower than some older published values, for pregnant women, physicians still often prescribe higher dosages.

Folate content varies among different potato cultivars, and recent work has shown a range between 1400 ng/g dry weight down to 550 ng, which represents an impressive difference of more than 60% (Goyer and Navarre, 2007). However, these amounts indicate that folate content in potato is not extremely high. This becomes more pronounced in processed potato where folate content is further reduced, and only reached 2.5% and 6.5% of the RDA in some boiled and baked potatoes, respectively (Goyer and Navarre, 2007, 2009; Goyer and Sweek, 2011). Still because of its high consumption in many countries worldwide, potato is overall considered a very good source of vitamin B$_9$ (Brussaard et al., 1997; Brevik et al., 2005; Alfthan et al., 2003). The wide variation in folate content in different potato varieties once again demonstrates the great potential for folate fortification through breeding, where a target of 20% of the RDA is not an

Published by Burleigh Dodds Science Publishing Limited, 2018.

unreasonable goal through conventional breeding. Alternatively, raising folate through metabolic engineering is also highly promising. This has been accomplished in different crop species including lettuce (Nunes et al., 2009), tomato (Diaz de la Garza et al., 2007), rice (Storozhenko et al., 2007) and white corn (Naqvi et al., 2009). In potato, levels were doubled in tubers to around 120 µg folate per 100 g fresh weight when compared to wild-type plants (Blancquaert et al., 2013).

3 Vitamin C

Vitamin C (ascorbic acid; Fig. 2) is likely the vitamin for which potatoes are best known. A 100 g portion (~3.5 ounces) of a red-skin potato provides 14% of the daily value (USDA SR21), but some high vitamin C varieties easily surpass that amount. Vitamin C is an enzyme cofactor that also detoxifies reactive oxygen species. It is associated with heart health,

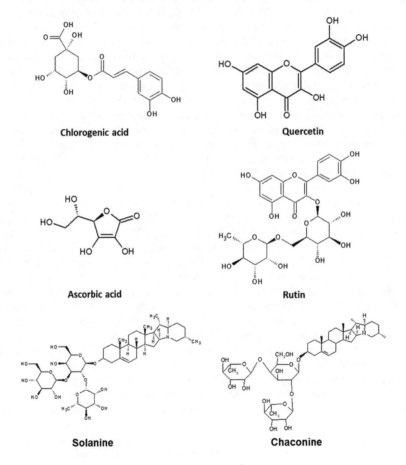

Figure 2 Structures of three phenylpropanoids (chlorogenic acid, quercetin and rutin), ascorbic acid and the two major glycoalkaloids (solanine and chaconine).

Published by Burleigh Dodds Science Publishing Limited, 2018.

has anti-inflammatory activity and may help prevent or manage diabetes and metabolic syndrome (Moser and Chun, 2016; Pearson et al., 2017). Vitamin C deficiency severe enough to cause scurvy, which is typified by loss of teeth, liver spots and bleeding, is rare in the developed world. In the third NHANES survey in the United States, 11–13% of young adults were vitamin C deficient or depleted (Hampl et al., 2004). A deficiency was found in one in seven Canadian young adults and the deficiency was associated with elevated chronic disease markers (Cahill et al., 2009); an even higher percentage (one in three) had suboptimal levels. In the developed world, deficiency is more common in people with advanced cancer, autism, diabetes, smokers, the obese and alcoholics (Mayland et al., 2005; Hampl et al., 2004; Gunton, 2015; Golriz et al., 2017; Ijaz et al., 2017). In the United Kingdom, 16% of women and 25% of men in the low-income population were deficient (Mosdøl et al., 2008). It is notable that potatoes were the least expensive vitamin C source in a study of 98 foods (Drewnowski and Rehm, 2013).

Plants are the main source of dietary vitamin C, with leaves and chloroplasts containing 5–25 mM (Wheeler et al., 1998). Tubers synthesize vitamin C and also import it from leaves and stems (Viola et al., 1998; Tedone et al., 2004). Plants have multiple vitamin C biosynthetic pathways, but the L-galactose pathway is the major route and all of the genes have been identified (Laing et al., 2007; Wolucka and Montagu, 2007; Bulley and Laing, 2016). Numerous transgenic approaches targeting different crops have been undertaken (Macknight et al., 2017), with some of the highest increases, up to six-fold, obtained in tomato (Bulley et al., 2012). Quantitative trait loci (QTLs) involved in ascorbate synthesis have also been identified and traditional breeding approaches can also be an effective way to increase vitamin C levels (Calafiore et al., 2016; Mellidou et al., 2012).

Analysis of vitamin C content in over 75 breeding lines developed and grown in the Pacific Northwest showed that it ranged from 11 to 40 mg/100 g FW, with both location and year affecting concentrations (Love and Pavek, 2008; Love et al., 2004). Vitamin C ranged from 13 to 31 mg/100 g FW in 33 cultivars grown in three European locations, assuming a tuber water content of 80% (Dale et al., 2003) and from 8 to 24 mg/100 g FW in 11 Indian potato varieties (Yamdeu Galani et al., 2017). Unusual levels, much higher than any previously reported, came from one study of Russet Burbank somaclones stated to contain from 31 to 139 mg/100 g FW (Nassar et al., 2014).

Vitamin C losses of up to 60% can occur after weeks to months of cold storage (Dale et al., 2003; Keijbets and Ebbenhorst-Seller, 1990). Whereas numerous studies report cold storage causes a loss of vitamin C, one group reported several-fold increases in vitamin C after storage of 11 Indian potato varieties (Yamdeu et al., 2017). Breeding programmes targeting vitamin C should also screen for lines that maintain levels in cold storage, which may be more easily accomplished if the biochemical mechanism leading to cold storage loss was better understood. All cultivars do not lose vitamin C at the same rate, as seen in a study showing some lines had no loss after 2 months of cold storage (Kulen et al., 2013), although all the lines had decreased vitamin C upon longer storage. One report suggested atmospheric oxygen levels had a larger effect than storage temperature and found that loss of vitamin C began during vine senescence (Blauer et al., 2013).

A high rate of nitrogen fertilization resulted in lower vitamin C amounts in fresh-cut products and in a more rapid loss after storage of the cut product (Licciardello et al., 2018). Commercial processing resulted in a 51% loss in vitamin C occurred during pre-freezing that could be avoided with optimal processing and flash-freezing of the cooked product (Tosun and Yücecan, 2008). Such results indicate that careful field management and processing can help maximize the amount in potatoes. Perhaps more than any other phytochemical discussed in this chapter, postharvest events have a large effect on tuber vitamin C content, emphasizing the need to breed for cultivars that retain higher amounts after storage.

4 Potassium

Potassium deficiency may be the most common and important dietary deficiency in the diet. A majority of Americans do not consume enough potassium, a problem that is listed as a 'public health concern' in the 2015–20 Dietary Guidelines for Americans. Adult intake averages 2640 mg/day, well below the 4700 mg/day recommended by the Institute of Medicine (Hoy and Goldman, 2012; Jackson et al., 2018). Potassium helps counter the effect of a high sodium diet; is required for muscle, heart, kidney, nerve and digestive system function as well as reduces risk of high blood pressure, stroke, inflammatory bowel disease, kidney stones and asthma.

Reduced dietary sodium and increased potassium intake for cardiovascular health is a top recommendation of many nutritionists. Potatoes have very low sodium and high potassium amounts, with the lowest Na/K ratio among examined vegetables (Pandino et al., 2011). The Dietary Guidelines for Americans recommends decreased sodium intake, yet many of the foods high in potassium are also high in sodium. In the developed world where cardiovascular disease is a greater threat to health than iron deficiency, foods with high potassium may be especially valuable.

We are not aware of any breeding programme specifically aiming to increase potassium in potatoes. Given its importance, the potassium content of potato germplasm has not been comprehensively surveyed, and the reported range is ~3.0–8.2 mg/g FW (Casanas et al., 2002; Rivero et al., 2003; Sanchez-Castillo et al., 1998; van Niekerk et al., 2016; Burrowes and Ramer, 2008). Even these limited reports make it clear that potatoes can provide even more potassium than commonly reported. The USDA nutrient database lists baked potatoes with skin to contain 5.5 mg/g FW, so a potato that contained 8.2 mg/g FW would provide ~50% more potassium. Given the large per capita consumption of potatoes, this would result in a significant uptake in potassium intake.

5 Carotenoids

Several phytonutrients are responsible for the flesh colour of potatoes, with anthocyanins being responsible for their red/blue/purple colour and carotenoids for yellow/orange. Zeaxanthin is the primary carotenoid responsible for orange colour and lutein for yellow (Brown et al., 1993). The Y locus encodes a β-carotene hydroxlase that is a primary determinant of tuber flesh colour (Wolters et al., 2010; Brown et al., 2006). Carotenoids have anti-inflammatory properties, and decrease the risk of cancer, cardiovascular disease, diabetes and age-related macular degeneration (Chucair et al., 2007; Tan et al., 2008; Abdel-Aal et al., 2013; Desmarchelier and Borel, 2017).

Carotenoids have been surveyed in a range of potato germplasm, including diploid potatoes from *S. stenotomum* and *S. phureja* that contained up to 20 µg carotenoids/g FW of zeaxanthin (Brown et al., 1993; Burmeister et al., 2011). Mainstream white potatoes contained 2.7–7.4 µg/g FW of carotenoids (Iwanzik et al., 1983), and Andean landraces ranged from 3 to 36 µg/g dry weight and up to 10 µg/g FW (Andre et al., 2007b; Burgos et al., 2012). Amounts up to 20 µg/g FW were found in a survey of 33 Andean cultivars (Brown et al., 2007). Total carotenoids ranged from trace amounts to 28 µg/g DW in the skin and 9 µg/g DW in the flesh in a study of 100 cultivars grown in Ireland or Spain (Valcarcel et al., 2015; Fernandez-Orozco et al., 2013).

Alleles affecting carotenoid amounts have been identified, including a QTL on chromosome 3 that accounted for 71% of the variation in carotenoids and was likely a β-carotene hydroxylase (Campbell et al., 2014; Sulli et al., 2017). The biochemical pathway of carotenoid synthesis is well understood, which creates possibilities for precision breeding approaches to increase amounts. Several transgenic approaches have successfully increased tuber carotenoid content (Diretto et al., 2006; Ducreux et al., 2005; Van Eck et al., 2007). The most successful approach overexpressed three bacterial carotenoid genes in 'Desiree' to produce a 'golden potato' that had a remarkable 3600-fold increase in β-carotene to 47 µg/g DW (Diretto et al., 2007) and has great potential for alleviating vitamin A deficiency in at-risk populations (Chitchumroonchokchai et al., 2017).

6 Potato phenylpropanoids

Phenylpropanoids are natural products produced by plants. They are small molecules also known as 'secondary metabolites'. Phenylpropanoids have many critical roles in the plant belaying their label as 'secondary metabolites'. This includes plant growth and development, cell wall synthesis, flowering, pigmentation, signal transduction and biotic and abiotic stress resistance, including a primary role in regulating plant inducible defences (Koes et al., 2005; Vogt, 2010). Plants produce thousands of different phenylpropanoids, with some of the better-known classes being polyphenols and anthocyanins.

These are valued by consumers who are familiar with phenylpropanoids due to popular media coverage of health benefits attributed to the phenylpropanoid content of food and beverages like coffee, tea, wine, kale, berries, along with coloured fruits and vegetables in general. Unlike the vitamins and minerals discussed above, phenylpropanoids do not have recommended daily intakes but are nevertheless important for health. They have a rather long list of health benefits that include conferring chemopreventive, anti-obesity and anti-inflammatory properties, reducing the risk of metabolic syndrome, strokes and diabetes, along with promoting gut health, cardiovascular health, longevity, mental acuity and eye health (Scalbert et al., 2005; Manach et al., 2004; Parr and Bolwell, 2000; Cardona et al., 2013; Tohge and Fernie, 2017; Wang et al., 2014; Tresserra-Rimbau et al., 2017).

Chlorogenic acid (CGA; Fig. 2) is typically the most abundant polyphenol in potatoes, can slow the entry of glucose into the bloodstream and may decrease the risk of type 2 diabetes (Bassoli et al., 2008; Legrand and Scheen, 2007). Interestingly, potatoes containing higher amounts of phenylpropanoids may have a lower glycaemic index (Ramdath et al., 2014). Flavonols found in potatoes like quercetin and rutin (Fig. 2) can reduce the risk of heart disease, asthma, bronchitis, emphysema and prostate and lung cancer (Kawabata et al., 2015).

Understanding the role of specific phenylpropanoids in health is a complex undertaking because there are thousands of different types and they are ingested within a complex matrix, resulting in cross interactions. Moreover, after ingestion, they are metabolized into products that can have different health-promoting properties than the parent compound (Hollman, 2014). Further complicating matters is that an individual's gut microbiota influences polyphenol bioefficacy, but people can differ in the composition of their gut microbiome; thus, the effect of phenylpropanoids may vary among the population (Cardona et al., 2013; Bolca et al., 2013; Espín et al., 2017).

Potatoes with high amounts of phenylpropanoids showed anticancer properties (Reddivari et al., 2007; Ombra et al., 2015; Charepalli et al., 2015; Sido et al., 2017) and reduced chromium toxicity (Zhao et al., 2011). Purple potato extracts lessened alcohol-induced hepatic injury in mice (Jiang et al., 2016), while in rats, on an obesity-promoting diet, purple potatoes provided metabolic and cardiovascular benefits that countered some of the effects of the diet (Ayoub et al., 2017). Adult human males fed 150 g of purple potatoes daily for 6 weeks had markers indicative of reduced DNA damage and inflammation (Kaspar et al., 2011). A decrease in postprandial glycaemia and insulinemia occurred in males fed purple potatoes (Linderborg et al., 2016), whereas after consuming purple potatoes, humans averaging 54 years in age had a significant drop in blood pressure without weight gain (Vinson et al., 2012).

6.1 Tuber phenylpropanoids

Potatoes were the third largest contributor of phenylpropanoids to the diet after apples and oranges in a study of 34 fruits and vegetables (Chun et al., 2005). The store-bought cultivars in this study were not specified, but were almost certainly white potatoes, which have lower amounts of phenylpropanoids than red- and purple-flesh potatoes (Brown, 2005; Navarre et al., 2011; Andre et al., 2007a). So even though potatoes are already a major source of dietary phenolics, cultivars are available with much higher amounts than the most common varieties. This shows the potential for consumers to choose varieties that provide the traits they value. However, grocery stores in the United States generally carry only a few types of potatoes, unlike in Europe, and specialty potatoes are not widely available outside of farmer's markets and perhaps an occasional medley carried in stores that is a mix of white, yellow and red/purple potatoes.

As with the vitamins and minerals discussed earlier, tuber phenylpropanoid content varies considerably among germplasm, illustrating the genetic diversity of potatoes. The most abundant phenylpropanoids in white potatoes are the hydroxycinnamic acids such as CGA that can constitute up to 90% of a tuber's total phenolics in some cultivars (Malmberg and Theander, 1985). The situation is different for red- and purple-flesh potatoes, which although they usually contain even higher amounts of CGA than white potatoes, also have high amounts of anthocyanins; therefore, CGA constitutes a smaller percentage of overall phenylpropanoid content (Navarre et al., 2011).

A 11-fold variation in phenolic acids and flavanols was reported in Andean potato landraces, which had a high correlation between total phenolics and total antioxidant capacity (Andre et al., 2007b; Valiñas et al., 2017; Ayvaz et al., 2016). Compared to the mainstream cultivars Desiree and Shepody, Chilean landraces had 8- and 11-fold more phenylpropanoids (Ah-Hen et al., 2012).

Anthocyanin levels in 50 coloured-fleshed cultivars varied from 0.5 to 7 mg/g FW in the skin and up to 2 mg/g FW in the flesh (Lewis et al., 1998; Jansen and Flamme, 2006). Among Andean cultivars, Guincho Negra, a dark purple potato from the Andigenum group, had the highest anthocyanin levels, at 16 mg/g DW (Andre et al., 2007c). A Phureja group genotype contained 41 mg/g DW (Pillai et al., 2013). We have screened over a thousand potato lines and observed over a 15-fold difference in phenylpropanoid amounts that include red- and purple-flesh breeding lines containing over 18 mg anthocyanins/g DW. Because the preferred culinary potatoes in many countries are white or yellow flesh varieties, the market for high-anthocyanin potatoes may be limited in the absence of consumer education campaigns.

To put the amounts of phenylpropanoids in potatoes in context relative to amounts provided by other vegetables, we compared various potatoes to 15 other vegetables, including broccoli, Brussels sprouts and spinach. Among the examined vegetables, the purple potato Magic Molly had the highest amount of total phenolics at ~1.7 mg/g FW (Navarre et al., 2011). Such data may come as a surprise to many who have the common misconception that potatoes provide little else besides starch.

6.2 Phenylpropanoid biosynthesis and regulation

Breeding for potatoes that contain the desired types and amounts of phenylpropanoids is feasible because the metabolic pathways of most major phenylpropanoids are well defined (Lovat et al., 2016; Deng and Lu, 2017). Transcription factors regulating expression of potato polyphenols, anthocyanins and flavonols are known. CGA in potatoes is synthesized via hydroxycinnamoyl CoA:quinate hydroxycinnamoyl transferase (Niggeweg et al., 2004; Payyavula et al., 2015). As will be discussed below, the MYB transcription factor StAN1 that regulates anthocyanin biosynthesis, also mediates CGA expression (Payyavula et al., 2013).

Three loci, D, P and R, which encode two structural genes and a transcription factor, control the amount of anthocyanins in potato skin (Jung et al., 2005, 2009; Zhang et al., 2009). Anthocyanins are regulated by MYB, bHLH and WD40 classes of transcription factors (Allan et al., 2008). In potatoes, an AN1 transcription factor complex regulates anthocyanins (Payyavula et al., 2013; Zhang et al., 2017). MYB transcription factors interact with their bHLH partners, which are essential for the activity of R2R3 MYBs. Compared to white and yellow potatoes, red- and purple-flesh potatoes that contained higher amounts of anthocyanins also more highly expressed *Anthocyanin1* (*StAN1*), *basic helix loop helix1* (*bHLH1*) and *StWD40* (Payyavula et al., 2013), supporting their key role in potato anthocyanin biosynthesis. Transcriptomic analysis of potatoes identified transcription factors variants, including an AN1 ten-amino acid C-terminal motif required for anthocyanin synthesis (Stushnoff et al., 2010; Liu et al., 2015, 2016; Cho et al., 2016).

StJAF13 was reported as a StAN1 co-regulator, and expression of StJAF13 with either StAN1 or StbHLH1 increased anthocyanins compared to overexpression of StAN1 alone (D'Amelia et al., 2014). MYBA1 is another MYB member related to anthocyanin biosynthesis in plants. Mutation of the MYBA1 promoter region results in a loss of anthocyanin production in grapevine (Kobayashi et al., 2004). An additional MYB, StMYBA1, which corresponds to the translated sequence of StAN3, shares 90% similarity to StAN1, and also regulated anthocyanin biosynthesis (Liu et al., 2016; Jung et al., 2009). Liu et al. (2017) transformed StMYBA1 into tobacco, and found that genes related to anthocyanin biosynthesis in tobacco were up-regulated. However, expression of those genes was down-regulated in the transgenic seedlings cultured in the dark, indicating that light is a key factor for anthocyanin accumulation regulated by this MYB. StbHLH appeared to be a limiting factor for anthocyanin biosynthesis, as it was differentially expressed in pigmented and white potatoes, whereas StAN1, StMYBA1 and StMYB113 were well expressed even in the white skin and white flesh tubers (Liu et al., 2016).

MYB12 is a transcription factor involved in flavonol biosynthesis (Mehrtens et al., 2005), with homologues in tomato (Adato et al., 2009). Specific overexpression of AtMYB12 in

Published by Burleigh Dodds Science Publishing Limited, 2018.

potato tubers increased the flavonol content (Li et al., 2016). Besides MYB12, other MYB members were also associated with flavonol biosynthesis in plants. MYB11 and MYB111 are MYBs involved in regulation of chalcone synthase, chalcone isomerase, flavonol-3-hydroxylase, flavonol-3'-hydroxylase and flavonol synthase (Stracke et al., 2007) in *Arabidopsis*. Whether there are homologous genes in potato remains to be investigated.

6.3 Environmental effects on phenylpropanoids

Biotic and abiotic environmental factors affect biosynthesis of phenylpropanoids, which have *in planta* roles relating to protection of the plant against stress. Photoperiod affects the biosynthesis of anthocyanin in many plants including potato. Potatoes grown in longer day length (14–15 h) contained higher anthocyanin than those grown in the shorter day lengths, but the total phenolic amounts were not notably influenced (Reyes et al., 2004). Light influences the biosynthesis of secondary metabolites (Jaakola and Hohtola, 2010). In addition, light can induce biosynthesis of flavonols in tubers, which of course are not normally exposed to light. Light-exposed potato strips accumulated two-fold more flavonols than those stored in the dark (Tudela et al., 2002).

Flavonols and anthocyanins can be induced by cold exposure (Schulz et al., 2016). Tubers in cold storage up to 6 days accumulated more flavonols (Tudela et al., 2002). Likewise, tuber growth and yield were associated with the cooler temperatures (Vayda, 1994). Reyes et al. (2004) found that anthocyanin content in tuber was enhanced in low temperatures. Drought also affects tuber phenylpropanoids. Anthocyanin content in the potato cultivars Guincho Nigra and Sullu decreased by 32% and 65% in drought conditions (Andre et al., 2009). In contrast, Huata Colorada produced 58% more anthocyanin in drought conditions compared to the control. Drought did not influence biosynthesis of rutin, but kaempferol was significantly reduced in Sullu and SS-2163, whereas it highly increased in Huata Colorada.

7 Glycoalkaloids

Potatoes, eggplants, tomatoes and some other plants primarily in the Solanaceae and Liliaceae plant families contain nitrogen-containing toxic compounds called GAs. More than 80 different GAs have been identified in various potato species (Ginzberg et al., 2009; Shakya and Navarre, 2008). Their primary role in plants is thought to be plant pest and pathogen resistance (Altesor et al., 2014). GAs occur in higher concentrations in leaves, sprouts and fruit than in tubers; concentrations of 18 g/kg FW are reported in sprouts (Valkonen et al., 1996). Environmental factors that increase GAs include drought (Bejarano et al., 2000), high temperature (Morris and Petermann, 1985; Lafta and Lorenzen, 2000), light exposure of tubers (Percival et al., 1994; Valkonen et al., 1996; Grunenfelder et al., 2006; Dale et al., 1993) and wounding (Bergenstråhle et al., 1992; Choi et al., 1994).

GAs are traditionally regarded as anti-nutrients. However, as will be discussed below, recent studies show a wide range of health benefits from GAs. If consumed in too high an amount, GAs can cause vomiting and other ill effects (Hopkins, 1995; McMillan and Thompson, 1979). In much of the world, potato breeders follow voluntary guidelines that new cultivars must contain less than 20 mg of total GAs/100 g FW (Wilson, 1959). Higher levels of GAs cause potatoes to have a bitter taste, but lower concentrations may positively effect flavour (Janskey, 2010). Bitterness does not necessarily indicate the presence of GAs, as phenylpropanoids can also be bitter.

Published by Burleigh Dodds Science Publishing Limited, 2018.

In contrast to several decades ago when it was widely thought that there were no positive effects of GAs on nutrition and that they were just toxins, studies showing health-promoting effects of GAs are now numerous. Summarizing all the recent GA health studies would take a stand-alone chapter, but the following synopsis is representative of the field.

Different GAs reduced metastasis in a human lung cancer cell line or inhibited malignant glioma growth (Shih et al., 2007; Shen et al., 2017; Wang et al., 2017). Another potato GA, solamargine, increased the susceptibility of human lung cancer cell lines to anticancer drugs (Liang et al., 2008). Tomatine, solanine and chaconine inhibited human colon and liver cancer cell growth in cell cultures with a similar efficacy to some chemotherapy drugs (Friedman et al., 2005; Lee et al., 2004). Concurrent use of multiple GAs showed synergistic and additive anticancer effects in cervical, lymphoma and stomach cancer cell studies (Friedman et al., 2005). GAs enhanced the susceptibility of breast and oesophageal cancer cells to anticancer drugs (Shiu et al., 2009; Zupko et al., 2014; Yelken et al., 2017; Wu et al., 2018). GAs induced apoptosis in human leukaemia cells and skin cell carcinoma (Cui et al., 2012; Sun et al., 2011). In cell culture studies, chaconine had efficacy against stomach, colon, liver, cervical cancer and prostate cancer (Reddivari et al., 2010; Friedman, 2015).

In cell culture studies, solanine was effective against cancers including pancreatic cancer, and also in mice where it suppressed proliferation, angiogenesis and metastasis (Lv et al., 2014). Creams containing GAs were effective against skin cancer in mice and in humans, including in a clinical trial in which low doses of GAs had 100% efficacy (Cham et al., 1991). Tomatine was effective against cancer in a feeding study using rainbow trout (Friedman et al., 2007), and suppressed growth of prostate cancer cells in mice (Lee et al., 2013). Mice treated with GAs that underwent a total remission of their cancer remained resistant to cancer cells subsequently injected, suggesting these GAs primed the immune system for long-term cancer protection (Cham and Chase, 2012).

It is possible that GAs at the concentrations found in potatoes may help protect against cancer, although epidemiological studies are needed to support this possibility (Friedman, 2015; Habli et al., 2017).

7.1 Glycoalkaloid biosynthesis and regulation

GA biosynthesis begins with the formation and condensation of C5 isoprenoid units from acetyl-CoA. Three isoprenoid units are condensed to form 2-trans,6-transfarnesyl diphosphate, which is condensed to form 2,3 oxidosqualene, leading to triterpene and sesquiterpenes biosynthesis, including cholesterol. Cholesterol is then used for GA synthesis. High expression HMG CoA reductase-1 and squalene synthase-encoding transcripts in potato are associated with high GA levels (Krits et al., 2007). The later steps of GA biosynthesis are controlled by a cluster of GLYCOALKALOID METABOLISM (GAME) genes that participate in the core pathway producing GAs in potato and tomato (Itkin et al., 2013). In potato, four GA-related genes GAME1 (glycosyltransferases 1), GAME2 (glycosyltransferases), GAME6/PGA2 (cytochrome P450 monooxygenase) and GAME11 (dioxygenase) are located on chromosome 7 and two (GAME4 and GAME12) on chromosome 12 (Nützmann et al., 2016; Cárdenas et al., 2016). GA biosynthesis in potato after cholesterol involves several hydroxylation, oxidation, transamination and glycosylation steps to generate GAs (Cárdenas et al., 2016; Nahar et al., 2017; Sawai et al., 2014; Umemoto et al., 2016). Finally, glycosyltransferases (GAME1, SGT2 and GAME2) are required for generating the sugar moieties that combined with the steroidal aglycone moiety make the final GA (McCue et al., 2005, 2006, 2007).

Published by Burleigh Dodds Science Publishing Limited, 2018.

GA accumulation is regulated by key genes at different steps in the pathway (Mekapogu et al., 2016). Transgenic potatoes overexpressing a heterologous soybean sterol methyltransferase or with altered expression of three genes responsible for glycosylating GAs had reduced amounts (McCue et al., 2005, 2006, 2007; Arnqvist et al., 2003). GAME4 involves an oxidation step required in the biosynthetic steps from cholesterol to the aglycone, solanidine. Silencing GAME4 in potato reduced GA levels up to 74-fold in both leaves and tubers (Itkin et al., 2013). Sawai et al. (2014) found down-regulation of cholesterol synthase gene, SSR2, decreased GAs by ~10%. Silencing of POTATO GLYCOALKALOID BIOSYNTHESIS1 (PGA1) and PGA2 resulted in a significant reduction in GAs (Umemoto et al., 2016). GA accumulation in response to environmental stresses appears to be regulated at different points in the pathway (Sinden et al., 1984; Valkonen et al., 1996; Ginzberg et al., 2009; Bartoszewski et al., 2000). When treated with light, different cultivars do not necessarily increase GAs to the same amount because GAs increased more in 'King Edward' tubers than in 'Bintje' (Nahar et al., 2017).

Because of these recent advances in GA biosynthesis and regulation, precision breeding approaches are likely to be highly effective in reducing the amount of GAs in potatoes, but whether eliminating GAs is a good idea is open to debate. Eliminating GAs from potatoes, or breeding them to unnecessarily low levels, could have adverse effects on flavour, result in the loss of potential health benefits and make it more difficult to use primitive germplasm as a source of superior new traits. This could also necessitate increased pesticide use, because GAs contribute to resistance against various pests and pathogens (Dahlin et al., 2017; Austin et al., 1988; Pehu et al., 1990; Percival et al., 1998; Sanford et al., 1997).

8 Conclusion and future trends

Potatoes contain an important array of vitamins, minerals and phytonutrients and are affordable, storable and high yielding. These traits combine to make potatoes one of the best options to provide the anticipated increase in crop yields needed to provide global food security in the coming decades. Some potato germplasm contain much higher amounts of phytonutrients than others, even though none of this germplasm was deliberately bred to have higher amounts. This fact highlights the potential of focused breeding for increased nutrition to make an already nutritious crop even more nutritious. The time for such a focus may have come given the rapidly expanding new knowledge about diet and health, biochemical pathways and the greatly increased interest in nutrition among much of the public. Logical targets for enhancement could be those phytonutrients that are not currently present in optimal amounts in the diet, and of which consuming more would be beneficial for health. Some potential targets have merit beyond their role in human health, such as multifunctional phenylpropanoids that not only influence nutritional value, but appearance, flavour, taste and environmental stress resistance.

9 Where to look for further information

Nutritional aspects of potatoes are an area of active research. Readers interested in future developments can follow potato-specific journals such as *The American Journal of Potato Research* and *Potato Research*. Food and nutrition journals such as *The Journal of Food* and *Agricultural Chemistry* and *Food Chemistry* often report such information. Academic, private industry and government researchers such as the United States Department of Agriculture – Agricultural Research Service work on potato nutrition. Extensive information about the nutritional composition of potatoes and potato products can be found in the USDA Food Composition Databases at https://ndb.nal.usda.gov/ndb/. This data is also presented in a different form at http://nutritiondata.self.com/. Information about the importance of potatoes can be found at http://www.fao.org/potato-2008/en/ and https://cipotato.org/.

10 References

Abdel-Aal, E.-S., Akhtar, H., Zaheer, K. and Ali, R. 2013. Dietary sources of lutein and zeaxanthin carotenoids and their role in eye health. *Nutrients*, 5, 1169.

Abdou, E. and Hazell, A. S. 2015. Thiamine deficiency: An update of pathophysiologic mechanisms and future therapeutic considerations. *Neurochemical Research*, 40, 353–61.

Adato, A., Mandel, T., Mintz-Oron, S., Venger, I., Levy, D., Yativ, M., Domínguez, E., Wang, Z., De Vos, R. C. and Jetter, R. 2009. Fruit-surface flavonoid accumulation in tomato is controlled by a SlMYB12-regulated transcriptional network. *PLoS Genetics*, 5, e1000777.

Ah-Hen, K., Fuenzalida, C., Hess, S., Contreras, A., Vega-Gálvez, A. and Lemus-Mondaca, R. 2012. Antioxidant capacity and total phenolic compounds of twelve selected potato landrace clones grown in Southern Chile. *Chilean Journal of Agricultural Research*, 72, 3–9.

Ajjawi, I., Rodriguez Milla, M. A., Cushman, J. and Shintani, D. K. 2007. Thiamin pyrophosphokinase is required for thiamin cofactor activation in Arabidopsis. *Plant Molecular Biology*, 65, 151–62.

Alfthan, G., Laurinen, M. S., Valsta, L. M., Pastinen, T. and Aro, A. 2003. Folate intake, plasma folate and homocysteine status in a random Finnish population. *European Journal of Clinical Nutrition*, 57, 81–8.

Allan, A. C., Hellens, R. P. and Laing, W. A. (2008). MYB transcription factors that colour our fruit. *Trends Plant Science* 13: 99–102.

Altesor, P., García, Á., Font, E., Rodríguez-Haralambides, A., Vilaró, F., Oesterheld, M., Soler, R. and González, A. 2014. Glycoalkaloids of wild and cultivated *Solanum*: Effects on specialist and generalist insect herbivores. *Journal of Chemical Ecology*, 40, 599–608.

Andre, C. M., Ghislain, M., Bertin, P., Oufir, M., Del Rosario Herrera, M., Hoffmann, L., Hausman, J.-F., Larondelle, Y. and Evers, D. 2007a. Andean potato cultivars (*Solanum tuberosum* L.) as a source of antioxidant and mineral micronutrients. *Journal of Agricultural and Food Chemistry*, 55, 366–78.

Andre, C. M., Ghislain, M., Bertin, P., Oufir, M., Herrera Mdel, R., Hoffmann, L., Hausman, J. F., Larondelle, Y. and Evers, D. 2007b. Andean potato cultivars (*Solanum tuberosum* L.) as a source of antioxidant and mineral micronutrients. *Journal of Agricultural and Food Chemistry*, 55, 366–78.

Andre, C. M., Oufir, M., Guignard, C., Hoffmann, L., Hausman, J. F., Evers, D. and Larondelle, Y. 2007c. Antioxidant profiling of native Andean potato tubers (*Solanum tuberosum* L.) reveals cultivars with high levels of beta-carotene, alpha-tocopherol, chlorogenic acid, and petanin. *Journal of Agricultural and Food Chemistry*, 55, 10839–49.

Andre, C. M., Schafleitner, R., Legay, S., Lefevre, I., Aliaga, C. A., Nomberto, G., Hoffmann, L., Hausman, J. F., Larondelle, Y. and Evers, D. (2009). Gene expression changes related to the production of phenolic compounds in potato tubers grown under drought stress. *Phytochemistry* 70: 1107–16.

Arnqvist, L., Dutta, P. C., Jonsson, L. and Sitbon, F. 2003. Reduction of cholesterol and glycoalkaloid levels in transgenic potato plants by overexpression of a type 1 sterol methyltransferase cDNA. *Plant Physiology*, 131, 1792–9.

Published by Burleigh Dodds Science Publishing Limited, 2018.

Arribas-Lorenzo, G. and Morales, F. J. 2009. Effect of pyridoxamine on acrylamide formation in a glucose/asparagine model system. *Journal of Agricultural and Food Chemistry*, 57, 901–9.

Ashihara, H., Stasolla, C., Yin, Y., Loukanina, N. and Thorpe, T. A. 2005. De novo and salvage biosynthetic pathways of pyridine nucleotides and nicotinic acid conjugates in cultured plant cells. *Plant Science*, 169, 107–14.

Austin, S., Lojkowska, E., Ehlenfeldt, K., Kelman, A. and Helgeson, J. P. 1988. Fertile interspecific somatic hybrids of *Solanum*: A novel source of resistance to *Erwinia* soft rot. *Phytopathology*, 78, 1216–20.

Ayoub, H. M., McDonald, M. R., Sullivan, J. A., Tsao, R., Platt, M., Simpson, J. and Meckling, K. A. 2017. The effect of anthocyanin-rich purple vegetable diets on metabolic syndrome in obese zucker rats. *Journal of Medicinal Food*, 20, 1240–9.

Ayvaz, H., Bozdogan, A., Giusti, M. M., Mortas, M., Gomez, R. and Rodriguez-Saona, L. E. 2016. Improving the screening of potato breeding lines for specific nutritional traits using portable mid-infrared spectroscopy and multivariate analysis. *Food Chemistry*, 211, 374–82.

Bartoszewski, G., Mujer, C. V., Smigocki, A. C. and Niemirowicz-Szczytt, K. 2000. A wound inducible cytochrome P450 from tomato. *Acta Physiologiae Plantarum*, 22, 269–71.

Bassoli, B. K., Cassolla, P., Borba-Murad, G. R., Constantin, J., Salgueiro-Pagadigorria, C. L., Bazotte, R. B., Da Silva, R. S. and De Souza, H. M. 2008. Chlorogenic acid reduces the plasma glucose peak in the oral glucose tolerance test: Effects on hepatic glucose release and glycaemia. *Cell Biochemistry and Function*, 26, 320–8.

Beaudin, A. E. and Stover, P. J. 2009. Insights into metabolic mechanisms underlying folate-responsive neural tube defects: A minireview. *Birth Defects Research Part A: Clinical and Molecular Teratology*, 85, 274–84.

Bejarano, L., Mignolet, E., Devaux, A., Espinola, N., Carrasco, E. and Larondelle, Y. 2000. Glycoalkaloids in potato tubers: The effect of variety and drought stress on the alpha-solanine and alpha-chaconine contents of potatoes. *Journal of the Science of Food and Agriculture*, 80, 2096–100.

Bender, D. A. 1999. Optimum nutrition: Thiamin, biotin and pantothenate. *Proceedings of the Nutrition Society*, 58, 427–33.

Bergenståhle, A., Tillberg, E. and Jonsson, L. 1992. Regulation of glycoalkaloid accumulation in potato tuber discs. *Journal of Plant Physiology*, 140, 269–75.

Birch, T. W., Gyorgy, P. and Harris, L. J. 1935. The vitamin B(2) complex. Differentiation of the antiblacktongue and the 'P.-P'. factors from lactoflavin and vitamin B(6) (so-called 'rat pellagra' factor). Parts I-VI. *Biochemical Journal*, 29, 2830–50.

Blancquaert, D., Storozhenko, S., Loizeau, K., De Steur, H., De Brouwer, V., Viaene, J., Ravanel, S., Rebeille, F., Lambert, W. and Van Der Straeten, D. 2010. Folates and folic acid: From fundamental research toward sustainable health. *Critical Reviews in Plant Sciences*, 29, 14–35.

Blancquaert, D., Storozhenko, S., Van Daele, J., Stove, C., Visser, R. G., Lambert, W. and Van Der Straeten, D. 2013. Enhancing pterin and para-aminobenzoate content is not sufficient to successfully biofortify potato tubers and Arabidopsis thaliana plants with folate. *Journal of Experimental Botany*, 64, 3899–909.

Blauer, J. N., Kumar, M. G. N., Knowles, L. O., Dingra, A. and Knowles, N. R. 2013. Changes in ascorbate and associated gene expression during development and storage of potato tubers. *Postharvest Biology and Technology*, 78, 76–91.

Bolca, S., Van De Wiele, T. and Possemiers, S. 2013. Gut metabotypes govern health effects of dietary polyphenols. *Current Opinion in Biotechnology*, 24, 220–5.

Brevik, A., Vollset, S. E., Tell, G. S., Refsum, H., Ueland, P. M., Loeken, E. B., Drevon, C. A. and Andersen, L. F. 2005. Plasma concentration of folate as a biomarker for the intake of fruit and vegetables: The Hordaland Homocysteine Study. *American Journal of Clinical Nutrition*, 81, 434–9.

Brown, C. R. 2005. Antioxidants in potato. *American Journal of Potato Research*, 82, 163–72.

Brown, C. R., Edwards, C. G., Yang, C. P. and Dean, B. B. 1993. Orange flesh trait in potato: Inheritance and carotenoid content. *Journal of the American Society for Horticultural Science*, 118, 145–50.

Brown, C., Kim, T., Ganga, Z., Haynes, K., De Jong, D., Jahn, M., Paran, I. and De Jong, W. 2006. Segregation of total carotenoid in high level potato germplasm and its relationship to beta-carotene hydroxylase polymorphism. *American Journal of Potato Research*, 83, 365–72.

Brown, C. R., Culley, D., Bonierbale, M. and Amorós, W. (2007). Anthocyanin, carotenoid content, and antioxidant values in native South American potato cultivars. *HortScience* 42: 1733–6.

Brussaard, J. H., Lowik, M. R., Van Den Berg, H., Brants, H. A. and Goldbohm, R. A. 1997. Folate intake and status among adults in the Netherlands. *European Journal of Clinical Nutrition*, 51(Suppl. 3), S46–50.

Bulley, S. and Laing, W. 2016. The regulation of ascorbate biosynthesis. *Current Opinion in Plant Biology*, 33, 15–22.

Bulley, S., Wright, M., Rommens, C., Yan, H., Rassam, M., Lin-Wang, K., Andre, C., Brewster, D., Karunairetnam, S. and Allan, A. C. 2012. Enhancing ascorbate in fruits and tubers through over-expression of the l-galactose pathway gene GDP-l-galactose phosphorylase. *Plant Biotechnology Journal*, 10, 390–7.

Burgos, G., Amoros, W., Salas, E., Munoa, L., Sosa, P., Diaz, C. and Bonierbale, M. 2012. Carotenoid concentrations of native Andean potatoes as affected by cooking. *Food Chemistry*, 133, 1131–7.

Burmeister, A., Bondiek, S., Apel, L., Kühne, C., Hillebrand, S. and Fleischmann, P. 2011. Comparison of carotenoid and anthocyanin profiles of raw and boiled *Solanum tuberosum* and *Solanum phureja* tubers. *Journal of Food Composition and Analysis*, 24, 865–72.

Burrowes, J. D. and Ramer, N. J. 2008. Changes in the potassium content of different potato varieties after cooking. *Journal of Renal Nutrition*, 18, 249.

Cahill, L., Corey, P. N. and El-Sohemy, A. 2009. Vitamin C deficiency in a population of young Canadian adults. *American Journal of Epidemiology*, 170, 464–71.

Calafiore, R., Ruggieri, V., Raiola, A., Rigano, M. M., Sacco, A., Hassan, M. I., Frusciante, L. and Barone, A. 2016. Exploiting genomics resources to identify candidate genes underlying antioxidants content in tomato fruit. *Frontiers in Plant Science*, 7, 397.

Campbell, R., Pont, S. D., Morris, J. A., McKenzie, G., Sharma, S. K., Hedley, P. E., Ramsay, G., Bryan, G. J. and Taylor, M. A. 2014. Genome-wide QTL and bulked transcriptomic analysis reveals new candidate genes for the control of tuber carotenoid content in potato (*Solanum tuberosum* L.). *Theoretical and Applied Genetics*, 127, 1917–33.

Cárdenas, P. D., Sonawane, P. D., Pollier, J., Bossche, R. V., Dewangan, V., Weithorn, E., Tal, L., Meir, S., Rogachev, I. and Malitsky, S. 2016. GAME9 regulates the biosynthesis of steroidal alkaloids and upstream isoprenoids in the plant mevalonate pathway. *Nature Communications*, 7, 10654.

Cardona, F., Andrés-Lacueva, C., Tulipani, S., Tinahones, F. J. and Queipo-Ortuño, M. I. 2013. Benefits of polyphenols on gut microbiota and implications in human health. *The Journal of Nutritional Biochemistry*, 24, 1415–22.

Casanas, R., Gonzalez, M., Rodriguez, E., Marrero, A. and Diaz, C. 2002. Chemometric studies of chemical compounds in five cultivars of potatoes from Tenerife. *Journal of Agricultural and Food Chemistry*, 50, 2076–82.

Chakauya, E., Coxon, K. M., Wei, M., Macdonald, M. V., Barsby, T., Abell, C. and Smith, A. G. 2008. Towards engineering increased pantothenate (vitamin B(5)) levels in plants. *Plant Molecular Biology*, 68, 493–503.

Cham, B. E. and Chase, T. R. 2012. Solasodine rhamnosyl glycosides cause apoptosis in cancer cells. Do they also prime the immune system resulting in long-term protection against cancer? *Planta Medica*, 78, 349–53.

Cham, B. E., Daunter, B. and Evans, R. A. 1991. Topical treatment of malignant and premalignant skin lesions by very low concentrations of a standard mixture (BEC) of solasodine glycosides. *Cancer Letters*, 59, 183–92.

Charepalli, V., Reddivari, L., Radhakrishnan, S., Vadde, R., Agarwal, R. and Vanamala, J. K. 2015. Anthocyanin-containing purple-fleshed potatoes suppress colon tumorigenesis via elimination of colon cancer stem cells. *The Journal of Nutritional Biochemistry*, 26, 1641–9.

Published by Burleigh Dodds Science Publishing Limited, 2018.

Chen, X., Song, B., Liu, J., Yang, J., He, T., Lin, Y., Zhang, H. and Xie, C. 2012. Modulation of gene expression in cold-induced sweetening resistant potato species *Solanum berthaultii* exposed to low temperature. *Molecular Genetics and Genomics*, 287, 411–21.

Chitchumroonchokchai, C., Diretto, G., Parisi, B., Giuliano, G. and Failla, M. L. 2017. Potential of golden potatoes to improve vitamin A and vitamin E status in developing countries. *PLoS ONE*, 12, e0187102.

Cho, K., Cho, K.-S., Sohn, H.-B., Ha, I. J., Hong, S.-Y., Lee, H., Kim, Y.-M. and Nam, M. H. 2016. Network analysis of the metabolome and transcriptome reveals novel regulation of potato pigmentation. *Journal of Experimental Botany*, 67, 1519–33.

Choi, D., Bostock, R. M., Avdiushko, S. and Hildebrand, D. F. 1994. Lipid-derived signals that discriminate wound-and pathogen-responsive isoprenoid pathways in plants: Methyl jasmonate and the fungal elicitor arachidonic acid induce different 3-hydroxy-3-methylglutaryl-coenzyme A reductase genes and antimicrobial isoprenoids in *Solanum tuberosum* L. *Proceedings of the National Academy of Sciences*, 91, 2329–33.

Chucair, A. J., Rotstein, N. P., Sangiovanni, J. P., During, A., Chew, E. Y. and Politi, L. E. 2007. Lutein and zeaxanthin protect photoreceptors from apoptosis induced by oxidative stress: Relation with docosahexaenoic acid. *Investigative Ophthalmology & Visual Science*, 48, 5168–77.

Chun, O. K., Kim, D. O., Smith, N., Schroeder, D., Han, J. T. and Lee, C. Y. 2005. Daily consumption of phenolics and total antioxidant capacity from fruit and vegetables in the American diet. *Journal of the Science of Food and Agriculture*, 85, 1715–24.

Copp, A. J. and Greene, N. D. 2010. Genetics and development of neural tube defects. *Journal of Pathology*, 220, 217–30.

Cui, C. Z., Wen, X. S., Cui, M., Gao, J., Sun, B. and Lou, H. X. 2012. Synthesis of solasodine glycoside derivatives and evaluation of their cytotoxic effects on human cancer cells. *Drug Discovery Therapy*, 6, 9–17.

Da Silva, R. P., Kelly, K. B., Al Rajabi, A. and Jacobs, R. L. 2014. Novel insights on interactions between folate and lipid metabolism. *Biofactors*, 40, 277–83.

Dahlin, P., Müller, M. C., Ekengren, S., McKee, L. S. and Bulone, V. 2017. The impact of steroidal glycoalkaloids on the physiology of phytophthora infestans, the causative agent of potato late blight. *Molecular Plant-Microbe Interactions*, 30, 531–42.

Dale, M., Griffiths, D., Bain, H. and Todd, D. 1993. Glycoalkaloid increase in Solarium tuberosum on exposure to light. *Annals of Applied Biology*, 123, 411–18.

Dale, M. F. B., Griffiths, D. W. and Todd, D. T. 2003. Effects of genotype, environment, and postharvest storage on the total ascorbate content of potato (*Solanum tuberosum*) tubers. *Journal of Agricultural and Food Chemistry*, 51, 244–8.

D'Amelia, V., Aversano, R., Batelli, G., Caruso, I., Castellano Moreno, M., Castro-Sanz, A. B., Chiaiese, P., Fasano, C., Palomba, F. and Carputo, D. (2014). High AN1 variability and interaction with basic helix-loop-helix co-factors related to anthocyanin biosynthesis in potato leaves. *The Plant Journal* 80: 527–40.

De Paiva, E. P., Marcella Melo Assis Costa, M. M. A. and De Azevedo Filho, C. A. 2015. Folate – Analytical properties, bioavailability and stability in foods. *Scientia Chromatographica*, 7, 199–222.

Delchier, N., Ringling, C., Maingonnat, J. F., Rychlik, M. and Renard, C. M. G. C. 2014. Mechanisms of folate losses during processing: Diffusion vs. heat degradation. *Food Chemistry*, 157, 439–47.

Deng, Y. and Lu, S. 2017. Biosynthesis and regulation of phenylpropanoids in plants. *Critical Reviews in Plant Sciences*, 36, 257–90.

Denslow, S. A., Rueschhoff, E. E. and Daub, M. E. 2007. Regulation of the Arabidopsis thaliana vitamin B6 biosynthesis genes by abiotic stress. *Plant Physiol Biochem*, 45, 152–61.

Depeint, F., Bruce, W. R., Shangari, N., Mehta, R. and O'Brien, P. J. 2006. Mitochondrial function and toxicity: Role of B vitamins on the one-carbon transfer pathways. *Chemico-Biological Interactions*, 163, 113–32.

Desmarchelier, C. and Borel, P. 2017. Overview of carotenoid bioavailability determinants: From dietary factors to host genetic variations. *Trends in Food Science & Technology*, 69, 270–80.

Devaux, A., Kromann, P. and Ortiz, O. 2014. Potatoes for sustainable global food security. *Potato Research*, 57, 185–99.

Diaz de la Garza, R. I., Gregory, J. F., 3rd and Hanson, A. D. 2007. Folate biofortification of tomato fruit. *Proceedings of the National Academy of Sciences USA*, 104, 4218–22.

Diretto, G., Tavazza, R., Welsch, R., Pizzichini, D., Mourgues, F., Papacchioli, V., Beyer, P. and Giuliano, G. 2006. Metabolic engineering of potato tuber carotenoids through tuber-specific silencing of lycopene epsilon cyclase. *BMC Plant Biology*, 6, 13.

Diretto, G., Al-Babili, S., Tavazza, R., Papacchioli, V., Beyer, P. and Giuliano, G. 2007. Metabolic engineering of potato carotenoid content through tuber-specific overexpression of a bacterial mini-pathway. *PLoS ONE*, 2, e350.

Dong, W., Stockwell, V. O. and Goyer, A. 2015. Enhancement of thiamin content in *Arabidopsis thaliana* by metabolic engineering. *Plant and Cell Physiology*, 56, 2285–96.

Donnelly, A. 2017. Wernicke-Korsakoff syndrome: Recognition and treatment. *Nursing Standard*, 31, 46–53.

Douce, R. and Neuburger, M. 1999. Biochemical dissection of photorespiration. *Current Opinion in Plant Biology*, 2, 214–22.

Drewnowski, A. and Eichelsdoerfer, P. 2010. Can low-income Americans afford a healthy diet? *Nutrition Today*, 44, 246–9.

Drewnowski, A. and Rehm, C. D. 2013. Vegetable cost metrics show that potatoes and beans provide most nutrients per penny. *PLoS ONE*, 8, e63277.

Ducreux, L. J. M., Morris, W. L., Hedley, P. E., Shepherd, T., Davies, H. V., Millam, S. and Taylor, M. A. 2005. Metabolic engineering of high carotenoid potato tubers containing enhanced levels of beta-carotene and lutein. *Journal of Experimental Botany*, 56, 81–9.

Espín, J. C., González-Sarrías, A. and Tomás-Barberán, F. A. 2017. The gut microbiota: A key factor in the therapeutic effects of (poly)phenols. *Biochemical Pharmacology*, 139, 82–93.

Fernandez-Orozco, R. B., Gallardo-Guerrero, L. and Hornero-Mendez, D. 2013. Carotenoid profiling in tubers of different potato (*Solanum* sp) cultivars: Accumulation of carotenoids mediated by xanthophyll esterification. *Food Chemistry*, 141, 2864–72.

Friedman, M. 2015. Chemistry and anticarcinogenic mechanisms of glycoalkaloids produced by eggplants, potatoes, and tomatoes. *Journal of Agricultural and Food Chemistry*, 63, 3323–37.

Friedman, M., Lee, K. R., Kim, H. J., Lee, I. S. and Kozukue, N. 2005. Anticarcinogenic effects of glycoalkaloids from potatoes against human cervical, liver, lymphoma, and stomach cancer cells. *Journal of Agricultural and Food Chemistry*, 53, 6162–9.

Friedman, M., McQuistan, T., Hendricks, J. D., Pereira, C. and Bailey, G. S. 2007. Protective effect of dietary tomatine against dibenzo[a,l]pyrene (DBP)-induced liver and stomach tumors in rainbow trout. *Molecular Nutrition & Food Research*, 51, 1485–91.

Fu, L. S., Doreswamy, V. and Prakash, R. 2014. The biochemical pathways of central nervous system neural degeneration in niacin deficiency. *Neural Regeneration Research*, 9, 1509–13.

Gardner, C. D., Trepanowski, J. F., Del Gobbo, L. C. Hauser, M. E., Rigdon, J., Ioannidis, J. P. A., Desai, M. and King, A. C. 2018. Effect of low-fat vs low-carbohydrate diet on 12-month weight loss in overweight adults and the association with genotype pattern or insulin secretion: The DIETFITS randomized clinical trial. *JAMA*, 319, 667–79.

GBDS. 2017. Global burden of disease study. *Lancet*, 390, http://www.thelancet.com/gbd.

Gehring, W. 2004. Nicotinic acid/niacinamide and the skin. *Journal of Cosmetic Dermatology*, 3, 88–93.

Ginzberg, I., Tokuhisa, J. G. and Veilleux, R. E. 2009. Potato steroidal glycoalkaloids: Biosynthesis and genetic manipulation. *Potato Research*, 52, 1–15.

Golriz, F., Donnelly, L. F., Devaraj, S. and Krishnamurthy, R. 2017. Modern American scurvy – experience with vitamin C deficiency at a large children's hospital. *Pediatric Radiology*, 47, 214–20.

Gorelova, V., Ambach, L., Rebeille, F., Stove, C. and Van Der Straeten, D. 2017. Folates in plants: Research advances and progress in crop biofortification. *Frontiers in Chemistry*, 5, 21.

Goyer, A. 2010. Thiamine in plants: Aspects of its metabolism and functions. *Phytochemistry*, 71, 1615–24.

Goyer, A. and Haynes, K. G. 2011. Vitamin B1 content in potato: Effect of genotype, tuber enlargement, and storage, and estimation of stability and broad-sense heritability. *American Journal of Potato Research*, 88, 374–85.

Goyer, A. and Navarre, D. A. 2007. Determination of folate concentrations in diverse potato germplasm using a trienzyme extraction and a microbiological assay. *Journal of Agricultural and Food Chemistry*, 55, 3523–8.

Goyer, A. and Navarre, D. A. 2009. Folate is higher in developmentally younger potato tubers. *Journal of the Science of Food and Agriculture*, 89, 579–83.

Goyer, A. and Sweek, K. 2011. Genetic diversity of thiamin and folate in primitive cultivated and wild potato (*Solanum*) species. *Journal of Agricultural and Food Chemistry*, 59, 13072–80.

Grunenfelder, L. A., Knowles, L. O., Hiller, L. K. and Knowles, N. R. 2006. Glycoalkaloid development during greening of fresh market potatoes (*Solanum tuberosum* L.). *Journal of Agricultural and Food Chemistry*, 54, 5847–54.

Gunton, J. 2015. Effects of vitamins C and D in type 2 diabetes mellitus. *Nutrition and Dietary Supplements*, 7, 21–8.

Gwirtz, J. A. and Garcia-Casal, M. N. 2014. Processing maize flour and corn meal food products. *Technical Considerations for Maize Flour and Corn Meal Fortification in Public Health*, 1312, 66–75.

Habli, Z., Toumieh, G., Fatfat, M., Rahal, O. and Gali-Muhtasib, H. 2017. Emerging cytotoxic alkaloids in the battle against cancer: Overview of molecular mechanisms. *Molecules*, 22, 250.

Hairong, W. 2015. Rediscovering the value of the potato. *Beijing Review*, 7, 1–2.

Hampl, J. S., Taylor, C. A. and Johnston, C. S. 2004. Vitamin C deficiency and depletion in the United States: The Third National Health and Nutrition Examination Survey, 1988 to 1994. *American Journal of Public Health*, 94, 870–5.

Hanson, A. D. and Gregory, J. F. 2011. Folate biosynthesis, turnover, and transport in plants. *Annual Review of Plant Biology*, 62, 105–25.

Havaux, M., Ksas, B., Szewczyk, A., Rumeau, D., Franck, F., Caffarri, S. and Triantaphylides, C. 2009. Vitamin B6 deficient plants display increased sensitivity to high light and photo-oxidative stress. *BMC Plant Biology*, 9, 130.

Hellmann, H. and Mooney, S. 2010. Vitamin B6: A molecule for human health? *Molecules*, 15, 442–59.

Herrero, S., Amnuaykanjanasin, A. and Daub, M. E. 2007. Identification of genes differentially expressed in the phytopathogenic fungus Cercospora nicotianae between cercosporin toxin-resistant and -susceptible strains. *FEMS Microbiology Letters*, 275, 326–37.

Hollman, P. C. H. 2014. Unravelling of the health effects of polyphenols is a complex puzzle complicated by metabolism. *Archives of Biochemistry and Biophysics*, 559, 100–5.

Hopkins, J. 1995. The glycoalkaloids: Naturally of interest (but a hot potato?). *Food and Chemical Toxicology*, 33, 323–8.

Horovitz, B. and Waggoner, J. 2014. McDonald's same-store sales get fried. *USA Today*, December 8. https://www.usatoday.com/story/money/business/2014/12/08/mcdonalds-same-store-sales-fried/20087793/ (accessed 30 May 2018).

Hoy, M. K. and Goldman, J. D. 2012. Potassium intake of the U.S. population: What we eat in America, national health and nutrition examination survey 2009–2012. Food surveys research group dietary data brief No. 10.

Ijaz, S., Jackson, J., Thorley, H., Porter, K., Fleming, C., Richards, A., Bonner, A. and Savović, J. 2017. Nutritional deficiencies in homeless persons with problematic drinking: A systematic review. *International Journal for Equity in Health*, 16, 71.

Itkin, M., Heinig, U., Tzfadia, O., Bhide, A. J., Shinde, B., Cardenas, P. D., Bocobza, S. E., Unger, T., Malitsky, S., Finkers, R., Tikunov, Y., Bovy, A., Chikate, Y., Singh, P., Rogachev, I., Beekwilder, J., Giri, A. P. and Aharoni, A. 2013. Biosynthesis of antinutritional alkaloids in solanaceous crops is mediated by clustered genes. *Science*, 341, 175–9.

Iwanzik, W., Tevini, M., Stute, R. and Hilbert, R. 1983. Carotinoidgehalt und -zusammensetzung verschiedener deutscher Kartoffelsorten und deren Bedeutung für die Fleischfarbe der Knolle. *Potato Research*, 26, 149–62.

Jaakola, L. and Hohtola, A. 2010. Effect of latitude on flavonoid biosynthesis in plants. *Plant, Cell & Environment*, 33, 1239–47.

Jackson, S. L., Cogswell, M. E., Zhao, L., Terry, A. L., Wang, C.-Y., Wright, J., Coleman King, S. M., Bowman, B., Chen, T.-C., Merritt, R. and Loria, C. M. 2018. Association between urinary sodium and potassium excretion and blood pressure among adults in the United States. *Circulation*, 137, 237.

Jansen, G. and Flamme, W. 2006. Coloured potatoes (*Solanum tuberosum* L.) – Anthocyanin content and tuber quality. *Genetic Resources and Crop Evolution*, 53, 1321–31.

Janskey, S. H. 2010. Potato flavor. In: Hui, Y. H. (Ed.), *Handbook of Fruit and Vegetable Flavors*. John Wiley & Sons, Hoboken, NJ, pp. 935–46.

Jiang, Z., Chen, C., Wang, J., Xie, W., Wang, M., Li, X. and Zhang, X. 2016. Purple potato (*Solanum tuberosum* L.) anthocyanins attenuate alcohol-induced hepatic injury by enhancing antioxidant defense. *Journal of Natural Medicines*, 70, 45–53.

Juguan, J. A., Lukito, W. and Schultink, W. 1999. Thiamine deficiency is prevalent in a selected group of urban Indonesian elderly people. *Journal of Nutrition*, 129, 366–71.

Jung, C. S., Griffiths, H. M., De Jong, D. M., Cheng, S., Bodis, M. and De Jong, W. S. 2005. The potato *P* locus codes for flavonoid 3′,5′-hydroxylase. *Theoretical and Applied Genetics*, 110, 269–75.

Jung, C. S., Griffiths, H. M., De Jong, D. M., Cheng, S., Bodis, M., Kim, T. S. and De Jong, W. S. 2009. The potato developer (*D*) locus encodes an R2R3 MYB transcription factor that regulates expression of multiple anthocyanin structural genes in tuber skin. *Theoretical and Applied Genetics*, 120, 45–57.

Kandutsch, A. A. and Baumann, C. A. 1953. Factors affecting the stability of thiamine in a typical laboratory diet. *Journal of Nutrition*, 49, 209–19.

Kaspar, K. L., Park, J. S., Brown, C. R., Mathison, B. D., Navarre, D. A. and Chew, B. P. 2011. Pigmented potato consumption alters oxidative stress and inflammatory damage in men. *Journal of Nutrition*, 141, 108–11.

Katoh, A., Uenohara, K., Akita, M. and Hashimoto, T. 2006. Early steps in the biosynthesis of NAD in Arabidopsis start with aspartate and occur in the plastid. *Plant Physiology*, 141, 851–7.

Kawabata, K., Mukai, R. and Ishisaka, A. 2015. Quercetin and related polyphenols: New insights and implications for their bioactivity and bioavailability. *Food & Function*, 6, 1399–417.

Keijbets, M. J. H. and Ebbenhorst-Seller, G. 1990. Loss of vitamin C (L-ascorbic acid) during long-term cold storage of Dutch table potatoes. *Potato Research*, 33, 125–30.

Kobayashi, S., Goto-Yamamoto, N. and Hirochika, H. 2004. Retrotransposon-induced mutations in grape skin color. *Science*, 304, 982.

Koes, R., Verweij, W. and Quattrocchio, F. 2005. Flavonoids: A colorful model for the regulation and evolution of biochemical pathways. *Trends Plant Science*, 10, 236–42.

Kohn, H. I. 1938. The concentration of coenzyme-like substance in blood following the administration of nicotinic acid to normal individuals and pellagrins. *Biochemical Journal*, 32, 2075–83.

Kossmann, J. and Lloyd, J. 2000. Understanding and influencing starch biochemistry. *Crit Rev Biochem Mol Biol*, 35, 141–96.

Krits, P., Fogelman, E. and Ginzberg, I. 2007. Potato steroidal glycoalkaloid levels and the expression of key isoprenoid metabolic genes. *Planta*, 227, 143–50.

Kulen, O., Stushnoff, C. and Holm, D. G. 2013. Effect of cold storage on total phenolics content, antioxidant activity and vitamin C level of selected potato clones. *Journal of the Science of Food and Agriculture*, 93, 2437–44.

Lafta, A. M. and Lorenzen, J. H. 2000. Influence of high temperature and reduced irradiance on glycoalkaloid levels in potato leaves. *Journal of the American Society for Horticultural Science*, 125, 563–6.

Laing, W. A., Wright, M. A., Cooney, J. and Bulley, S. M. 2007. The missing step of the L-galactose pathway of ascorbate biosynthesis in plants, an L-galactose guanyltransferase, increases leaf ascorbate content. *Proceedings of the National Academy of Sciences of the United States of America.*, 104(22), 9534–9.

Lee, K. R., Kozukue, N., Han, J. S., Park, J. H., Chang, E. Y., Baek, E. J., Chang, J. S. and Friedman, M. 2004. Glycoalkaloids and metabolites inhibit the growth of human colon (HT29) and liver (HepG2) cancer cells. *Journal of Agricultural and Food Chemistry*, 52, 2832–9.

Lee, S. T., Wong, P. F., He, H., Hooper, J. D. and Mustafa, M. R. 2013. Alpha-tomatine attenuation of in vivo growth of subcutaneous and orthotopic xenograft tumors of human prostate carcinoma PC-3 cells is accompanied by inactivation of nuclear factor-kappa B signaling. *PLoS ONE*, 8, e57708.

Legrand, D. and Scheen, A. J. 2007. Does coffee protect against type 2 diabetes? *Revue Medicale De Liege*, 62, 554–9.

Lewis, C. E., Walker, J. R. L., Lancaster, J. E. and Sutton, K. H. 1998. Determination of anthocyanins, flavonoids and phenolic acids in potatoes. I. Coloured cultivars of *Solanum tuberosum* L. *Journal of the Science of Food and Agriculture*, 77, 45–57.

Li, G. M., Presnell, S. R. and Gu, L. 2003. Folate deficiency, mismatch repair-dependent apoptosis, and human disease. *Journal of Nutritional Biochemistry*, 14, 568–75.

Li, Y., Tang, W., Chen, J., Jia, R., Ma, L., Wang, S., Wang, J., Shen, X., Chu, Z., Zhu, C. and Ding, X. (2016). Development of marker-free transgenic potato tubers enriched in caffeoylquinic acids and flavonols. *Journal of Agricultural Food Chemistry* 64: 2932–40.

Liang, C. H., Shiu, L. Y., Chang, L. C., Sheu, H. M., Tsai, E. M. and Kuo, K. W. 2008. Solamargine enhances HER2 expression and increases the susceptibility of human lung cancer H661 and H69 cells to trastuzumab and epirubicin. *Chemical Research in Toxicology*, 21, 393–9.

Licciardello, F., Lombardo, S., Rizzo, V., Pitino, I., Pandino, G., Strano, M. G., Muratore, G., Restuccia, C. and Mauromicale, G. 2018. Integrated agronomical and technological approach for the quality maintenance of ready-to-fry potato sticks during refrigerated storage. *Postharvest Biology and Technology*, 136, 23–30.

Linderborg, K. M., Salo, J. E., Kalpio, M., Vuorinen, A. L., Kortesniemi, M., Griinari, M., Viitanen, M., Yang, B. and Kallio, H. 2016. Comparison of the postprandial effects of purple-fleshed and yellow-fleshed potatoes in healthy males with chemical characterization of the potato meals. *International Journal of Food Sciences and Nutrition*, 67, 581–91.

Liu, Y., Lin-Wang, K., Deng, C., Warran, B., Wang, L., Yu, B., Yang, H., Wang, J., Espley, R. V., Zhang, J., Wang, D. and Allan, A. C. 2015. Comparative transcriptome analysis of white and purple potato to identify genes involved in anthocyanin biosynthesis. *PLoS ONE*, 10, e0129148.

Liu, Y., Lin-Wang, K., Espley, R. V., Wang, L., Yang, H., Yu, B., Dare, A., Varkonyi-Gasic, E., Wang, J., Zhang, J., Wang, D. and Allan, A. C. 2016. Functional diversification of the potato R2R3 MYB anthocyanin activators AN1, MYBA1, and MYB113 and their interaction with basic helix-loop-helix cofactors. *Journal of Experimental Botany*, 67, 2159–76.

Liu, Y., Wang, L., Zhang, J., Yu, B., Wang, J. and Wang, D. 2017. The MYB transcription factor StMYBA1 from potato requires light to activate anthocyanin biosynthesis in transgenic tobacco. *Journal of Plant Biology*, 60, 93–101.

Lojzova, L., Riddellova, K., Hajslova, J., Zrostlikova, J., Schurek, J. and Cajka, T. 2009. Alternative GC-MS approaches in the analysis of substituted pyrazines and other volatile aromatic compounds formed during Maillard reaction in potato chips. *Analytica Chimica Acta*, 641, 101–9.

Lovat, C., Nassar, A. M. K., Kubow, S., Li, X.-Q. and Donnelly, D. J. 2016. Metabolic biosynthesis of potato (*Solanum tuberosum* L.) antioxidants and implications for human health. *Critical Reviews in Food Science and Nutrition*, 56, 2278–303.

Love, S. L. and Pavek, J. J. 2008. Positioning the potato as a primary food source of vitamin C. *American Journal of Potato Research*, 83, 171–80.

Love, S. L., Salaiz, T., Shafii, B., Price, W. J., Mosley, A. R. and Thornton, R. E. 2004. Stability of expression and concentration of ascorbic acid in North American potato germplasm. *HortScience* 39, 156–60.

Lv, C., Kong, H., Dong, G., Liu, L., Tong, K., Sun, H., Chen, B., Zhang, C. and Zhou, M. 2014. Antitumor efficacy of alpha-solanine against pancreatic cancer in vitro and in vivo. *PLoS ONE*, 9, e87868.

Macknight, R. C., Laing, W. A., Bulley, S. M., Broad, R. C., Johnson, A. A. T. and Hellens, R. P. 2017. Increasing ascorbate levels in crops to enhance human nutrition and plant abiotic stress tolerance. *Current Opinion in Biotechnology*, 44, 153–60.

Malmberg, A. and Theander, O. 1985. Determination of chlorogenic acid in potato tubers. *Journal of Agricultural and Food Chemistry*, 33, 549–51.

Manach, C., Scalbert, A., Morand, C., Remesy, C. and Jimenez, L. 2004. Polyphenols: Food sources and bioavailability. *American Journal of Clinical Nutrition*, 79, 727–47.

Manzetti, S., Zhang, J. and Van Der Spoel, D. 2014. Thiamin function, metabolism, uptake, and transport. *Biochemistry*, 53, 821–35.

Mayland, C. R., Bennett, M. I. and Allan, K. 2005. Vitamin C deficiency in cancer patients. *Palliative Medicine*, 19, 17–20.

McCue, K. F., Shepherd, L. V. T., Allen, P. V., Maccree, M. M., Rockhold, D. R., Corsini, D. L., Davies, H. V. and Belknap, W. R. 2005. Metabolic compensation of steroidal glycoalkaloid biosynthesis in transgenic potato tubers: Using reverse genetics to confirm the in vivo enzyme function of a steroidal alkaloid galactosyltransferase. *Plant Science*, 168, 267–73.

McCue, K. F., Allen, P. V., Shepherd, L. V. T., Blake, A., Whitworth, J., Maccree, M. M., Rockhold, D. R., Stewart, D., Davies, H. V. and Belknap, W. R. 2006. The primary in vivo steroidal alkaloid glucosyltransferase from potato. *Phytochemistry*, 67, 1590–7.

McCue, K. F., Allen, P. V., Shepherd, L. V. T., Blake, A., Malendia Maccree, M., Rockhold, D. R., Novy, R. G., Stewart, D., Davies, H. V. and Belknap, W. R. 2007. Potato glycosterol rhamnosyltransferase, the terminal step in triose side-chain biosynthesis. *Phytochemistry*, 68, 327–34.

McKillop, D. J., Pentieva, K., Daly, D., McPartlin, J. M., Hughes, J., Strain, J. J., Scott, J. M. and McNulty, H. 2002. The effect of different cooking methods on folate retention in various foods that are amongst the major contributors to folate intake in the UK diet. *British Journal of Nutrition*, 88, 681–8.

McMillan, M. and Thompson, J. C. 1979. An outbreak of suspected solanine poisoning in schoolboys: Examinations of criteria of solanine poisoning. *Quarterly Journal of Medicine*, 48, 227–43.

Mehrtens, F., Kranz, H., Bednarek, P. and Weisshaar, B. 2005. The Arabidopsis transcription factor MYB12 is a flavonol-specific regulator of phenylpropanoid biosynthesis. *Plant physiology*, 138, 1083–96.

Mekapogu, M., Sohn, H.-B., Kim, S.-J., Lee, Y.-Y., Park, H.-M., Jin, Y.-I., Hong, S.-Y., Suh, J.-T., Kweon, K., Jeong, J.-C., Kwon, O.-K. and Kim, Y.-H. (2016). Effect of light quality on the expression of glycoalkaloid biosynthetic genes contributing to steroidal glycoalkaloid accumulation in potato. *American Journal of Potato Research* 93: 264–77.

Mellidou, I., Chagné, D., Laing, W. A., Keulemans, J. and Davey, M. W. 2012. Allelic variation in paralogs of GDP-l-galactose phosphorylase is a major determinant of vitamin C concentrations in apple fruit. *Plant Physiology*, 160, 1613–29.

Mooney, S. and Hellmann, H. 2010. Vitamin B6: Killing two birds with one stone? *Phytochemistry*, 71, 495–501.

Mooney, S., Chen, L., Kuhn, C., Navarre, R., Knowles, N. R. and Hellmann, H. 2013. Genotype-specific changes in vitamin B6 content and the PDX family in potato. *BioMed Research International*, 2013, 389723.

Morris, S. and Petermann, J. 1985. Genetic and environmental effects on levels of glycoalkaloids in cultivars of potato (*Solanum tuberosum* L.). *Food Chemistry*, 18, 271–82.

Mosdøl, A., Erens, B. and Brunner, E. J. 2008. Estimated prevalence and predictors of vitamin C deficiency within UK's low-income population. *Journal of Public Health*, 30, 456–60.

Moser, M. and Chun, O. 2016. Vitamin C and heart health: A review based on findings from epidemiologic studies. *International Journal of Molecular Sciences*, 17, 1328.

Mozaffarian, D., Hao, T., Rimm, E. B., Willett, W. C. and Hu, F. B. 2011. Changes in diet and lifestyle and long-term weight gain in women and men. *New England Journal of Medicine*, 364, 2392–404.

Nahar, N., Westerberg, E., Arif, U., Huchelmann, A., Olarte Guasca, A., Beste, L., Dalman, K., Dutta, P. C., Jonsson, L. and Sitbon, F. 2017. Transcript profiling of two potato cultivars

Published by Burleigh Dodds Science Publishing Limited, 2018.

during glycoalkaloid-inducing treatments shows differential expression of genes in sterol and glycoalkaloid metabolism. *Scientific Reports*, 7, 43268.

Naqvi, S., Zhu, C., Farre, G., Ramessar, K., Bassie, L., Breitenbach, J., Perez Conesa, D., Ros, G., Sandmann, G., Capell, T. and Christou, P. 2009. Transgenic multivitamin corn through biofortification of endosperm with three vitamins representing three distinct metabolic pathways. *Proceedings of the National Academy of Sciences USA*, 106, 7762–7.

Nassar, A. M. K., Kubow, S., Leclerc, Y. and Donnelly, D. J. 2014. Somatic mining for phytonutrient improvement of 'Russet Burbank' potato. *American Journal of Potato Research*, 91, 89–100.

Navarre, D. A., Pillai, S., Shakya, R. and Holden, M. J. 2011. HPLC profiling of phenolics in diverse potato genotypes. *Food Chemistry*, 127, 34–41.

Niggeweg, R., Michael, A. J. and Martin, C. 2004. Engineering plants with increased levels of the antioxidant chlorogenic acid. *Nature Biotechnology*, 22, 746–54.

Nunes, A. C., Kalkmann, D. C. and Aragao, F. J. 2009. Folate biofortification of lettuce by expression of a codon optimized chicken GTP cyclohydrolase I gene. *Transgenic Research*, 18, 661–7.

Nunn, N. and Qian, N. 2011. The potato's contribution to population and urbanization: Evidence from a historical experiment. *Quarterly Journal Economics*, 126, 593–650.

Nützmann, H. W., Huang, A. and Osbourn, A. 2016. Plant metabolic clusters–from genetics to genomics. *New Phytologist*, 211, 771–89.

Oakley, G. P., Jr. 2009. The scientific basis for eliminating folic acid-preventable spina bifida: A modern miracle from epidemiology. *Annals of Epidemiology*, 19, 226–30.

Ombra, M. N., Fratianni, F., Granese, T., Cardinale, F., Cozzolino, A. and Nazzaro, F. 2015. In vitro antioxidant, antimicrobial and anti-proliferative activities of purple potato extracts (*Solanum tuberosum* cv *Vitelotte noire*) following simulated gastro-intestinal digestion. *Natural Product Research*, 29, 1087–91.

Ottenhof, H. H., Ashurst, J. L., Whitney, H. M., Saldanha, S. A., Schmitzberger, F., Gweon, H. S., Blundell, T. L., Abell, C. and Smith, A. G. 2004. Organisation of the pantothenate (vitamin B5) biosynthesis pathway in higher plants. *Plant Journal*, 37, 61–72.

Pandino, G., Lombardo, S. and Mauromicale, G. 2011. Mineral profile in globe artichoke as affected by genotype, head part and environment. *Journal of the Science of Food and Agriculture*, 91, 302–8.

Parr, A. J. and Bolwell, G. P. 2000. Phenols in the plant and in man. The potential for possible nutritional enhancement of the diet by modifying the phenols content or profile. *Journal of the Science of Food and Agriculture*, 80, 985–1012.

Payyavula, R. S., Singh, R. K. and Navarre, D. A. 2013. Transcription factors, sucrose, and sucrose metabolic genes interact to regulate potato phenylpropanoid metabolism. *Journal of Experimental Botany*, 64, 5115–31.

Payyavula, R. S., Shakya, R., Sengoda, V. G., Munyaneza, J. E., Swamy, P. and Navarre, D. A. 2015. Synthesis and regulation of chlorogenic acid in potato: Rerouting phenylpropanoid flux in HQT-silenced lines. *Plant Biotechnology Journal*, 13, 551–64.

Pearson, J., Pullar, J., Wilson, R., Spittlehouse, J., Vissers, M., Skidmore, P., Willis, J., Cameron, V. and Carr, A. 2017. Vitamin C status correlates with markers of metabolic and cognitive health in 50-year-olds: Findings of the Chalice cohort study. *Nutrients*, 9, 831.

Pehu, E., Gibson, R., Jones, M. and Karp, A. 1990. Studies on the genetic basis of resistance to potato leaf roll virus, potato virus Y and potato virus X in *Solanum brevidens* using somatic hybrids of *Solanum brevidens* and *Solanum tuberosum*. *Plant Science*, 69, 95–101.

Percival, G., Dixon, G. and Sword, A. 1994. Glycoalkaloid concentration of potato tubers following continuous illumination. *Journal of the Science of Food and Agriculture*, 66, 139–44.

Percival, G., Karim, M. and Dixon, G. 1998. Influence of light-enhanced glycoalkaloids on resistance of potato tubers to *Fusarium sulphureum* and *Fusarium solani* var. *coeruleum*. *Plant Pathology*, 47, 665–70.

Pillai, S., Navarre, D. A. and Bamberg, J. B. 2013. Analysis of polyphenols, anthocyanins and carotenoids in tubers from *Solanum tuberosum* group Phureja, Stenotomum and Andigena. *American Journal of Potato Research*, 90, 440–50.

Ramdath, D. D., Padhi, E., Hawke, A., Sivaramalingam, T. and Tsao, R. 2014. The glycemic index of pigmented potatoes is related to their polyphenol content. *Food and Function*, 5, 909–15.

Raschke, M., Boycheva, S., Crevecoeur, M., Nunes-Nesi, A., Witt, S., Fernie, A. R., Amrhein, N. and Fitzpatrick, T. B. 2011. Enhanced levels of vitamin B(6) increase aerial organ size and positively affect stress tolerance in Arabidopsis. *Plant Journal*, 66, 414–32.

Ray, D. K., Mueller, N. D., West, P. C. and Foley, J. A. 2013. Yield trends are insufficient to double global crop production by 2050. *PLoS ONE*, 8, e66428.

Reddivari, L., Vanamala, J., Chintharlapalli, S., Safe, S. H. and Miller, J. C., Jr. 2007. Anthocyanin fraction from potato extracts is cytotoxic to prostate cancer cells through activation of caspase-dependent and caspase-independent pathways. *Carcinogenesis*, 28, 2227–35.

Reddivari, L., Vanamala, J., Safe, S. H. and Miller, J. C., Jr. 2010. The bioactive compounds alpha-chaconine and gallic acid in potato extracts decrease survival and induce apoptosis in LNCaP and PC3 prostate cancer cells. *Nutrition and Cancer*, 62, 601–10.

Rees, E. and Gowing, L. R. 2013. Supplementary thiamine is still important in alcohol dependence. *Alcohol and Alcoholism*, 48, 88–92.

Reyes, L. F., Miller, J. C. and Cisneros-Zevallos, L. 2004. Environmental conditions influence the content and yield of anthocyanins and total phenolics in purple- and red-flesh potatoes during tuber development. *American Journal of Potato Research*, 81, 187–93.

Riaz, M., Asif, M. and Ali, R. 2009. Stability of vitamins during extrusion. *Critical Reviews in Food Science and Nutrition*, 49, 361–8.

Rivero, R. C., Hernández, P. S., Rodríguez, E. M. R. G., Martín, J. D. and Romero, C. D. A. 2003. Mineral concentrations in cultivars of potatoes. *Food Chemistry*, 83, 247–53.

Roje, S. 2007. Vitamin B biosynthesis in plants. *Phytochemistry*, 68, 1904–21.

Rolfe, H. M. 2014. A review of nicotinamide: Treatment of skin diseases and potential side effects. *Journal of Cosmetic Dermatology*, 13, 324–8.

Sanchez-Castillo, C. P., Dewey, P. J. S., Aguirre, A., Lara, J. J., Vaca, R., Barra, P. L. D. L., Ortiz, M., Escamilla, I. and James, W. P. T. 1998. The mineral content of Mexican fruits and vegetables. *Journal of Food Composition Analysis*, 11, 340–56.

Sanford, L., Kobayashi, R., Deahl, K. and Sinden, S. 1997. Diploid and tetraploid *Solanum chacoense* genotypes that synthesize leptine glycoalkaloids and deter feeding by Colorado potato beetle. *American Potato Journal*, 74, 15–21.

Sansano, M., Juan-Borras, M., Escriche, I., Andres, A. and Heredia, A. 2015. Effect of pretreatments and air-frying, a novel technology, on acrylamide generation in fried potatoes. *Journal of Food Science*, 80, T1120–8.

Sawai, S., Ohyama, K., Yasumoto, S., Seki, H., Sakuma, T., Yamamoto, T., Takebayashi, Y., Kojima, M., Sakakibara, H., Aoki, T., Muranaka, T., Saito, K. and Umemoto, N. 2014. Sterol side chain reductase 2 is a key enzyme in the biosynthesis of cholesterol, the common precursor of toxic steroidal glycoalkaloids in potato. *The Plant Cell*, 26, 3763–74.

Scalbert, A., Manach, C., Morand, C., Remesy, C. and Jimenez, L. 2005. Dietary polyphenols and the prevention of diseases. *Critical Reviews in Food Science and Nutrition*, 45, 287–306.

Schulz, E., Tohge, T., Zuther, E., Fernie, A. R. and Hincha, D. K. 2016. Flavonoids are determinants of freezing tolerance and cold acclimation in Arabidopsis thaliana. *Scientific Reports*, 6. doi:10.1038/srep34027.

Shakya, R. and Navarre, D. A. 2008. LC-MS analysis of solanidane glycoalkaloid diversity among tubers of four wild potato species and three cultivars (*Solanum tuberosum*). *Journal of Agricultural and Food Chemistry*, 56, 6949–58.

Shen, K.-H., Hung, J.-H., Chang, C.-W., Weng, Y.-T., Wu, M.-J. and Chen, P.-S. 2017. Solasodine inhibits invasion of human lung cancer cell through downregulation of miR-21 and MMPs expression. *Chemico-Biological Interactions*, 268, 129–35.

Shih, Y. W., Chen, P. S., Wu, C. H., Jeng, Y. F. and Wang, C. J. 2007. Alpha-chaconine-reduced metastasis involves a PI3K/Akt signaling pathway with downregulation of NF-kappaB in human lung adenocarcinoma A549 cells. *Journal of Agricultural and Food Chemistry*, 55, 11035–43.

Published by Burleigh Dodds Science Publishing Limited, 2018.

Shiu, L. Y., Liang, C. H., Chang, L. C., Sheu, H. M., Tsai, E. M. and Kuo, K. W. 2009. Solamargine induces apoptosis and enhances susceptibility to trastuzumab and epirubicin in breast cancer cells with low or high expression levels of HER2/neu. *Bioscience Reports*, 29, 35–45.

Sido, A., Radhakrishnan, S., Kim, S. W., Eriksson, E., Shen, F., Li, Q., Bhat, V., Reddivari, L. and Vanamala, J. K. 2017. A food-based approach that targets interleukin-6, a key regulator of chronic intestinal inflammation and colon carcinogenesis. *The Journal of Nutritional Biochemistry*, 43, 11–17.

Sinden, S. L., Sanford, L. L. and Webb, R. E. 1984. Genetic and environmental control of potato glycoalkaloids. *American Potato Journal*, 61, 141–56.

SR28. 2016. *USDA, Agricultural Research Service, Nutrient Data Laboratory* [Online]. Available:/nea/bhnrc/ndl (accessed December 2017).

Storozhenko, S., De Brouwer, V., Volckaert, M., Navarrete, O., Blancquaert, D., Zhang, G. F., Lambert, W. and Van Der Straeten, D. 2007. Folate fortification of rice by metabolic engineering. *Nature Biotechnology*, 25, 1277–9.

Stracke, R., Ishihara, H., Huep, G., Barsch, A., Mehrtens, F., Niehaus, K. and Weisshaar, B. 2007. Differential regulation of closely related R2R3-MYB transcription factors controls flavonol accumulation in different parts of the Arabidopsis thaliana seedling. *The Plant Journal*, 50, 660–77.

Stushnoff, C., Ducreux, L. J., Hancock, R. D., Hedley, P. E., Holm, D. G., McDougall, G. J., McNicol, J. W., Morris, J., Morris, W. L., Sungurtas, J. A., Verrall, S. R., Zuber, T. and Taylor, M. A. 2010. Flavonoid profiling and transcriptome analysis reveals new gene-metabolite correlations in tubers of *Solanum tuberosum* L. *Journal of Experimental Botany*, 61, 1225–38.

Sulli, M., Mandolino, G., Sturaro, M., Onofri, C., Diretto, G., Parisi, B. and Giuliano, G. 2017. Molecular and biochemical characterization of a potato collection with contrasting tuber carotenoid content. *PLoS ONE*, 12, e0184143.

Sun, L., Zhao, Y., Yuan, H., Li, X., Cheng, A. and Lou, H. 2011. Solamargine, a steroidal alkaloid glycoside, induces oncosis in human K562 leukemia and squamous cell carcinoma KB cells. *Cancer Chemotherapy and Pharmacology*, 67, 813–21.

Tahiliani, A. G. and Beinlich, C. J. 1991. Pantothenic-Acid in Health and Disease. *Vitamins and Hormones-Advances in Research and Applications*, 46, 165–228.

Tan, J. S., Wang, J. J., Flood, V., Rochtchina, E., Smith, W. and Mitchell, P. 2008. Dietary antioxidants and the long-term incidence of age-related macular degeneration: The Blue Mountains Eye Study. *Ophthalmology*, 115, 334–41.

Tedone, L., Hancock, R. D., Alberino, S., Haupt, S. and Viola, R. 2004. Long-distance transport of L-ascorbic acid in potato. *BMC Plant Biology*, 4, 16.

Tibbetts, A. S. and Appling, D. R. 2010. Compartmentalization of mammalian folate-mediated one-carbon metabolism. *Annual Review of Nutrition*, 30, 57–81.

Titiz, O., Tambasco-Studart, M., Warzych, E., Apel, K., Amrhein, N., Laloi, C. and Fitzpatrick, T. B. 2006. PDX1 is essential for vitamin B6 biosynthesis, development and stress tolerance in Arabidopsis. *Plant Journal*, 48, 933–46.

Tohge, T. and Fernie, A. R. 2017. An overview of compounds derived from the shikimate and phenylpropanoid pathways and their medicinal importance. *Mini Reviews in Medicinal Chemistry*, 17, 1013–27.

Tosun, B. N. and Yücecan, S. 2008. Influence of commercial freezing and storage on vitamin C content of some vegetables. *International Journal of Food Science and Technology*, 43, 316–21.

Tresserra-Rimbau, A., Arranz, S. and Vallverdu-Queralt, A. 2017. New insights into the benefits of polyphenols in chronic diseases. *Oxidative Medicine and Cellular Longevity*, 2017.

Tudela, J. A., Cantos, E., Espín, J. C., Tomás-Barberán, F. A. and Gil, M. I. 2002. Induction of antioxidant flavonol biosynthesis in fresh-cut potatoes. Effect of domestic cooking. *Journal of Agricultural and Food Chemistry*, 50, 5925–31.

Umemoto, N., Nakayasu, M., Ohyama, K., Yotsu-Yamashita, M., Mizutani, M., Seki, H., Saito, K. and Muranaka, T. 2016. Two cytochrome P450 monooxygenases catalyze early hydroxylation steps in the potato steroid glycoalkaloid biosynthetic pathway. *Plant Physiology*, 171(4), 2458–67.

Published by Burleigh Dodds Science Publishing Limited, 2018.

Valcarcel, J., Reilly, K., Gaffney, M. and O'Brien, N. 2015. Total carotenoids and l-ascorbic acid content in 60 varieties of potato (*Solanum tuberosum* L.) grown in Ireland. *Potato Research*, 58, 29–41.

Valiñas, M. A., Lanteri, M. L., Ten Have, A. and Andreu, A. B. 2017. Chlorogenic acid, anthocyanin and flavan-3-ol biosynthesis in flesh and skin of Andean potato tubers (*Solanum tuberosum* subsp. *andigena*). *Food Chemistry*, 229, 837–46.

Valkonen, J. P. T., Keskitalo, M., Vasara, T. and Pietila, L. 1996. Potato glycoalkaloids: A burden or a blessing? *Critical reviews in plant sciences*, 15 (1), 1–20.

Van Eck, J., Conlin, B., Garvin, D. F., Mason, H., Navarre, D. A. and Brown, C. R. 2007. Enhancing beta-carotene content in potato by RNAi-mediated silencing of the beta-carotene hydroxylase gene. *American Journal of Potato Research*, 84(4), 331–42.

van Niekerk, C., Schönfeldt, H., Hall, N. and Pretorius, B. 2016. The role of biodiversity in food security and nutrition: A potato cultivar case study. *Food and Nutrition Sciences*, 7, 371.

Vayda, M. 1994. *Environmental Stress and its Impact on Potato Yield*. CABI, Wallingford, UK.

Vinson, J. A., Demkosky, C. A., Navarre, D. A. and Smyda, M. A. 2012. High-antioxidant potatoes: Acute in vivo antioxidant source and hypotensive agent in humans after supplementation to hypertensive subjects. *Journal of Agricultural and Food Chemistry*, 60, 6749–54.

Viola, R., Vreugdenhil, D., Davies, H. V. and Sommerville, L. 1998. Accumulation of L-ascorbic acid in tuberising stolon tips of potato (*Solanum tuberosum* L). *Journal of Plant Physiology*, 152, 58–63.

Vogt, T. 2010. Phenylpropanoid biosynthesis. *Molecular Plant*, 3, 2–20.

Wahlberg, G., Adamson, U. and Svensson, J. 2000. Pyridine nucleotides in glucose metabolism and diabetes: A review. *Diabetes/Metabolism Research and Reviews*, 16, 33–42.

Walczak-Jedrzejowska, R., Wolski, J. K. and Slowikowska-Hilczer, J. 2013. The role of oxidative stress and antioxidants in male fertility. *Central European Journal of Urology*, 66, 60–7.

Wang, S., Moustaid-Moussa, N., Chen, L., Mo, H., Shastri, A., Su, R., Bapat, P., Kwun, I. and Shen, C.-L. 2014. Novel insights of dietary polyphenols and obesity. *The Journal of Nutritional Biochemistry*, 25, 1–18.

Wang, X., Zou, S., Lan, Y.-L., Xing, J.-S., Lan, X.-Q. and Zhang, B. 2017. Solasonine inhibits glioma growth through anti-inflammatory pathways. *American Journal of Translational Research*, 9, 3977–89.

Webb, M. E., Smith, A. G. and Abell, C. 2004. Biosynthesis of pantothenate. *Natural Product Reports*, 21, 695–721.

Wheeler, G. L., Jones, M. A. and Smirnoff, N. 1998. The biosynthetic pathway of vitamin C in higher plants. *Nature*, 393, 365–9.

Williams, A. and Ramsden, D. 2005. Nicotinamide: A double edged sword. *Parkinsonism & Related Disorders*, 11, 413–20.

Wills, R. B. H., Lim, J. S. K. and Greenfield, H. 1984. Variation in nutrient composition of Australian retail potatoes over a 12-month period. *Journal of the Science of Food and Agriculture*, 35, 1012–17.

Wilson, G. S. 1959. A small outbreak of solanine poisoning. *Monthly Bulletin of the Ministry of Health and the Public Health Laboratory Service*, 18, 207–10.

Wojtczak, L. and Slyshenkov, V. S. 2003. Protection by pantothenic acid against apoptosis and cell damage by oxygen free radicals – The role of glutathione (Reprinted from Thiol Metabolism and Redox Regulation of Cellular Functions). *Biofactors*, 17, 61–73.

Wolters, A.-M. A., Uitdewilligen, J. G. A. M. L., Kloosterman, B. A., Hutten, R. C. B., Visser, R. G. F. and Van Eck, H. J. 2010. Identification of alleles of carotenoid pathway genes important for zeaxanthin accumulation in potato tubers. *Plant Molecular Biology*, 73, 659–71.

Wolucka, B. A. and Montagu, M. V. 2007. The VTC2 cycle and the de novo biosynthesis pathways for vitamin C in plants: An opinion. *Phytochemistry*, 68, 2602–13.

Wu, J., Wang, L., Du, X., Sun, Q., Wang, Y., Li, M., Zang, W., Liu, K. and Zhao, G. 2018. α-solanine enhances the chemosensitivity of esophageal cancer cells by inducing microRNA-138 expression. *Oncology Reports*, 39, 1163–72.

Yamdeu, G., Joseph Hubert, Mankad, P. M., Shah, A. K., Patel, N. J., Acharya, R. R. and Talati, J. G. 2017. Effect of storage temperature on vitamin C, total phenolics, UPLC phenolic acid profile and antioxidant capacity of eleven potato (*Solanum tuberosum*) varieties. *Horticultural Plant Journal*, 3, 73–89.

Yamdeu Galani, J. H., Mankad, P. M., Shah, A. K., Patel, N. J., Acharya, R. R. and Talati, J. G. 2017. Effect of storage temperature on vitamin C, total phenolics, UPLC phenolic acid profile and antioxidant capacity of eleven potato (*Solanum tuberosum*) varieties. *Horticultural Plant Journal*, 3, 73–89.

Yasmin, A., Jalbani, A. A. and Raza, S. 2011. Effect of growth regulators on meristem tip culture of local potato CVS desiree and patrones. *Pakistan Journal of Agriculture, Agricultural Engineering & Veterinary Sciences*, 27, 143–9.

Yelken, B. Ö., Balcı, T., Süslüer, S. Y., Kayabaşı, Ç., Avcı, Ç. B., KırmıZıbayrak, P. B. and Gündüz, C. 2017. The effect of tomatine on metastasis related matrix metalloproteinase (MMP) activities in breast cancer cell model. *Gene*, 627, 408–11.

Zeng, X., Kong, R. P., Cheng, K. W., Du, Y., Tang, Y. S., Chu, I. K., Lo, C., Sze, K. H., Chen, F. and Wang, M. 2010. Direct trapping of acrylamide as a key mechanism for niacin's inhibitory activity in carcinogenic acrylamide formation. *Chemical Research in Toxicology*, 23, 802–7.

Zhang, Y., Cheng, S., De Jong, D., Griffiths, H., Halitschke, R. and De Jong, W. 2009. The potato *R* locus codes for dihydroflavonol 4-reductase. *Theoretical and Applied Genetics*, 119, 931–7.

Zhang, Y. F., Liu, B., Li, X. H., Ouyang, Z. G., Huang, L., Hong, Y. B., Zhang, H. J., Li, D. Y. and Song, F. M. 2014. The de novo biosynthesis of vitamin B6 is required for disease resistance against Botrytis cinerea in tomato. *Molecular Plant-Microbe Interactions*, 27, 688–99.

Zhang, Y., Jin, X., Ouyang, Z., Li, X., Liu, B., Huang, L., Hong, Y., Zhang, H., Song, F. and Li, D. 2015. Vitamin B6 contributes to disease resistance against Pseudomonas syringae pv. tomato DC3000 and Botrytis cinerea in Arabidopsis thaliana. *Journal of Plant Physiology*, 175, 21–5.

Zhang, H., Yang, B., Liu, J., Guo, D., Hou, J., Chen, S., Song, B. and Xie, C. 2017. Analysis of structural genes and key transcription factors related to anthocyanin biosynthesis in potato tubers. *Scientia Horticulturae*, 225, 310–16.

Zhao, X., Sheng, F., Zheng, J. and Liu, R. 2011. Composition and stability of anthocyanins from purple *Solanum tuberosum* and their protective influence on Cr(VI) targeted to bovine serum albumin. *Journal of Agricultural and Food Chemistry*, 59, 7902–9.

Zupko, I., Molnar, J., Rethy, B., Minorics, R., Frank, E., Wolfling, J., Molnar, J., Ocsovszki, I., Topcu, Z., Bito, T. and Puskas, L. G. 2014. Anticancer and multidrug resistance-reversal effects of solanidine analogs synthetized from pregnadienolone acetate. *Molecules*, 19, 2061–76.

Chapter 10

Improving the breeding, cultivation and use of sweetpotato in Africa

Putri Ernawati Abidin and Edward Carey, International Potato Center (CIP), Ghana

1 Introduction

Sweetpotato, *Ipomoea batatas* (L.) Lam., is a New World crop. Austin (1988) postulated its origin as being somewhere between the Yucatan Peninsula of Mexico and the mouth of the Orinoco River in Venezuela. The crop was spread widely by Portuguese explorers to Africa, India and the East Indies in the sixteenth century (Yen, 1974; Huaman and Zhang, 1997). In many parts of sub-Saharan Africa (SSA), sweetpotato is an important low-input crop (Ewell and Mutuura, 1994; Tayo, 2000; Abidin, 2004; Abidin, et al., 2013). In addition, Bashaasha et al. (1995) found that in most places in Uganda, women play a major role in cultivating sweetpotato and they are able to distinguish the varieties (Hakiza et al., 2000).

Sweetpotato has a tendency to mutate, and this has served as important source of genetic variation (Hernandez et al., 1964; Collins and Cannon, 1983) along with its outcrossing nature and capacity to flower and set seed (Miller, 1937, 1939; Hernandez and Miller, 1964; Jones, 1965a,b). This results in new genotypes which have given rise to large numbers of landraces grown by farmers in several areas of Asia, Africa and Oceania (Yen, 1974; Carey, 1996; Carey et al., 1998; Abidin and Carey, 2001).

http://dx.doi.org/10.19103/AS.2017.0016.12

Carey et al. (1998) reported that sweetpotato production in SSA is mainly based on large numbers of landrace varieties. In recent years, as sweetpotato has become more commercially important, and as its public health value is recognised, resources have begun to be allocated to the crop. In contrast, until recently, sweetpotato was among the so-called 'orphan crops'. In China, sweetpotato crop is mostly grown for food, animal feed and processed products. In SSA, sweetpotato fulfils a food security role and has tremendous potential to contribute to reducing vitamin A deficiency (VAD) and serving as a wheat flour substitute in baked products. Orange-fleshed sweetpotato (OFSP) is an example of a bio-fortified crop, in which the micronutrient status of staple foods has been enhanced through breeding and has reached the point where impact on micronutrient status can be achieved (Bouis, 2002).

Woolfe (1992) described sweetpotato as having a low status due to its image as a subsistence crop, a 'poor man's food' or something to be eaten only in times of famine or war. This status may have been a limiting factor in its exploitation as a food of high nutritional quality. Pre- and post-harvest losses, resulting in excessive waste, have increased prices to levels unattractive to those searching for a low-cost nutritious substitute for more prestigious but expensive foods.

In many areas, the lack of cultivars with characteristics catering to consumer preferences for colour, texture, flavour and low fibre levels, combined with the difficulties of handling and storing a highly perishable commodity under tropical conditions, frequently results in inferior quality sweetpotatoes. The high levels of sweetness and the strong flavour associated with many cultivars may also limit its potential as a staple food and make it difficult to combine with other foods in a variety of dishes. Sweetpotato leaves and tips are often considered to be tough and too strongly flavoured in comparison with other green leafy vegetables. However, poor nutrition and VAD are high among children under five years of age in many countries (Black et al., 2008; Liu et al., 2012) and the OFSP has been recognised as having significant potential to contribute to alleviating the problems of hidden hunger and malnutrition. While contributing to health and food security, where there is a surplus in production, farmers can generate income from growing OFSP, thus contributing significantly to the alleviation of poverty. Just 100 g (1/2 cup) of OFSP can supply the daily vitamin A needs of young children under five years of age and vulnerable women, the groups most at risk of VAD (Hotz et al., 2011).

2 Programmes for improving sweetpotato as a crop

2.1 Breeding and crop development

A good number of OFSP farmers and bred OFSP varieties are found in SSA. One OFSP variety, Ejumula, collected from Katakwi, Uganda, in 1999 has become a favourite variety for farmers in Uganda and was reported by Abidin in 2004. She suggested this variety could be potentially used in the breeding programme to genetically improve the beta-carotene levels in new cultivars bred for the rural poor in Uganda (Abidin, 2004). All types of sweetpotatoes are good sources of vitamins C, E, K and several B vitamins but only OFSP has provitamin A (Jaarsveld et al., 2005; Nastel et al., 2006; Low et al., 2007). Research in South Africa (Jaarsveld et al., 2005) has demonstrated the efficacy of OFSP as a bio-available source of vitamin A, and community-level research in Mozambique (Low et al., 2007) demonstrated that an integrated approach using OFSP can reduce VAD in a resource-poor population. Yamakawa and Yoshimoto (2002) claimed that sweetpotato can play an important role in activating some physiological functions of the human body in order to prevent serious diseases such as diabetes, cancers, liver injury and blood pressure problems. The International Potato Center (CIP) has been involved in integrated, food-based approaches to reducing VAD since 2002 in Mozambique.

The Sweetpotato for Profit and Health Initiative (SPHI), co-led by the CIP with the Forum for Agricultural Research in Africa aims to reposition sweetpotato in the food economies of SSA. It is a multi-partner and multi-donor initiative which seeks to improve the lives of 10 million households targeting 17 countries in SSA by 2020. It aims to enhance production, improve varieties through accelerated breeding programmes, strengthen seed systems and expand the use of, and demand for, sweetpotato among both urban and rural consumers.

In Malawi, sweetpotato is ranked as the third food crop after maize and cassava. A massive distribution of vines of improved materials underlies the transformation in estimated per capita consumption of sweetpotato from 12 kg/capita in 1970, 88 kg/capita in 2003 and 240 kg/capita in 2012 (Ministry of Agriculture, unpublished). In Ghana, the area under sweetpotato cultivation is estimated to be increasing at 9% annually (Ministry of Agriculture, unpublished). The crop provides more food (194 MJ) per hectare per day than maize (145 MJ), cassava (138 MJ) or yam (94 MJ) (Scott et al., 2000). Most sweetpotato varieties have a short production cycle compared to other staple root and tuber crops such as cassava and yam (3–5 months vs. 9–24 for cassava and 8–11 for yam), although it can also be a perennial crop. The ability of sweetpotato to grow under marginal conditions, lower soil fertility requirements whilst benefiting from flexible planting and harvesting times, also contribute to its expansion. Sweetpotato requires little input and has relatively low labour requirements, making it particularly suitable for households threatened by migration, civil disorder or diseases such as HIV and AIDS (Jayne, et al., 2004). However, attention to good agricultural practices (GAP) is important to ensure good yields in many environments where soil fertility is declining.

2.2 Key research developments

The following section introduces key approaches to sweetpotato research and development and the findings of two projects under the SPHI: the Rooting Out Hunger Project in Malawi and Jumpstarting Orange-Fleshed Sweetpotato in West Africa through Diversified Markets, with activities in Ghana, Nigeria and Burkina Faso. The OFSP project in Malawi was funded by Irish Aid and effectively created demand for the crop through a social safety net-based approach where vulnerable households (e.g. those with children under five and pregnant women) were targeted through a voucher scheme. This voucher system was used to create a system of decentralised vine multipliers to effectively deliver OFSP vines to the targeted producers. The project, which is aligned with the agricultural policy of Malawi, provided a model for contributing to food and nutrition security in the country. Farmers generated incomes from, and improved their livelihoods by, cultivating OFSP.

An ongoing project in Ghana, Burkina Faso and Nigeria, West Africa, takes a market-oriented approach in considering the commercialisation of OFSP vines where diversified markets are created to provide farmers with a reason to produce. The OFSP is used as an entry point to improve the nutrition status of households by linking agriculture, health, nutrition and wealth creation. In both projects, local knowledge was considered and included in action research. Post-harvest trials on OFSP, intercropping OFSP and maize, OFSP and soya beans, OFSP and onion, and other crops have been conducted. The findings are contributing to enhancement of food security and nutrition security with the dual function of reducing VAD while improving household income (CIP, 2014).

2.3 Farmer-participatory selection approach in sweetpotato breeding

In 1999, a total of 206 accessions of farmers' varieties from five districts of north-eastern Uganda were collected along with knowledge about the varieties. Morphological diversity and duplication among the accessions were assessed and 188 were classified as distinct accessions (Abidin and Carey, 2001; Abidin 2004). Farmer participation was initially used to select the best and most promising varieties based on the farmers' preferences. This was followed by on-farm and multi-locational testing. The findings from this research work were published by Abidin et al. in 2002 and by Abidin in 2004.

Farmer-managed on-farm testing was carried out in Soroti District in north-eastern Uganda to assess the selected varieties and compared them with cultivars from the Ugandan breeding programme and local check varieties (Abidin, 2004). Multi-locational testing and stability analysis was undertaken in 20 environments over three seasons. Farmers' knowledge of sweetpotato varieties from north-eastern Uganda proved useful in selecting varieties for the breeding programme. Further discussion was presented in Abidin (2004) and Abidin et al. (2005). It was suggested that farmer participation in variety selection can contribute to accelerated selection of new varieties, taking into account the preferred attributes of farmers and end users. This was further confirmed with the release of a variety started from seed and selected by farmers in Uganda (Gibson et al., 2011).

2.4 Sweetpotato utilisation and culinary attributes

Between 1997 and 2000, CIP in Uganda worked to reposition sweetpotato from a poor man's crop to a higher status crop by improving its utilisation and processing. Food scientists made doughnuts, chapattis (tortillas), buns, bread and cakes from 30% sweetpotato flour as a substitute for wheat flour (Hagenimana and Owori, 1997; Owori et al., 1997; CIP, 1998; Owori and Hagenimana, 2000).

More recently, scientists from national programmes in SSA and CIP have developed various OFSP processed products using sweetpotato puree (mashed sweetpotato). It was noted that drying OFSP to make flour resulted in higher losses of micronutrients such as provitamin A than was the case in production of products using puree (Bechoff, 2010). Therefore, sweetpotato puree is currently promoted by CIP as a nutritionally superior ingredient. Various recipes using OFSP puree made from fresh steamed roots have been introduced and encouraged by CIP through local recipes from Malawi and Ghana in Africa (available at https://www.researchgate.net/publication/279886348 and/or/279886197). Golden bread prepared from OFSP puree is being sold by roadside bakeries in urban markets in southern, great Accra, and northern Ghana. In southern Malawi, fried bread ('mandazi') and doughnuts made from fresh steamed roots of OFSP were sold in the urban markets (the CIP blog is available at
  https://www.researchgate.net/publication/280111108_Changing_lives_in_Malawi_ with_orange-fleshed_sweetpotatoes).

2.5 Provitamin A-rich OFSP for communities in Africa

Grüneberg et al. (2009) described current challenges and a way forward in sweetpotato breeding. They proposed an accelerated breeding scheme (ABS) to resolve the challenges posed by the need to produce results under time-limited grant-funded projects. Carey et al. (2016) reviewed progress on adoption and adaptation of the ABS approach in SSA and in Peru (CIP headquarters). Following the initial description, four varieties were released in Peru in 2010. According to Carey et al. (2016), accelerated breeding has led to the release and/or registration of 51 varieties, many of them OFSP. Grüneberg et al. (2015) have reviewed advances in sweetpotato breeding from 1992 to 2012. Nevertheless, using disease-free planting material is also considered to be important in boosting the yield. The work on quality-declared planting material (QDPM) is currently introduced at the farmers' level to maintain the performance of the foundation seeds of preferred varieties by the end users (available at https://www.researchgate.net/publication/310548161 and/ or 310547055). In-depth information on linkage between breeding programmes and seed systems is presented in Section 3.

A number of multi-donor projects have been funded, including those under the SPHI: 'Sweetpotato Action for Security and Health in Africa' (SASHA), 'Rooting Out Hunger in Malawi with Nutritious Orange-fleshed Sweetpotato', 'Jumpstarting orange-fleshed sweetpotato in West Africa through diversified markets', 'Breaking post-harvest bottlenecks: long-term sweetpotato storage in adverse climates' and so on (see the Sweetpotato Knowledge Portal (www.sweetpotatoknowledge.org) for the portfolio of completed and ongoing sweetpotato projects). The projects were developed to help address the challenges described by Woolfe (1992) in achieving the full potential of sweetpotato.

A multi-partner approach is being applied to reach 10 million households in 17 countries in SSA by 2020, improving their nutritional status and incomes. Breeding programmes have been established at sweetpotato support platforms in Ghana, Mozambique and Uganda. In Ghana, the focus is on development of non-sweet quality types and drought-resistant varieties for the savanna agricultural areas. In Central and East Africa, the focus is on developing resistance to sweetpotato virus diseases (SPVD) and sweetpotato weevils (*Cylas* spp). In Southern Africa, the focus is on drought-resistant varieties. Integrated efforts, including linkages between breeding efforts and seed systems, agronomy, utilisation and marketing are a hallmark of the SPHI.

A number of research and trials were conducted to complement breeding efforts as part of an integrated crop improvement effort. Recently, a comparative study was undertaken in Ghana on sources of sweetpotato planting material for quality vines and root yield (Abidin, et al., 2016a). Another study evaluating the potential of some sweetpotato cultivars to yield both roots and vines in drought-prone parts of Ghana was done by Akansake et al. (2016). End-user attributes and yield are key traits at each sweetpotato breeding platform. A two-year case study on consumer participation in Ghana's breeding programme was recently reported by Dery et al. (2016). Many more studies on consumer preferences can also be found in a number of publications. Research bringing local knowledge into practice by farmers has also been featured. This includes intercropping sweetpotato and maize (Abidin et al., 2015), sweetpotato and onion, and sweetpotato and soya bean (Abidin et al., 2016b), and using sand storage for extending the shelf life of fresh sweetpotato roots for home consumption and market sales (Abidin, et al., 2016c). Local knowledge on pests and diseases in Uganda has been investigated and followed by research on the damage done to sweetpotato by millipedes and weevils, piecemeal versus one-time harvesting, studies of crop rotation and cultural practices to prevent pest damage (Ebregt, 2007a). Ebregt et al. (2004a,b; 2005; 2007a–c) found that millipedes can cause considerable damage in sweetpotato production and several other crops, such as maize, groundnut and cassava. They noticed that damage typically occurred at the early stages of crop development.

3 Developments in breeding and seed dissemination

Sweetpotato genotypes are propagated clonally following initial breeding from 'true' seeds. The main challenge in identifying new varieties is to select those possessing the required end-use attributes and the ability to perform well across production environments in the target agro-ecology. Using the traditional method, the sweetpotato breeding schemes anticipated a lengthy selection process of up to eight years from hybridisation to variety release. During 2004–13, a total of 89 sweetpotato varieties were released in sub-Saharan countries such as Ghana, Burkina Faso, Nigeria, Burundi, Rwanda, Uganda, Tanzania, Kenya, Mozambique, Malawi and South Africa. Sixty-two of these variety releases were OFSP (Grüneberg et al., 2015). Improvement of storage root yield is a high priority in all countries where the average yields are relatively low (<12t/ha).

Varieties resistant to drought, SPVD and sweetpotato weevil (Cylas spp.) are also needed for the region and are among breeding programme priorities. Since 2009, an ABS has been introduced through the SPHI. ABS has led to the release and/or registration of up to 51 varieties in under five years from crossing to release. This has resulted in 22 varieties in Mozambique, six varieties in Rwanda, five varieties in Malawi, five varieties in Zambia, two varieties in Uganda and five in Burkina Faso (Carey et al., 2016).

A multinational, multi-locational test of elite sweetpotato germplasm from countries in Southern and Eastern Africa was carried out in 1999 separately from those described above in Section 2.3. The research aimed at understanding selection efficiency to help improve resource allocation and accelerate the selection of new varieties (Grüneberg, 2004; Abidin, 2004). In Uganda, Mwanga et al. (2007a,b) reported that a number of sweetpotato varieties were released, and the orange-fleshed farmers' variety Ejumula collected by Abidin in 1999 (Abidin, 2004) was included among others. Although this variety is relatively susceptible to SPVD, it has been used as a parent in the breeding programme in Mozambique, improving the beta-carotene quality of the next generation of sweetpotato varieties. Other local farmer varieties have also proved to be good progenitors, giving rise to new varieties, including OFSP varieties in a number of countries.

3.1 Linking breeding programmes and seed systems for sustainability

A number of research have been conducted. The reports from these research works have indicated that there has been a large potential impact of getting beta-carotene-rich orange-fleshed varieties into the diets of rural and urban consumers, especially into those of young children. However, both white- and yellow-fleshed sweetpotatoes are commonly used by farmers in SSA. There was therefore a need to determine elements for rapid and effective dissemination of planting material (also called 'seed'). Figure 1 details elements required for a sustainable and functional seed system based on lessons learnt from Malawi and Ghana. It explains the roadmap to achieving a sustainable sweetpotato seed system through an actor-centred approach. Key elements include (i) an established 1-2-3 seed system closely linked with sweetpotato breeding efforts; (ii) improved knowledge and empowerment through training; (iii) monitoring and tracking systems; (iv) advocacy, awareness campaign, sensitisation, visits, agricultural trade fairs, meetings, action research and so on; and (v) value chains, market development and funding resources that will empower these programmes.

Figure 2 provides a schematic presentation of the so-called '1-2-3 sweetpotato seed system' used under the Rooting Out Hunger in Malawi project (1 is read as primary, 2 as secondary and 3 as tertiary). Figure 3 provides additional details with respect to these elements and their sustainable functioning within both commercial and regulatory frameworks.

In many countries of SSA, farmers are used to free planting material received from neighbours, families and friends. Typically, the quality of the vines may be poor as the cuttings often come from SPVD-infected plants (either symptomatic or asymptomatic) and/or plants infested by sweetpotato weevils (*Cylas* spp.), or from degenerated varieties. However, such informal seed systems do provide selection pressure for varieties that are able to perform reliably and provide planting material of apparently decent quality. Formal seed systems, including the use of QDPM has the potential to maintain quality seed in varieties that might otherwise degenerate due to disease

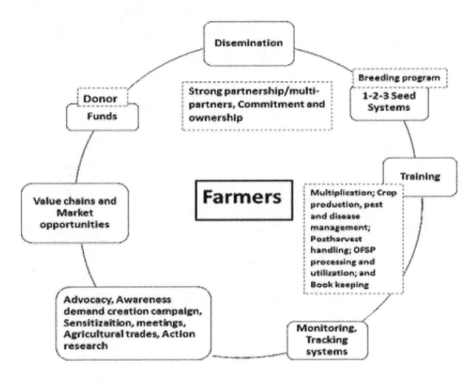

Figure 1 Elements required for functioning seed systems for sustainability.

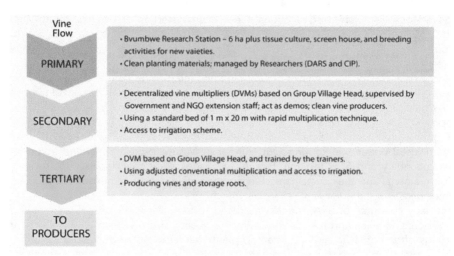

Figure 2 Schematic representation of the 1-2-3 vine multiplication system implemented in Malawi introduced by CIP-led project Rooting Out Hunger in Malawi during its OFSP intervention (2010–14).

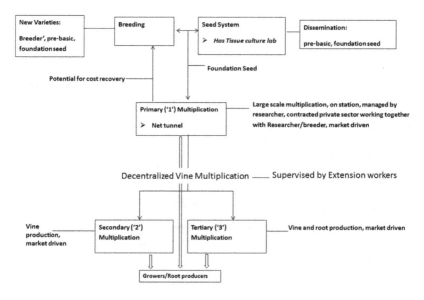

Figure 3 Linkage of seed system and breeding programme initially used in Malawi and further developed in Ghana.

and pest problems. It should be noted that provision of free planting material is not restricted to farmers, but is also frequently used by development and relief projects, and makes sense when the objective of the dissemination effort is related to food and nutrition security. However, it is difficult to sustain these subsidised schemes following the completion of projects. This is leading to consideration of the development of seed systems sustained through commercial demand for planting material, but capable of responding to humanitarian needs when necessary. Figure 3 details schematic linkages between breeding programmes and seed systems in Malawi and Ghana which are based on market-driven approaches.

3.2 The Rooting Out Hunger in Malawi with Nutritious Orange-fleshed Sweetpotato project

In October 2009, the CIP launched the 'Rooting Out Hunger in Malawi with Nutritious Orange-fleshed Sweetpotato' project after intensive consultations with various stakeholders. This was a 4.5-year programme funded by Irish Aid in Malawi. The project sought to use OFSP to improve vitamin A and energy intake for at least 70,000 households with young children, to assist 20% of the households to earn at least US$40 per year from OFSP sales, and to increase average sweetpotato yields by at least 50%. A multi-partner approach was taken to achieve project objectives. The implementing partners were Catholic Development Commission (CADECOM) in Chikwawa District; Concern Universal (CU) in Dedza, Phalombe and Mulanje Districts; and the Millennium Villages Project in Zomba District. Various relevant government agencies assisted in project implementation, including the Department of Agricultural Research Services; the Department of Agricultural Extension Services (both under the Ministry of Agriculture); the Department of Nutrition and Health, Gender, HIV & AIDS in the office of the President; and the Scaling Up Nutrition (SUN) movement coordinator at the presidential office.

The conceptual framework introduced by Low et al. (2007) (Fig. 4) was used to implement the project. Aligning the project with the millennium development goals also contributed to success in Malawi. Establishing the seed system, working through a multi-partner approach including relevant government agencies, building the capacity of farmers and implementing actors, and fostering a sense of ownership by project partners were key elements for its success.

Together with implementing partners, the sweetpotato seed system was defined to guide the project on how to implement the dissemination of OFSP vines. In the case of the OFSP project in Malawi, the market was stimulated through a voucher scheme targeting vulnerable populations, the benefits of which were well distributed through a decentralised vine multiplication (DVM) system set up so that all producers were within walking distance of their seed source.

Training, sensitisation and field days were essential elements in ensuring increased demand and adoption of OFSP. Three training modules were developed and used extensively: (i) sweetpotato planting material multiplication and root production, (ii) sweetpotato post-harvest handling, and (iii) OFSP processing and utilisation. As the work was done under low-input conditions, drip irrigation was introduced and the farmers were trained on how to use and maintain simple drip kits. A gender lens was also applied at all project stages. The implementing non-governmental organisation (NGO) partners and agricultural extension staff (4,131 people, 45% women as accumulative data after four years) were trained to be the master trainers, and were directly trained by the CIP project staff. Step-down training was managed by the NGOs to train lead farmers (24,815 farmers, 52% women; accumulative data after four years). To ensure a wider reach, a farmer-to-farmer approach was introduced, in which each farmer had to train another 10 for crop production management and another five for OFSP processing and

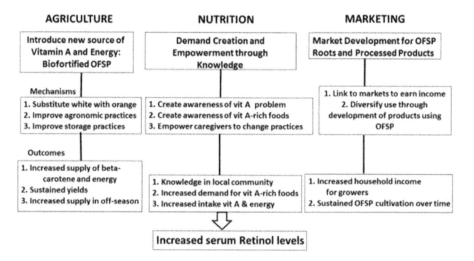

Figure 4 Conceptual framework for an integrated OFSP-led food-based approach (source: Low et al., 2007).

utilisation. Figure 5 shows the increasing number of districts and partners reached from 2009 to 2012. Training, visits and sensitisation improved knowledge on GAP which led to increased yields. As a result, the yield in 2014 was increased to 16.5 t/ha compared to 8 t/ha recorded in 2005 and 2006 (Source: Sarah Tione, Spokesperson of MoAFS, Malawi, 2013, unpublished).

In the 2013/14 rainy season, the average income per DVM was US$158 (Mk 66,360) through vouchers and US$152 from other sources, bought by international and national NGOs and by individual farmers. Table 1 shows details on number of DVM, gender and area of vine production in each rainy season in every district involved. Table 2 gives some information on the number of household beneficiaries who received vines and the revenue by DVM. The household beneficiaries also generated incomes from root and leaf sales, and OFSP processed products. Gender disaggregated data are presented as women were the important targets.

By the end of the project, in 2014, the total household of Malawians who benefited from OFSP intervention was 191,092. Of that total, 106,478 households benefited through the subsidised voucher scheme (Table 2). There has been a rule from the project that the beneficiary received 300 vine cuttings only one time from the project. The beneficiaries got some trainings on GAP including multiplication management, and OFSP processing and utilisation. The objective of the training was the beneficiaries were able to maintain the OFSP vines by their own. Furthermore, of that total, 79,617 households received OFSP vine cuttings from other projects and sources, such as USAID projects in Mulanje District; rural development programme in Balaka District; Agriculture Development Division in Machinga District; a number of NGOs working in Zomba, Nsanje, Chiradzulu Thyolo, Salima districts; Food Agriculture Organization (FAO)'s food security and nutrition project in Mzimba and Kasungu districts; Farmers Union of Malawi; farmer-based organisation; and Health Care Center in Salima. This covered all three regions of

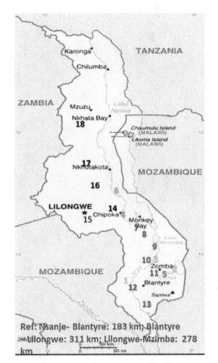

Implementing Partners and project area since Oct 2009
1. Chikhwawa under Cadecom
2. Phalombe under CU
3. Zomba under MVP
4. Dedza under CU

Implementing Partner and project area since Jan 2011
5. Mulanje under CU

Implementing Partner and project area since Sep 2011
6. Salima under Kachele Club (farmers club)

Implementing Partners and project area since Nov 2011
7. Balaka under WALA
8. Machinga under WALA
9. Zomba under WALA
10. Chiradzulu under WALA
11. Thyolo under WALA
12. Chikhwawa under WALA
13. Nsanje under WALA

Implementing Partners and project area since Jan 2012 and Dec 2013
14. Dedza under FUM
15. Lilongwe under FUM
16. Dowa under FUM
17. Kasungu under FICA-FAO, CRS-Mzuzu CADECOM
18. Mzimba under FICA-FAO, CRS-Mzuzu CADECOM

Ref: Nsanje- Blantyre: 183 km; Blantyre – Lilongwe: 311 km; Lilongwe-Mzimba: 278 km

Figure 5 Map showing where OFSP vine dissemination in Malawi has taken place. Adoption of OFSP in Malawi.

Malawi. The implementing partners (IPs) helped the multipliers (DVMs) find other NGOs to market their vines when they had surplus vines in the garden. The revenue written in Table 2 was not only from the voucher scheme but also from other organisations or individual farmers who bought the vine cuttings. They strictly followed the standard measurement, for instance, one bundle contained 300 vine cuttings (67.5 m²) and the price per bundle was the same with the price sold through the voucher scheme. From the yield estimate annually measured by farmers, it has been noted to have an improved yield with an average of 16.5 t/ha. This is a great improvement that has been accomplished by the Rooting Out Hunger project.

Awareness and demand creation campaigns through radio and TV, sensitisation efforts, meetings, training and advocacy all contributed to the success of the projects. As indicated above, many organisations joined the OFSP intervention (Fig. 5), allowing project efforts to directly or indirectly reach more than 190,000 households in just four years. The cost of vine dissemination was relatively low, at US$10/household as a result of multi-partner working. The area covered by the sweetpotato crop increased from 9.1% (baseline survey) to 18.4% (endline survey). The area covered by OFSP also increased from 2.8% (baseline) to 14.4% at endline. The endline survey showed that the dietary diversity of children under five was higher compared to the baseline survey, an indication that there is an improvement in the quality of the children's diet (CIP, 2014).

Table 1 Number and gender of DVMs providing clean OFSP planting material (M = male; F = female) and the area of vine production (Source: CIP, 2014)

District	2010/11 rainy season (Y1)				2011/12 rainy season (Y2)				2012/13 rainy season (Y3)				2013/14 rainy season (Y4)			
	DVM	% M	% F	Area (ha)	DVM	% M	% F	Area (ha)	DVM	% M	% F	Area (ha)	DVM	% M	% F	Area (ha)
Dedza	17	52	48	3.4	26	32	68	16	21	64	36	16	36	64	36	27
Phalombe and Mulanje	39	47	53	0.8	92	93	107	5	30	57	43	2.7	54	48	52	4.9
Balaka	–	–	–	–	–	–	–	–	2	63	37	3.6	2	50	50	3.6
Zomba	44	53	47	0.9	16	36	64	2.5	45	42	58	2.3	58	34	66	3
Chikwawa	33	100	0	0.7	12	92	8	3	12	100	0	3.8	21	95	5	3
Total	**133**	**52**	**48**	**5.8**	**146**	**62**	**38**	**26.5**	**110**	**63**	**37**	**28.4**	**171**	**53**	**47**	**41.5**

Table 2 Number of household beneficiaries receiving vines from DVMs and the DVMs' revenues in the period of 2010/11 to 2013/14 rainy seasons (Source: CIP, 2014)

District	2010/11 Rainy season		2011/12 Rainy season		2012/13 Rainy season		2013/14 Rainy season	
	No. of HH	Revenue by DVM (US$)	No. of HH	Revenue by DVM (US$)	No. of HH	Vine sales by DVM	No. of HH	Revenue by DVM (US$)
Dedza	4733	4406.1	3000	3159.2	3693	2363.5	3991	2433.5
Phalombe and Mulanje	859	799.6	6727	10041.1	7053	4513.9	21700	13231.7
Balaka	–	–	80	74.4	1000	640	500	305
Zomba	3250	3025.5	8000	26923.2	8058	5395.1	10000	6097.6
Chikwawa	2126	1984.8	6208	7897.9	7500	17715.2	8000	4878.1
6 districts	**10968**	**10216**	**24015**	**26566.6**	**27304**	**30627.7**	**44191**	**26945.9**

Notes: the price of vine cuttings in 2010/11 and 2011/12 was Mk155/bundle or 93 cents US$; in 2012/13 was Mk190/bundle or 64 cents; and in 2013/14 Mk250/bundle or 56 cents US$. IPs: Concern Universal (Dedza, Phalombe, Mulanje), Millennium Villages Project (Zomba), Catholic Development Commission (Chikwawa). In 2010/11 rainy season: women 51% and men 49%; In 2011/12 rainy season: women 63% and men 37%. In 2012/13 rainy season: women 46% and men 54%. In 2013/14 rainy season: women 49% and men 51%.

4 Improvements in cultivation and post-harvest handling

4.1 Intercropping OFSP with maize, soya beans and onions

Studies on the intercropping of OFSP and three other important crops were conducted in Malawi. The research was inspired by the need for intensification of land use and for farmers to feed their families and sustain their livelihoods. Malawi is a land-locked country with limited agriculture land available. It is one the poorest countries in the world, having a per capita income below the estimated cost of the daily recommended calorie intake (CIP, 2013). Through its agricultural policy, the Government of Malawi focused on both crop and food diversification. The SUN movement in Malawi has supported this policy to improve the diets of people at the household level. Intercropping trials were set up to examine OFSP intercropped with maize, onion and soya beans. The selection of maize and onions for intercropping with OFSP was based on the findings of previous research in Uganda reporting the potential of these non-host crops for reducing sweetpotato weevil (*Cylas* spp.) damage to the sweetpotato crop (Nampeera et al., 2011). Soya bean was chosen as a nutritious crop which was listed in the nutrition policy of Malawi. The OFSP-maize experiment was conducted during the 2010/11 and 2011/12 rainy seasons (Abidin et al., 2015a). The OFSP-soya bean research was conducted during rainy seasons of 2012–13 and 2013–14. The opinions of farmers were collected twice.

Sweetpotato crop was left in the ground for two months beyond its normal harvest date. The OFSP-onion experiment was conducted only during the winter/dry season of 2013. Various intercrop patterns were evaluated in alternating mono-cropped strips or within-row intra-cropping. The trial was a randomised block design with three replicates. The land equivalent ratio (LER) and economic analysis was calculated. The LER for all intercrop combinations was above 1, meaning that intercropping had an advantage over the mono-crops (Mead and Willey 1980). The best intercrop pattern for sweetpotato/maize was in strips of two rows of OFSP with one row of maize; the best for sweetpotato/onion was four rows of OFSP/one row of onions, and for soya bean the pattern was three OFSP/six soya bean plants within the row. The best intercrop pattern was evaluated after considering the farmers' opinions (Abidin et al., 2016e). Again, it has been demonstrated from this research that Maize and/ or onion intercrops with OFSP, or sweetpotato in general has significantly reduced weevil damage (Nampeera, et al., 2011; Abidin et al., 2015a; Kapalamula, 2014).

The economic analysis showed that intercropping OFSP with maize, onion or soya bean resulted in better profits than mono-cropping (Tables 3, 4 and 5). The findings showed

Table 3 Cost and returns per unit area (US$) from intercropping maize and OFSP Zondeni averaged two rainy seasons of 2010/11 and 2011/12 in Malawi (Abidin et al., 2015)

	2:1 OFSP/ maize row	1:1 OFSP/ maize row	1:2 OFSP/ maize rows	3:1 OFSP/maize plant*	Sole maize	Sole OFSP
Gross return	2805	2619	2030	2843	1146	6061
Total variable cost	1457	2939	1484	1450	1221	1144
Gross margins	1348	−320	546	1393	−75	4917

Note: unit area = 90 m²; *intra-cropping: two crops were planted in the same row, in this case a high gross margin was contributed by OFSP because of its popularity. Maize did not do well with this type of intercropping (1:1 OFSP/Maize row).

Table 4 Cost and returns per unit area (US$) from intercropping onion and OFSP Zondeni during the dry season in 2013 in Malawi (Kapalamula, 2014)

	Sole OFSP	Sole onion	2:1 OFSP/onion row	4:1 OFSP/onion row
Gross return	2920	2726	2592	2424
Total variable costs	362	5478	2136	1674
Gross margins	2558	−2752	456	750

Note: unit area = 90 m²; onion was too expensive to be grown during the dry season.

Table 5 Cost and returns per unit area (US$) from intercropping soya beans and OFSP Zondeni averaged two rainy seasons of 2012/13 and 2013/14 in Malawi

	2:1 OFSP/ soya bean row	1:1 OFSP/ soya bean row	1:2 OFSP/ soya bean rows	3:6 OFSP/ soya bean plants*	Sole OFSP	Sole soya bean
Gross return	3109	2545	1564	10038	1324	707
Total variable costs	604	745	550	1340	387	621
Gross margins	2505	1800	1015	8697	937	86

Note: unit area = 90 m²; * intra-cropping – 2 crops were planted in the same row.

that intecrop systems could sustainably increase agricultural productivity and incomes with respect to food and nutrition security, adapt and build resilience to climate change in food production systems and contribute to climate-smart agriculture defined by FAO, United Nations (Abidin et al., 2016e).

4.2 Post-harvest handling through various storage facilities

The short shelf life of sweetpotato roots is a major constraint. In tropical environments where refrigerated storage is not economically feasible, particularly in developing countries, roots are generally kept for only a few weeks. For food security, farmers have developed various methods to extend sweetpotato shelf life. For example, in-ground storage, under grass or covered by ash, storage in pits as in Malawi (Abidin et al., 2013) and many other countries in southern and eastern Africa, or simply left in the ground and harvested piecemeal when required as in Uganda (Abidin, 2004; Ebregt et al., 2007) and elsewhere in East Africa. To avoid the need for fresh storage, sweetpotato is also processed into sun-dried chips in drier regions such as in north-eastern Uganda (Abidin, 2004) and in many other countries in SSA. Normally, these indigenous fresh storage methods are only effective for a month or two, after which quality declines due to rotting, pest infestations and physiological deterioration.

Recently, post-harvest handling trials employing two indigenous methods from Malawi, pit storage and traditional raised granary storage were evaluated. The storage pit was modified with steps to allow easy access to its lower levels. The inside of the traditional raised granary structure was plastered with clay cement. Dry sand was used in both pit and granary stores to cover layers of stored roots and prevent them from coming into contact with each other.

An indigenous method from Ghana, the moistened heap, in which roots are stored indoors in heaps and covered with moistened straw, was compared to a newly introduced sandbox. The sandbox was made from dried mud (adobe). The roots were packed in layers separated by dry sand as described above in raised granary structures. A thatch roof was used to cover sandbox storage rooms. The detailed methodology is described by Abidin et al. (2016c). The research concluded that by using sand-storage methods, the shelf life of sweetpotato roots could be extended beyond the normal period reported by farmers. Roots were kept up to 6.5 months in Malawi and up to four months in Ghana. The most important finding was that at 6.5 months, the beta-carotene of the storage roots from the OFSP Zondeni variety remained as high as when freshly harvested. It was also found that farmers gained a higher price when selling stored roots, as their price increased out of season. Women favoured the pit method with steps for its convenience. In Ghana, the sandbox was found superior to the moistened heap method. Abidin et al. (2016c) suggested that the methods designed for this trial were suitable for home consumption, but would require modification for commercial applications.

5 Nutritional quality and its improvement

5.1 Nutrient content of OFSP storage roots

Sweetpotato storage roots are ranked among the crops highest in nutritional value due to their fibre content, complex carbohydrates, protein, two main vitamins A and C, iron and

Table 6 Comparison of sweetpotato storage roots to other food staples (in raw form)

Staple	Maize/ corn	Rice	Wheat	Potato	Cassava	Soya bean	Sweet potato	Sorghum	Yam	Plantain
Component (per 100g portion)						Amount				
Water (g)	76	12	11	79	60	68	77	9	70	65
Energy (kJ)	360	1528	1419	322	670	615	360	1419	494	511
Protein (g)	3.2	7.1	13.7	2.0	1.4	13.0	1.6	11.3	1.5	1.3
Fat (g)	1.18	0.66	2.47	0.09	0.28	6.8	0.05	3.3	0.17	0.37
Carbohydrates (g)	19	80	71	17	38	11	20	75	28	32
Fibre (g)	2.7	1.3	10.7	2.2	1.8	4.2	3	6.3	4.1	2.3
Sugar (g)	3.22	0.12	0	0.78	1.7	0	4.18	0	0.5	15
Calcium (mg)	2	28	34	12	16	197	30	28	17	3
Iron (mg)	0.52	4.31	3.52	0.78	0.27	3.55	0.61	4.4	0.54	0.6
Magnesium (mg)	37	25	144	23	21	65	25	0	21	37
Phosphorus (mg)	89	115	508	57	27	194	47	287	55	34
Potassium (mg)	270	115	431	421	271	620	337	350	816	499
Sodium (mg)	15	5	2	6	14	15	55	6	9	4
Zinc (mg)	0.45	1.09	4.16	0.29	0.34	0.99	0.3	0	0.24	0.14
Copper (mg)	0.05	0.22	0.55	0.11	0.10	0.13	0.15	–	0.18	0.08
Manganese (mg)	0.16	1.09	3.01	0.15	0.38	0.55	0.26	–	0.40	–
Selenium (mg)	0.6	15.1	89.4	0.3	0.7	1.5	0.6	0	0.7	1.5
Vitamin C (mg)	6.8	0	0	19.7	20.6	29	2.4	0	17.1	18.4
Thiamin (mg)	0.20	0.58	0.42	0.08	0.09	0.44	0.08	0.24	0.11	0.05
Riboflavin (mg)	0.06	0.05	0.12	0.03	0.05	0.18	0.06	0.14	0.03	0.05
Niacin (mg)	1.70	4.19	6.74	1.05	0.85	1.65	0.56	2.93	0.55	0.69
Pantothenic acid (mg)	0.76	1.01	0.94	0.30	0.11	0.15	0.80	–	0.31	0.26
Vitamin B6 (mg)	0.06	0.16	0.42	0.30	0.09	0.07	0.21	–	0.29	0.30
Folate (total mcg)	46	231	43	16	27	165	11	0	23	22
Vitamin A (IU)*	208	0	0	2	13	180	14187	0	138	1127
Vitamin E, alpha-tocopherol (mg)	0.07	0.11	0	0.01	0.19	0	0.26	0	0.39	0.14
Vitamin K (mg)	0.3	0.1	0	1.9	1.9	0	1.8	0	2.6	0.7
Beta-carotene (mcg)*	52	0	0	1	8	0	8509	0	83	457
Lutein+Zeaxanthin (mcg)	764	0	0	8	0	0	0	0	0	30
Saturated fatty acids (g)	0.18	0.18	0.45	0.03	0.07	0.79	0.02	0.46	0.04	0.14
Monounsaturated fatty acids (g)	0.35	0.21	0.34	0.00	0.08	1.28	0.00	0.99	0.01	0.03
Polyunsaturated fatty acids	0.56	0.18	0.98	0.04	0.05	3.20	0.01	1.37	0.08	0.07

*Only found in OFSP varieties.

calcium. The crop is a good source of dietary fibre (2.5–3.3 g/100 g) and is classified as having a much lower glycaemic index than most other sources of carbohydrates (grains, potatoes). Despite the name 'sweet', this can be a beneficial food for diabetics, as preliminary studies on animals have revealed it helps to stabilise blood sugar levels and to lower insulin resistance (http://www.whfoods.com/genpage.php?tname=foodspice&dbid=64; the world's healthiest foods: Sweetpotatoes, 2001–12, the George Mateljan Foundation). The need to avoid consuming foods that increase the risk of diabetes is reflected in high glycaemic index values. Table 6 shows some nutritional values of sweetpotato storage roots and other food staples.

5.2 Nutrient content of sweetpotato leaves

Sweetpotato leaves may help in the fight against age-related macular degeneration (AMD) (Kachathrian et al., 2003). This is because its leaf is an excellent source of lutein (3,3'-dihydroxy-β-carotene). Lutein has been identified as a dietary component that can delay the onset of AMD. Major sources of lutein are green vegetables and marigold flowers. Kachathrian et al. (2003) found that sweetpotato leaves rank second in lutein content after marigold flowers. Information given by USDA SR-21 shows that sweetpotato leaves are low in saturated fat and sodium, and are very low in cholesterol. They are also a good source of protein, niacin, calcium and iron, and a very good source of dietary fibre, vitamin A, vitamin C, thiamin, riboflavin, vitamin B6, folate, magnesium, phosphorus, potassium and manganese. Table 7 provides nutritional information on sweetpotato leaves (raw) at serving size of 100 g (accessed 20 November 2016 at http://nutritiondata.self.com/facts/vegetables-and-vegetable-products/2664/2#ixzz1uyfrdSXq).

5.3 Vitamin A content of OFSP, nutrition and market development

Vitamin A is an essential micronutrient necessary for growth and development, good eyesight, immunity and foetal development. VAD is usually a result of dietary deficiency, or disorders which interfere with its absorption, storage or transport in the body. VAD during pregnancy may lead to poor foetal development and birth defects. VAD in children may also lead to stunted growth and an increase in the duration and severity of childhood illnesses such as measles, diarrhoea, pneumonia and common cold. Furthermore, other VAD can cause disorders including night blindness, drying and damage of the eyes, and an increased incidence of illness and death (West, 2003).

Johnson (2016) indicated that the liver normally stores between 80 and 90% of the body's vitamin A. It is released as needed, bound to pre-albumin (transthyretin) and retinol-binding protein. Retinol activity equivalents were developed because provitamin carotenoids have less vitamin A activity than preformed vitamin A. The conversion ratio of provitamin carotenoids to vitamin A (retinol) varies by carotenoid and source, but the accepted conversion ratio from β-carotene to retinol in sweetpotato storage roots is 12:1; 1 µg retinol = 3.33 IU. In contrast, as Johnson (2016) pointed out, vitamin A toxicity can occur. This can be acute or chronic. Both types usually cause headache and increased intracranial pressure.

β-carotene and other provitamin carotenoids, contained in green leafy and yellow vegetables and deep-coloured or bright-coloured fruits, are converted to vitamin A

Table 7 Nutrition facts of sweetpotato leaves at serving size of 100 grams (provided by USDA SR-21) compiled by the principal author of this chapter from http://nutritiondata.self.com/facts/vegetables-and-vegetable-products/2664/2#ixzz1uyfrdSXq; accessed 20 November 2016)

Items	At serving size of 100 g sweetpotato leaves	% DV
Calories from	147 KJ	2
a) Carbohydrate	95 KJ	–
b) Fat	10.5 KJ	–
c) Protein	41 KJ	–
d) Alcohol	0 KJ	–
Carbohydrate	6.4 g	2
Dietary Fibre	2 g	8
Starch	–	–
Sugar	–	–
Protein and Amino Acids	4 g	8
Fat and Fatty Acids		
a) Total Fat	0.3 g	0
b) Saturated fat	0.1 g	0
c) Mono-saturated fat	0.0 g	–
d) Poly-saturated fat	0.1 g	–
e) Total omega-3 fatty acid	21 mg	–
f) Total omega-6 fatty acid	113 mg	–
Vitamins		–
a) Vit A	1028 IU	21
b) Vit C	11 mg	18
c) Vit D	–	–
d) Vit E	–	–
e) Vit K	–	–
f) Thiamin	0.2 mg	10
g) Riboflavin	0.3 mg	20
h) Niacin	1.1 mg	6
i) Vit B6	0.2 mg	9
j) Folate	80 mg	20
k) Vit B12	0 mg	0
l) Pantothenic acid	0.2 mg	2
Minerals		
a) Calcium	37 mg	4

Items	At serving size of 100 g sweetpotato leaves	% DV
b) Iron	1 mg	6
c) Magnesium	61 mg	15
d) Phosphorus	94 mg	9
e) Potassium	518 mg	15
f) Sodium	9 mg	0
g) Zinc	0.3 mg	2
h) Copper	0 mg	–
i) Manganese	0.3 mg	13
j) Selenium	0.9 mg	1
Cholesterol (Steroids)	0 mg	–
Water	87.9 g	–
Alcohol	0 g	–
Ash	1.4 g	–

Note: Per cent daily values (% DV) are for adult or children aged 4 or older, and are based on 2,000 calorie reference diet. Everybody has different individual need of the daily values; it can be higher or lower than that shown in the above table.

(Johnson, 2016). West (2003) indicated that vitamin A is found in the form of retinol in animal foods, such as liver, poultry, eggs and fish. Carotenoids are absorbed better from vegetables when they are cooked or homogenised and served with some fat (e.g. oils). According to Bechoff (2010), ß-carotene is the carotenoid with the highest provitamin A activity (100%), because it can be entirely converted into two molecules of vitamin A (retinol). Doering et al. (1995) described that carotenoids are found in nature as *trans*-carotenoids and Bengsston et al. (2008) found that all-*trans* ß-carotene represents about 80–90% of the total carotenoid in OFSP. Kósambo et al. (1998) clarified that in the OFSP crop, initial levels of carotenoids are influenced by variety, root maturation and location. In addition, Simonne et al. (1993) grouped sweetpotato varieties into four general categories based on their ß-carotene content on a dry weight basis: non-detectable (<1 µg/g); low ß-carotene (1–39 µg/g) – pale orange; moderate ß-carotene (40–129 µg/g) – orange; and high ß-carotene (>130 µg/g) – dark orange. According to Bechoff et al. (2010), the body can convert the ß-carotene into retinol (vit A) at the ratio of 12:1.

The introduction, adoption and utilisation of OFSP provides a simple, cost-effective tool to improve nutrition, empower women and increase earning opportunities, even for the poorest households. Bechoff et al. (2011) suggested that a serving of porridge (one mug), boiled root (half a root of approximately 100–150 g), OFSP mandazis (two) or OFSP chapatti (one) would provide a significant part of the daily vitamin A requirements of a child under six years; one mug of porridge being 20% of daily vitamin A requirements, half boiled root 46%, two mandazis 75% or one chapatti 100%. Therefore, when consuming OFSP, there is no cause for concern about vitamin A toxicity. Food diversification in the daily diet is important, however, to combat malnutrition. Even at low yields of six tons/ha, just 500 square metres can generate an adequate annual supply of vitamin A for a family of five. Clearly, OFSP is a vitamin A powerhouse. Various recipes from sweetpotato storage roots and leaves have been compiled by Abidin et al. (2015b,c) from local recipes in East and West Africa are available at https://www.researchgate.net/.

Low et al. (2007) reported significant improvements in vitamin A intake and serum retinol concentrations (a proxy for vitamin A status) from an integrated agriculture–nutrition market intervention promoting OFSP in a very resource-poor setting in Mozambique. An illustration of the integrated agriculture, nutrition and marketing approach is presented in Fig. 4 (Low et al., 2007). The approach introduces new sources of vitamin A and energy on the agricultural side (i.e. the OFSP crop). This is complemented by demand creation and empowerment through increasing popular knowledge to ensure the desired improvements in consumption and hence in nutrition.

Finally, it is necessary to think about market development for OFSP roots and processed products. The end result of this integrated effort is to obtain increased serum retinol levels in the target population, specifically in children under five years of age. In 2011, the International Food Policy Research Institute (IFPRI), the Ministry of Agriculture of Malawi, Irish Aid and USAID organised a conference in Malawi, 'Unleashing agriculture's potential for improved nutrition and health in Malawi' (http://malawi2011.ifpri.info/program/). During the conference, the project using the nutritious OFSP, 'How agriculture can contribute to nutrition and health, a lesson learnt from the Rooting Out Hunger project in Malawi', was presented (Abidin, 2011). The goal was to use the nutritious OFSP to combat hunger and reduce VAD among children under the age of five. The project used a voucher scheme to ensure poor rural communities could benefit from the programme with respect to food and nutrition security. The use of vouchers also ensured that the decentralised vine multipliers were paid for their efforts.

6 Crop diversification for new uses

6.1 Commercialisation of vines in the sweetpotato value chain

Marketing a new crop, such as OFSP, can be challenging. A number of studies were undertaken on the market for sweetpotatoes. A study by Mukundi et al. (2013) concluded that over 70% of sweetpotatoes produced in Kenya are sold in informal markets where farmers get a low price for their crop. They suggested that collective action by farmers will improve their marketing opportunities as they will be able to share transaction costs and obtain improved access to critical market information. In addition, Etumnu (2015, pers. comm.) also studied the contributing factors of bio-fortified food selection, using OFSP as his entry point in Ghana. He found that presenting information to consumers on the nutritional benefits of OFSP exerts a substantial, positive and significant impact on willingness to pay for the produce. He concluded that providing nutritional information appears to be essential for the successful introduction of OFSP and other bio-fortified foods to stimulate market demand.

The CIP-led three-year project 'Jumpstarting orange-fleshed sweetpotato in West Africa through Diversified Markets' is funded by the Bill and Melinda Gates Foundation. The project began in April 2014 and will end in March 2017. In contrast to eastern and southern Africa, the promotion of OFSP in West Africa is relatively new, and the sweetpotato crop is not typically as important in this region. The pilot locations of the project are in Ghana, Nigeria and Burkina Faso. In Ghana, Northern and Upper East Regions, were chosen, where sweetpotato ranged from being an important cash crop (districts around Bawku), to being of moderate (districts around Navrongo), to only minor importance in the farming system (Tolon and Kumbungu Districts near Tamale). In Nigeria, the project area is in Osun State, where a successful school feeding programme is piloting OFSP, and in adjacent Kwara State, which is traditionally a major producer of sweetpotato. In Burkina Faso, the project is targeting communities near Orodara in Kénédougou Province in the southwest of the country, which are major suppliers of fresh sweetpotato to markets in Burkina Faso and Mali. While sweetpotato is a well-known crop in West Africa, and is often commercially important, OFSP varieties are not widely available and the nutritional value of OFSP is not widely recognised. As a result, it is not yet in demand from consumers and marketers. There is a need to capitalise on the potential of OFSP through communication and promotion in order to contribute effectively to combating VAD and improving farmer incomes.

The jumpstarting project is focused on developing market opportunities for OFSP, so that market demand will drive OFSP adoption and the development of a commercialised seed system. This will ensure fulfilment of the crop's potential in contributing to both wealth and health. The project seeks to determine distinct formal markets such as school feeding programmes, restaurants, boarding schools and informal market channels including conventional wholesale and retail markets, as well as products for fresh consumption and agro-processing at various scales. An integrated approach of the public and private sector partnership is also used to match into the four jumpstarting OFSP project outcomes (Fig. 6), including both income and nutrition (Abidin et al., 2016d).

The four major outcomes were assigned at each target location: (i) formal and informal diversified OFSP market opportunities developed in pilot areas in Ghana, Nigeria and Burkina Faso; (ii) viable QDPM seed systems in target areas capable of expansion in response to increased demand; (iii) households including women and children, in target areas increased vitamin A consumption from OFSP; and (iv) commercial sweetpotato planting material and OFSP producers, including women, increased their income through participation in OFSP value chains. Again, CIP works with multiple partners, that is, relevant government agencies and NGOs.

By the final year of project implementation, market-driven approaches showed a succeeding trend, demonstrating that these approaches to OFSP had led to larger incomes and increased consumption of vitamin A-rich foods including OFSP, especially in women and children. 'Theory of Change' workshops held early in the project helped the development of actor-centred impact pathways. During the first two years of the project, major emphasis was placed on the establishment of vine production through DVM, developing groups of root producers, strengthening the capacity of extension and the training of trainers, including farmers, on GAP including the management of pests and diseases, OFSP utilisation and processing. Multifaceted promotional efforts included radio programmes, visits, awareness/demand creation campaigns and advocacy from high-level policy makers and politicians down to the local level of village leaders and chiefs. Faith-based organisations, international organisations and NGOs were also targeted. OFSP value chains were targeted and linkages between producers and buyers as well as market segments were identified. By the third year, the production of storage roots began to boom at farm gates, and simultaneously, efforts to develop and support these diversified markets continued, so as not to disappoint producers who were taking the risk of responding to new and still uncertain market opportunities.

6.2 Markets in Ghana, Nigeria and Burkina Faso

Results from Ghana, Nigeria and Burkina Faso showed that considerable progress was made during the first two and a half years of the project. DVMs and farmers' groups for OFSP storage root production were firmly established. For the 2016 planting season, a total of 54 DVMs comprising 110 vine producers and 76 groups of root producers consisting of 1,423 farmers (32% women) grew OFSP for markets. A mid-line survey conducted in Ghana showed that 32% of the sweetpotato area grown by respondents was covered by OFSP in the area of project implementation. In Burkina Faso, value chain committees comprising storage root producers and buyers have been established, and commitments made by buyers to purchase up to 2,500 tonnes of OFSP storage roots, based on their newly acquired knowledge of the nutritional value of OFSP, and of customer demand for the product.

In Nigeria, the O-MEALS school feeding in Osun State put OFSP on the menu in January of 2015 and a pilot programme was initiated in eight schools. The number of schools was increased to 17 in August of 2015, serving one OFSP meal weekly to a total of 8,157 pupils. In August 2016, the O-MEALS programme expanded to an additional 100 schools with 31,395 students in grades 1 to 4. Numerous producers have entered the effort to meet the weekly demand for OFSP, which is served as a soup. Additionally, the O-MEALS programme includes bread on the menu one day a week, and is promoting the use of nutritionally enhanced bread made with OFSP puree as a partial substitute for wheat flour.

In all three countries, quality management protocols for QDPM and GAP have been developed. Pretesting of QDPM has been done in each country. Teams comprising a lead farmer for each group, an extension officer and the inspected farmer have been created in project intervention areas to assure that farmers bought clean planting materials for

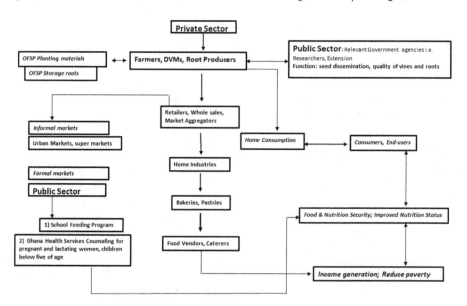

Figure 6 An integrated approach: public and private sector partnership matching into the four jumpstarting OFSP project outcomes.

Table 8 Number of households benefiting either directly or indirectly from orange-fleshed sweetpotato vitamin A powerhouse across the three countries since April 2014 through July 2016

Indicator variables	Year 1 (4 April 2014 to 31 March 2015)			Year 2 (1 April 2015 to 31 March 2016)			Year 3 (1 April 2016 to 31 July 2016)									Grand Total
	Ghana	Nigeria	Burkina Faso	Ghana	Nigeria	Burkina Faso	Ghana			Nigeria			Burkina Faso			
				direct			direct	% M	% F	direct	% M	% F	direct	% M	% F	
Households reached with OFSP vines	2466	6334	784	2461	5000	1883	3362	31	69	–	–	–	193	78	22	22483
No of people reached[1]	12577	19860	–	9953	13238	4170	4181	80	20	8684	52	48	91	95	5	72754
No of pregnant women reached[2]	390	834	–	4412	5081	–	2000	–	100	–	–	–	–	–	–	12717
No of breastfeeding women reached[2]	641	1013	–	2674	–	–	–	–	–	–	–	–	–	–	–	4328
Pregnant, breastfeeding and/or lactating women through Ghana Health Service intervention	–	–	–	–	–	–	4745	–	100	–	–	–	–	–	–	4745
No of women reached[3]	–	–	–	–	–	–	–	–	–	133	100	0	–	–	–	133
DVMs	–	–	–	–	–	–	15	100	0	8	100	0	48	92	8	71
Total	16074	28041	784	19500	23319	6053	14303	31	69	8825	51	49	332	84	16	117231

[1] Number of people reached include training beneficiaries, vine recipients including direct from the project and indirect beneficiaries (leveraged). Indirect beneficiaries are defined as the beneficiaries recorded through the joint programme partners, such as MEDA (2000 hhs), Trax Ghana (1 350 hhs), World Vision (2 700 hhs), MOFA activities from regions (3 243 hhs) receiving services from jumpstarting project recorded particularly only through (Year 1 of project activities).
[2] Pregnant/Lactating mothers counselled through Health Services.
[3] Women were not pregnant/breastfeeding.

root production. Dry season storage root production has been introduced, particularly for providing the OFSP roots needed by the School Feeding Programmes piloted in Nigeria and Ghana. In Ghana, two schools in Kumbungu District, Northern Region, began a pilot effort in June 2016. The programme, which served over 800 students, will be expanded to other schools and there is a need to develop mechanisms to ensure timely payment of producers.

In addition to school food, other formal and informal markets have been identified and developed in each country, including urban and local markets, Ghana Health Services counselling and bakeries in four regions of Ghana. Radio programmes were designed by partners as part of the marketing strategy. Monitoring and evaluation tools have been developed to monitor the DVMs, root production, market prices and volumes, using open data kit forms. Strong partnerships have been established to help the project achieve its goals more quickly. More than 117,230 households benefited either directly or indirectly from OFSP across the three countries from April 2014 through to July 2016. The detailed figures can be seen in Table 8.

In July 2016, 22 vine multipliers (DVMs) in Ghana sold 1,128,030 clean planting materials and grossed US$14,650 of which 42% was from subsidised vouchers through the Ghana Health Service, 34% from individual farmers and 24% from the NGOs. In Nigeria, six DVMs sold 1,264,700 clean planting materials to 13 root producer groups (13% women) generating an incomes of US$12,647. In Burkina Faso, 37 DVMs sold 1,210,000 clean planting materials to root producers in two project areas and a neighbouring community, and also generated sales of vines to Mali. From the dry season root production, eight small producers (three women) and one large producer (male) in Northern Region of Ghana sold 8,015 kg of OFSP storage roots, and earned a total of US$1,655. This includes (i) 4,655 kg to retailers and consumers; (ii) 895 kg to two caterers for the Ghana School Feeding Programme; and (iii) 2,465 kg to a buyer in Accra for various markets. In Nigeria, the school feeding programme has involved 17 schools in 12 local government areas with a total of 8,157 pupils through the O-MEALS School Feeding Programme in Osun State since January 2016. The food vendors bought the storage roots from 13 groups of storage root producers comprising 30% women who earned at least US$16,000. It appears that the concept of OFSP commercialisation laid out in Fig. 6 is meeting with success, and should ensure both its income and nutrition outcomes by the end of the project.

The following individual cases illustrate the use of research to improve cultivation and living standards and were generated from the experiences of two projects in Malawi and Ghana.

7 Case studies: Malawi

7.1 Widowed mother

Lucia Graleta from the Group Villages Head Maguleje, Mbwewe Extension Planning Area (EPA) in Chikwawa District, a widow with seven children under the age of five, was a household beneficiary who received 300 OFSP vine cuttings from the project 'Rooting out Hunger' through CADECOM. These were planted in an area of 67.5 m² in the 2010/11 rainy season. After four years, she had expanded the area of cultivation for OFSP to 0.2 ha. Annually, she produces 31 bags of 50 kg and sells the OFSP roots in the local and urban markets. The price of the OFSP storage roots are double that of the white and/ or yellow-fleshed varieties. Two kg of roots sold for Mk200 in September–October when sweetpotato is scarce, and for Mk100 in June–July during the peak harvesting period. Mrs Graleta used the money from selling the roots to pay the school fees of her children and to buy household items such as salt, sugar and soap. She mainly grew sweetpotato during winter and maize, sorghum, millet and cotton during the summer or rainy season. According to Christopher Masamba, the Agriculture Extension Development Officer of Mbwewe EPA, the OFSP Zondeni variety performed well during the winter season but yielded poorly in the rainy season. Farmers, therefore, intend to expand their areas of growing OFSP during winter.

7.2 Promotion of OFSP by Michesi Club in Phalombe District

Phalombe is the most heavily populated and impoverished district of Malawi. Through a number of donors working on development aid, the United Nations has given close attention to building development research capacity in this district. Various donor-funded projects, including the Irish Aid project 'Rooting out Hunger in Malawi with Nutritious Orange-fleshed Sweetpotato' have chosen this district for implementing research and development programmes.

The members of Michesi Club were among the beneficiaries targeted by this OFSP project. The club is situated in Thomu Village, Kulambe EPA. It began with a group of people living with HIV and AIDS, with the purpose of providing mutual support to members. The club was established in June 2009 and has 90 members (12 males and 68 females). The objective was to reduce the day-to-day problem of people lacking incentives for farming. A number of activities were created by the members, such as working with extension and NGOs on composting and conservation agriculture, and taking care of orphans whose parents had died, mostly because of HIV and AIDS. Later, the club accepted anyone wanting to join. It does not discriminate against people living with HIV/AIDS, older people, widowers, single mothers, married couples and others. They do not have a specific meeting room and usually gather under a mango tree. Activities are focused on teaching and sharing information. In the rainy season of 2010/2011, the members became involved in OFSP vine multiplication introduced by CU, under the USAID-MOBI LISE project, based in Mulanje. They received training on vine multiplication, sweetpotato production, pest and disease management and OFSP processing. Most of their vine cuttings were sold through a voucher scheme introduced by the Rooting Out Hunger project.

To become a member, a farmer initially pays Mk100. If there is a need for communication, each farmer contributes an additional Mk20. In the fourth year of being involved in this OFSP project, the club has 11 plots of 20 m² for an OFSP nursery. Four plots were bought for Mk80,000 and seven plots rented for Mk2,000/plot. All members initially received free OFSP vine cuttings though the club which sold replacement vines to members for Mk250/bundle if the originals dried out in individual plots. Other farmers who wanted to grow OFSP also bought the vines from the club. One member reported realising Mk210,000, and another Mk500,000 from vine sales, mostly through the voucher scheme. They used the money to renovate their houses; bought land (upland and/or dambo land), household items and corrugated iron; and paid school fees, among other things.

During the dry season, vines were planted on low dambo land near a source of water. When the vines were ready for further multiplication, they were brought to the upland garden near the houses. The club helped members whose vines had dried out from lack of management skills by sharing knowledge from skilled farmers with them. In 2012, the skilled members trained two groups: the first group comprised five women and two men and the second 11 women and five men. Members, therefore, have a strong sense of ownership of the programme and learn to take care of the crop, storing sliced sun-dried roots at home to avoid theft. Most of the roots were for home consumption, but the surplus was sold to traders at Mk2,500 for a 50 kg bag. Those involved in the project reported feeling more healthy, having better eyesight, being better off financially and happier in their marriages.

7.3 Home-grown school feeding programme

Nkhulambe EPA is amongst the poorest districts in Malawi, and financial assistance is still needed to help improve livelihoods in the community. Of the 12 primary schools in this district, nine were involved in the school meals programme funded by WFP. Through its Social Cash Transfer Scheme programme, UNICEF further supported 46 households with school-age children (five to 16 years of age) to purchase books, school uniforms, pencils and pens. The Rooting Out Hunger project also chose this community for its OFSP intervention. Innocent Kaponya, an Agriculture Extension Development Coordinator at Nkhulambe EPA in Phalombe District reported that this community used the initiative to introduce the OFSP crop to school menus through the home-grown school feeding programme. As there were three schools not in the feeding programme supported by the WFP, the community decided to choose these schools for a special programme which they named 'Home Grown School Feeding'. Vegetables were grown at these schools to ensure the children received more nutritious meals. This began in the rainy season of 2013/14 (the 4th year of the Rooting Out Hunger project) and included the OFSP Zondeni. The community contributed the OFSP vines, land and labour while the District Council provided fertilisers.

8 Case studies: Ghana

8.1 Commercial vine and root producer

Dauda Zakaria lives in Bontanga, Northern Region. After being involved for two years with the Jumpstarting OFSP through Diversified Markets programme, he has earned GHc55,595 (US$13,894) from vine and root sales. He sold his vines to individual farmers, NGOs and Ghana Health Service, reaching pregnant and lactating women. The OFSP storage roots were sold in local and urban markets. He made some contacts with caterers, farmers, food vendors, restaurants and farmer organisations through telephone calls and announcements on local radio. He also brought his produce to the local markets not far from the village if customers were unable to come to his farm. The vine and root sales enabled him to help in paying the school fees for his younger brother, and at the beginning of 2016, he bought two acres of land, of which he has used 1.5 acres for root production and 0.5 acre for vine production. He was able to marry and build a concrete (cement) house with a metal roof. He continues growing OFSP and plans to buy a motor tricycle to transport sweetpotato roots and vines to customers.

8.2 OFSP bread

In Ghana, bread is popular as a breakfast and mid-afternoon snack. Three main types of bread are typically sold in markets or at roadsides: tea bread, sugar bread and butter bread. Recently, brown bread has been introduced and is mainly sold in supermarkets, its buyers being largely from the upper middle class. Drawing on experience from eastern and southern Africa, the Jumpstarting OFSP project introduced OFSP puree as a partial (up to 50%) substitute for wheat flour. Beside the nutritional benefits from the vitamin A-rich OFSP, the use of puree can reduce the cost of bread production, depending on the relative cost of OFSP storage roots and wheat flour. Locally produced OFSP becomes competitive with imported wheat flour at roughly 75% of the cost, a price which is easily achieved in the present market. However, the nutritional value and attractive colour of the 'golden bread' made from OFSP allow it to be a sold at a premium, providing pricing flexibility.

OFSP bread made from puree using an affordable and locally fabricated motorised grating machine was introduced to artisanal bakers during training exercises in various locations. The Vekon Bakery in Sogakope, in the Volta Region, served as one entry point. The bakery, which is owned by a woman and has more than 100 employees (>75% women), runs wood-fired ovens, uses about 50 bags of flour per day and sells its butter bread at a well-patronised roadside rest stop owned by the bakery on a major highway which passes through the town.

In August 2016, the bakery began producing and selling bread made with puree from one 60 kg bag of fresh OFSP and one 50 kg bag of wheat flour daily. There was a stable demand for the bread and a stable supply of roots from the producers. The project helped the bakery to design labels for the bread, emphasising its nutritional value, and worked with them to give guidance in understanding consumer acceptance, improving the shelf life of the bread and increasing production. Other bakeries in the Volta, Greater Accra and Northern Regions have also been trained and demand for golden bread and the OFSP needed to produce it is increasing consistently. It is anticipated that the market opportunity for golden bread is very large as it is well accepted by consumers, is nutritious and easy to produce, both at an artisanal bakery level and by larger, commercial bakeries and coordination of a supply of affordable OFSP can be managed easily.

9 Conclusion and future trends

The future of bio-fortified OFSP in helping to combat VAD appears to be bright. Awareness of the effectiveness of OFSP in combating VAD is stimulating ever more investment to help scale up this approach. OFSP has the potential for use as an entry point in diverse interventions including those concerned with social safety nets, early childhood and schooling. Action research will be required, both in the applied setting of project implementation and at the more basic research level of crop and cropping systems improvement. Complementary bio-fortified crops is another promising area, including the nutritious food basket approach. For example, producing bio-fortified crops such as OFSP and high iron beans has already been tried in Malawi. In addition to genetic improvement, agronomic approaches to bio-fortification, including the use of zinc, fertilisation have shown promise in potato crops and may have similar application for sweetpotatoes (Kromann et al., 2016).

Efforts are underway to increase iron and zinc contents in OFSP through conventional breeding. In cases of iron or zinc bio-fortification, studies will be required to determine the degree of bioavailability of these elements. Interdisciplinary approaches are being undertaken to combat micronutrient deficiency such as the SUN movement, which embraced OFSP in Malawi, and will continue to develop this approach. It is anticipated that breeding efforts will be more driven by private sector demand based on specific end use and the additional nutritional characteristics of provitamin A, as new types such as the purple fleshed sweetpotato, rich in healthy anthocyanins, are developed. Sustainable models for seed systems which contribute to consistency and increases in production will be needed, and will certainly need to involve breeding linkages. Knowledge of the nutritional value of sweetpotato will stimulate demand for the crop and may be a key element to sustainable scaling. Optimum involvement of various actors, including government and the private sector, will vary depending on circumstances. Ongoing research on adoption and the economic impact of research investments will help to justify further investment. Effective models for scaling/extension will be necessary, with an emphasis on empowerment of women in order to ensure the optimum nutritional impact in target demographics.

10 Where to look for further information

Literature cited in this chapter provides a number of relevant references. The paper by Gruneberg et al. (2015) is noteworthy, providing an in-depth review of sweetpotato breeding. Ruel et al. (2013) discussed nutrition-sensitive interventions including the use of bio-fortified crops and indicated that OFSP was the only case in which bio-fortification was proved to be effective at the community level. They further pointed out that the potential of school feeding and its associated nutritional education is very effective in improving nutritional knowledge at the household level.

Public health interventions, particularly those introducing the OFSP in antenatal care and infant and young child feeding efforts in Western Kenya have provided some evidence of beneficial nutritional impact. The Sweetpotato Knowledge Portal (www.sweetpotatoknowledge.org) provides information on numerous former and current sweetpotato project activities in SSA.

11 Acknowledgements

The authors wish to acknowledge the support of Irish Aid, USAID-Office of Foreign Disaster Assistance and the Bill & Melinda Gates Foundation for supporting efforts in Malawi and Ghana. IPs, including governments of Malawi and Ghana, NGOs and farmers are also recognised. The authors also acknowledge the reviewers who offered constructive suggestions on this chapter.

12 References

Abidin, P. E. and E. E. Carey, 2001. Sweetpotato genetic diversity in north-eastern Uganda: germplasm collection, farmer knowledge, and morphological characterization. Presented at the American

Society for Horticultural Science (ASHS) International Symposium in Sacramento, the USA: 22nd to 25th July 2001, *HortScience* 36(3): 487.

Abidin, P. E., F. A. van Eeuwijk, P. Stam, P. C. Struik, D. P. Dapeng, M. Hermann and E. E. Carey, 2002. Evaluation of sweetpotato (Ipomoea batatas (L.) Lam.) germplasm from North-eastern Uganda through a farmer participatory approach. In: T. Ames (ed.). Proceedings 1st International Symposium on Sweetpotato, the International Society for Horticultural Science (ISHS) in Lima, Peru: 26th till 29th November 2001. *Acta Holticulturae* 583: 61–8.

Abidin, P. E., 2004. Sweetpotato breeding for northeastern Uganda: Farmer varieties, farmer-participatory selection, and stability of performance. PhD Thesis, Wageningen University, The Netherlands, 152pp. ISBN:90–8504-033-7. Available at https://www.researchgate.net/publication/40125709

Abidin, P. E., 2011. How agriculture can contribute to nutrition and health outcomes: Experience to date from the rooting out hunger in Malawi with nutritious orange-fleshed sweetpotato project. A key-note. In C. Swann (ed.). *Unleashing Agriculture's Potential for Improved Nutrition and Health in Malawi*, pp. 28–31. Min of Agriculture and Food Security of Malawi, IFPRI, Irish Aid and USAID. 74pp. Available at http://malawi2011.ifpri.info/program/

Abidin, P. E., T. Nyekanyeka, S. Heck, S. McLean, G. Mnjengezulu, F. Chipungu, R. Chimsale and B. Botha, 2013. Less hunger, better health and more wealth: the benefits of knowledge sharing in Malawi's Orange-Fleshed Sweet Potato project. In Annom. (ed.). *Hunger-Nutrition-Climate Justice*. A new dialogue: putting people at the heart of global development. Conference papers, 15–16 April 2013, pp. 55–8, Irish Presidency of the Council of the European Union, Irish Aid, Mary Robinson Foundation Climate Justice, CGIAR-Climate Change, Agriculture and Food Security (CCAFS), World Food Programme (WFP), Dublin, Ireland. Available at http://cgspace.cgiar.org/handle/10568/27890.

Abidin, P. E., F. Chipungu, T. Nyekanyeka, T. Chilanga, O. Mwenye, J. Kazembe, B. Botha and E. E. Carey, 2015a. Maize-orange-fleshed sweetpotato (OFSP) intercropping: potential for use to enhance food security and the scaling-up nutrition effort in Malawi. In J. Low (ed.). *Transforming Potato and Sweetpotato Value Chains for Food and Nutrition Security*, pp. 405–13, CABI Pub, Cambridge.

Abidin, P. E., E. Dery, F. K. Amagloh, K. Asare, E. F. Amoaful and E. E. Carey, 2015b. Training of trainers' module for orange-fleshed sweetpotato (OFSP). Utilization and processing. International Potato Center (CIP); Nutrition Department of the Ghana Health Service, Tamale (Ghana). 32pp.

Abidin, P. E., E. Dery, F. K. Amagloh, K. Asare, E. F. Amoaful and E. E. Carey, 2015c. Golden sweetpotato dishes. International Potato Center (CIP); Nutrition Department of the Ghana Health Service, Tamale (Ghana). 24pp.

Abidin, P. E., D. A. Akansake, K. B. Asare, K. Acheremu and E. E. Carey, 2016a. Comparative studies on sources of sweetpotato planting material for quality vines and root yield in Ghana,. Presented at the 10th Triennial Conference of African Potato Association (APA) in Addis Ababa, Ethiopia, 9–13 October 2016. Journal De Gruyter. (submitted).

Abidin, P. E., J. Kazembe, C. Kapalamula, C. Takagi and E. E. Carey, 2016b. Intercropping orange-fleshed sweetpotato with maize, soybeans or onion in Malawi for sustainable intensification (Accepted Abstract). Presented in the World Congress on Root and Tuber Crops (WCRTC) in Nanning, China, 18–22 January 2016. Available at http://www.gcp21.org/wcrtc/S08.html

Abidin, P. E., J. Kazembe, R. A. Atuna, F. K. Amagloh, K. Asare, E. K. Dery and E. E. Carey, 2016c. Sand storage, extending the shelf-life of fresh sweetpotato roots for home consumption and market sales. *Journal of Food Science and Engineering* 6: 227–36; doi:10.17265/2159-5828/2016.04.005

Abidin, P. E., S. Adekambi, J. Njoku, K. Acheremu, S. Koussao, J. Nchor, J. Trucker, I. Koara, D. Yesseh, B. Kiger, E. F. Amoaful, D. Akansake, E. Dery and E. E. Carey, 2016c. Jumpstarting orange-fleshed sweetpotato through diversified markets: An approach to accelerating adoption, and increasing production to benefit many end users (Accepted abstract). The 13th International Symposium of the International Society for Tropical Root Crops – Africa Branch (ISTRC-AB). Dar Salam, Tanzania, 5–10 March 2017.

Abidin, P. E., J. Kazembe, C. Kapalamula, C. Takagi and E. E. Carey, 2016e. Intercropping orange-fleshed sweetpotato with maize, soybeans or onion in Malawi for sustainable intensification (Accepted abstract for oral presentation). Presented in the World Congress on Root and Tuber Crops (WCRTC) in Nanning, China, 18–22 January 2016.

Akansake, A. D., P. E. Abidin, K. B. Asare, K. Acheremu and E. E. Carey, 2016.Evaluation the potential of some sweetpotato cultivars to yield both roots and vines in drought prone parts of Ghana. (Abstract submitted). 10th Triennial Conference of African Potato Association (APA) in Addis Ababa, Ethiopia, 9–13 October 2016.

Austin, D. F., 1988. The taxonomy, evolution and genetic diversity of sweetpotatoes and related wild species. In P. Gregory (ed.). *Exploration, Maintenance, and Utilization of Sweetpotato Genetic Resources*, pp. 27–60. CIP, Lima, Peru.

Bashaasha, B., R. O. M. Mwanga, C. Ocitti p'Obwoya and P. T. Ewell, 1995. Sweetpotato in the farming and food systems of Uganda. A farm survey report. CIP-NARO. Kenya, Uganda.

Bechoff, A., 2010. Investigating carotenoid loss after drying and storage of orange-fleshed sweetpotato. PhD thesis. University of Greenwich. UK. 330pp.

Bechoff, A., M. Poulaert, K. I. Tomlins, E. Westby, G. Menya, S. Young and C. Dhuique-Mayer, 2011. Retention and bioaccessibility of B-Carotene in blended foods containing orange-fleshed sweetpotato flour. *Journal of Agricultural and Food Chemistry* 59(18): 10373–80.

Bengtsson, A., A. Namutebi, M. L. Alminger and U. Svanberg, 2008. Effects of various traditional processing methods on the all-trans-b-carotene content of orangefleshed sweet potato. *Journal of Food Composition and Analysis* 21: 134–43.

Black, R. E., Allen, L. H., Bhutta, Z. A., Caulfield, L. E., de Onis, M., Ezzati, M., Mathers, C. and Rivera, J., 2008.Maternal and child undernutrition: global and regional exposures and health consequences. *Lancet* 371: 243–60.

Bouis, H., 2002. Plant breeding: A new tool for fighting micronutrient malnutrition. The American Society for Nutritional Sciences. *Journal of Nutrition* 132: 491–4S.

Carey, E. E., 1996. Not every clone is sacred: Arguments for the increased use of seed populations for conservation and utilization of sweetpotato germplasm. In V. Ramanatha Rao (ed.). Proceedings of the workshop on the formation of a network for the conservation of sweetpotato biodiversity in Asia held in Bogor, Indonesia, 1–5 May 1996, pp. 100–6. International Plant Genetic Resources Institute, Rome, Italy.

Carey, E. E., H. M. Kidanemariam and P. T. Ewell, 1998. Farmers' varieties and farmer based seed systems: Aspects of CIP's work on sweetpotato and potato improvement in eastern, central and southern Africa. In: Proceedings of the USC Canada/Seeds of Survival (SoS), 9th International Training Workshop On Sustainable Management, Development and Utilization of Plant Genetic Resources, 9–18 February 1998, pp. 86–97. Lesotho Agricultural College, Maseru, Lesotho. Unitarian Service Committee of Canada, Ottawa, Canada.

Carey, E. E., R. O. M. Mwanga, M. Andrade, G. Makunde, M. Chiona, E. Baafi, K. Some, J. Ndirigwe, G. Ssemakula, F. Chipungu, J. W. Low, P. E. Abidin and W. Grüneberg, 2016. Uptake, Adaptation and Output from Accelerated Breeding Programs for Sweetpotato in SSA (Keynote presentation). APA Conference, Addis Ababa, Ethiopia, October 2016 (in-press).

CIP 1998. Food security in East Africa: a battle on many fronts. International Potato Center (CIP) Annual Report 1998: 10–14.

CIP 2013. Rooting out hunger in Malawi with nutritious orange-fleshed sweetpotato. Annual Project Report (1 October 2012 to 31 October 2013) submitted to Irish Aid by International Potato Center (CIP), prepared by P. E. Abidin (project leader). 36pp.

CIP 2014. Rooting out hunger in Malawi with nutritious orange-fleshed sweetpotato. End of Phase I Project Report (1 October 2009 to 30 April 2014) submitted to Irish Aid by International Potato Center (CIP), prepared by P. E. Abidin (project leader). 38pp.

Collins, W. W. and J. M. Cannon, 1983. Sweetpotato production and problems in the United States. In D. I. Dolly (ed.). *Root Crops in the Caribbean*. Proceedings of the Caribbean Regional Workshop on Tropical Root Crops, Kingston, Jamaice, 10–16 April 1983, pp. 251–4. West Indies Univ. St. Augustine (Trinidad and Tobago).

Dery, E. K., K. B. Asare, K. Adofo, E. Obeng-Bio, E. Owusu-Mensah, I. Oduro, P. E. Abidin and E. E. Carey, 2016. Consumer participation in Ghana's sweetpotato breeding program: a two year case study. (Abstract submitted). 10th Triennial Conference of African Potato Association (APA) in Addis Ababa, Ethiopia, 9–13 October 2016.

Doering, W. von E., C. Sotiriou-Leventis and W. R. Roth 1995. Thermal interconversions among 15-cis, 13-cis, and all-trans-b-carotene: kinetics, Arrhenius parameters, thermochemistry, and potential relevance to anticarcinogenicity of all-trans-b-carotene. *Journal of American Chemical Society* 117: 2747–57.

Ebregt, E., P. C. Struik, P. E. Abidin and B. Odongo 2004a. Farmers' information on sweetpotato production and millipede infestation in north-eastern Uganda. I. Associations between spatial and temporal crop diversity and the level of pest infestation. *NJAS – Wageningen Journal of Life Sciences* 52–1:47–68.

Ebregt, E., P. C. Struik, P. E. Abidin and B. Odongo 2004b. Farmers' information on sweetpotato production and millipede infestation in north-eastern Uganda. II. Pest incidence and indigenous control strategies. *NJAS – Wageningen Journal of Life Sciences* 52–1:69–84.

Ebregt, E., P. C. Struik, B. Odongo and P. E. Abidin 2005. Pest damage in sweet potato, groundnut and maize in north-eastern Uganda with special reference to damage by millipedes (Diplopoda). *NJAS – Wageningen Journal of Life Sciences* 53–1:49–69.

Ebregt, E. 2007a. Are millipedes a pest in low-input crop production in north-eastern Uganda? Farmers' perception and experimentation. PhD thesis, Wageningen University, The Netherlands. With summaries in English and Dutch, 168pp.

Ebregt, E., P. C. Struik, B. Odongo and P. E. Abidin 2007b. Feeding activity of the East-African millipede Omopyge sudanica Kraus on different crop products in laboratory experiments. *NJAS – Wageningen Journal of Life Sciences* 54(3): 313–23. https://doi.org/10.1016/s1573-5214(07)80022-4

Ebregt, E., P. C. Struik, B. Odongo and P. E. Abidin 2007c. Piecemeal versus one-time harvesting of sweet potato in north-eastern Uganda with special reference to pest damage *NJAS – Wageningen Journal of Life Sciences* 55: 75–92.

Etumnu, C. E. 2015. Behavioral determinants of Bio-fortified food selection: The case of Orange-fleshed sweetpotato in Ghana. Study funded by University of San Fransisco and International Potato Center in Ghana (personal communication).

Ewell, P. T. and J. N. Mutuura 1994. Sweet potato in the food systems of eastern Africa. In F. Ofori and S. K. Hahn (eds.). *Tropical Root Crops in a Developing Economy*. Proceedings of the ninth symposium of the international society for tropical root crops, 20–26 October 1991, pp. 405–20. Accra, Ghana. IITA, Ibadan, Nigeria.

Gibson, R. W., I. Mpembe and R. O. M. Mwanga. 2011. Benefits of participatory plant breeding (PPB) as exemplified by the first-ever officially released PPB-bred sweet potato cultivar. *Journal of Agricultural Science* 149: 625–32. doi:10.1017/S0021859611000190

Grüneberg, W. J., E. Abidin, P. 'Ndolo, C. A. Pereira and M. Hermann 2004. Variance component estimations and allocation of resources for breeding sweetpotato under East African conditions. *Plant Breeding* 123: 311–15.

Grüneberg, W., R. Mwanga, M. Andrade and H. Dapaah 2009. Challenge Theme Paper 1: Sweet Potato Breeding. In Andrade, M., Barker, I., Cole, D., Fuentes, S., Gruneberg, W., Kapinga, R., Kroschel, J., Labarta, R., Lemaga, B., Loechl, C., Low, J., Ortiz, O.,Oswald, A. and Thiele, G. (eds). *Unleashing the Potential of Sweetpotato in Sub-Saharan Africa: Current Challenges and Way Forward*. International Potato Center (CIP). 208pp.

Grüneberg, W. J., D. Ma, R. O. M. Mwanga, E. E. Carey, K. Huamani, K. Huamani, F. Diaz, R. Eyzaguirre, E. Guaf, M. Jusuf, A. Karuniawan, K. Tjintokohadi, Y.-S. Song, S. R. Anil, M. Hossain, E. Rahaman, S. I. Attaluri, K. Somé, S. O. Afuape, K. Adofo, E. Lukonge, L. Karanja, J. Ndirigwe, G. Ssemakula, S. Agili, J. M. Randrianaivoarivony, M. Chiona, F. Chipungu, S. M. Laurie, J. Ricardo, M. Andrade, F. Rausch Fernandes, A. S. Mello, M. A. Khan, D. R. Labonte and G. C. Yencho 2015. Advances in sweetpotato breeding from 1992 to 2012. In: J. Low (eds). *Transforming Potato and Sweetpotato Value Chains for Food and Nutrition Security*, pp. 3–68, CABI Pub, Cambridge.

Hagenimana, V. and C. Owori 1997. Feasibility, acceptability and production costs of sweetpotato baked products in Lira municipality, Uganda. *Africa Potato Association (APA) Symposium Proceedings* 4: 209–13.

Hakiza, J. J., G. Turyamureeba, R. M. Kakuhenzire, B. Odongo, R. M. Mwanga, R. Kanzikwera and E. Adipala 2000. Potato and sweetpotato improvement in Uganda: A historical perspective. *African Potato Association (APA) Conference Proceedings* 5: 47–58.

Hernandez, T. P., T. Hernandez and J. C. Miller 1964. Frequency of somatic mutations in several sweetpotato varieties. *Proceedings of the American Society for Horticultural Science* 85: 430–3.

Hernandez, T. P. and J. C. Miller 1964. Further studies on incompatibility in the sweetpotato. *Proceedings of the American Society for Horticultural Science* 85: 426–9.

Hotz, C., C. Loechl, A. d. Brauw, P. Eozenou, D. Gilligan, M. Moursi, B. Munhaua, P. v. Jaarsveld, A. Carriquiry and J. V. Meenakshi 2011. A large-scale intervention to introduce orange sweet potato in rural Mozambique increases vitamin A intakes among children and women. *British Journal of Nutrition* 1–14.

Huaman, Z. and D. P. Zhang 1997. Sweetpotato. In D. Fuccilo, L. Sears and P. Stapleton (eds). *Biodiversity in Trust.* Conservation and use of plant genetic resources in CGIAR centres, Cambridge University Press, Cambridge, pp. 29–38.

Jaarsveld, P. V., D. W. Marais, E. Harmse and D. Rodriguez-Amaya 2006. Retention of β-carotene in boiled, mashed orange-fleshed sweet potato. *Journal of Food Composition Analysis* 19: 321–9.

Jayne, T., M. Villareal, M. Pingali and G. Hemrich 2004 Interactions between the agricultural sector and the HIV/AIDS pandemic: Implications for agricultural policy. In International Development Working Paper. pp. http://www.aec.msu.edu/agecon/fs2/index.htm. East Lansing, Michigan.

Johnson, L. E. 2016. Vitamin A (Retinol). Merck and the Merck Manuals. Professional Version. Accessible 19 November 2016 at http://www.merckmanuals.com/professional/nutritional-disorders/vitamin-deficiency,-dependency,-and-toxicity/vitamin-a.

Jones, A. 1965a. Cytological observations and fertility measurements of sweetpotato (Ipomoea batatas (L.) Lam.). *Proceedings of American Society for Horticultural Science* 86: 527–37.

Jones, A. 1965b. A proposed breeding procedure for sweetpotato. *Crop Science* 5: 191–2.

Khachatryan, A., R. R. Bansode, D. R. Labonte and J. N. Losso (2003). Age-related macular degeneration (AMD) is a pathological condition with real cure. Institute of Food Technologists (IFT) Annual Meeting, Chigago, USA. (http://ift.confex.com/ift/2003/techprogram/paper_20401.htm)

Kromann, P., F. Valverde, S. Alvarado, R. Vélez, J. Pisuña, B. Potosí, A. Taipe, D. Caballero, A. Cabezas and A. Devaux. 2016. Can Andean potatoes be agronomically biofortified with iron and zinc fertilizers? *Plant Soil.* DOI 10.1007/s11104-016-3065-0.

Liu, L., Johnson, H. L., Cousens, S., Perin, J., Scott, S., Lawn, J. E., Rudan, I., Campbell, H., Cibulskis, R., Li, M., Mathers, C., Black, R. E., for the Child Health Epidemiology Reference Group of WHO and UNICEF. Global, regional, and national causes of child mortality: an updated systematic analysis for 2010 with time trends since 2000. *Lancet* 379: 2151–61.

Low, J. W., M. Arimond, N. Osman, B. Cunguara, F. Zano and D. Tschirley 2007. A food-based approach introducing orange-fleshed sweet potatoes increased vitamin A intake and serum retinol concentrations in young children in rural Mozambique. *Journal of Nutrition* 137: 1320–7.

Kapalamula, C. 2014. Effects of intercropping 'orange-fleshed' sweetpotato with onion on the Level of weevil damage. Master Thesis, National Chung Hsing University International Master Program of Agriculture, Taiwan. 83pp.

Mead, R. and Willey R. 1980. The concept of 'Land Equivalent Ratio' and advantages in yields from intercropping. *Experimental Agriculture* 16: 217–28.

Miller, J. C. 1937. Inducing the sweetpotato to bloom and set seed. *The Journal of Heredity* 28: 347–9.

Miller, J. C. 1939. Further studies and technic used in sweetpotato breeding in Louisiana. *The Journal of Heredity* 30: 485–92.

Mukundi, E., M. Mathenge and M. Ngigi 2013. Sweet Potato Marketing Among Smallholder Farmers: The role of collective action. Invited paper presented at the 4th International Conference of the African Association of Agricultural Economists, 22–25 September 2013, Hammamet, Tunisia.

Mwanga, R. O. M., B. Odongo, C. Niringiye, R. Kapinga, S. Tumwegamire, P. E. Abidin, E. E. Carey, B. Lemaga, J. Nsumba and D. Zhang 2007a. Sweetpotato Selection Releases: lessons learnt from Uganda. *African Crop Science Journal* 15(1): 11–23.

Mwanga, R. O. M., B. Odongo, C. Niringiye, A. Alajo, P. E. Abidin, R. Kapinga, S. Tumwegamire, B. Lemaga, J. Nsumba and E. E. Carey 2007b. Release of two orange-fleshed sweetpotato cultivars, SPK 004 (kakamega) and ejumula in 2004 in Uganda. *HortScience* 42(7): 1728–30.

Nampeera, E. L., H. Talwana and M. Potts 2011. Effects of nonhost crop barriers on Cylas spp. sweetpotato infestation and damage (abstract). In Annon (ed.). *From Soil to Soul: Crop Production for Improved African Livelihoods and a Better Environment for Future Generation*, Entomology Session D3. 10th African Crop Science Society Conference in Maputo, Mozambique, pp. 112, 338.

Nastel, P., H. E. Bouis, J. V. Meenakshi and W. Pfeiffer 2006. Biofortification of staple food crops. *Journal of Nutrition* 136: 1064–7.

Owori, C., J. Kigozi and A. Mwesigye 1997. Standardisation and development of recipes for processing sweetpotato products. PRAPACE/CIP/NARO Annual progress report (June 1996 to February 1997).

Owori, C. and V. Hegenimana 2000. Quality evaluation of sweetpotato flour processed in different agro-ecological sites using small scale processing technologies. *African Potato Association Conference (APA) Proceedings* 5: 483–90.

Ruel, M. T., H. Alderman and the Maternal and Child Nutrition Study Group. 2013. Nutrition-sensitive interventions and programmes: how can they help to accelerate progress in improving maternal and child nutrition? *Lancet* 382: 536–51. http://dx.doi.org/10.1016/S0140-6736(13)60843-0

Scott G. J., R. Best, M. Rosegrant and M. Bokanga 2000. Roots and tubers in the global food system: A vision statement to the year 2020 (including Annex). A co-publication of CIP, CIAT, IFPRI, IITA, and IPGRI. Printed in Lima Peru: International Potato Center.

Simonne, A. H., Kays, S. J., Koehler, P. E. and Eitenmiller, R. R. 1993. Assessment of β-carotene content in sweetpotato breeding lines in relation to dietary requirements. *Journal of Food Composition and Analysis* 6: 336–45.

Tayo, T. O. 2000. Opportunities for increased potato and sweetpotato production in West Africa. *African Potato Association (APA) Conference Proceedings* 5: 27–30.

West, K. P. Jr. 2003. Vitamin A deficiency disorders in children and women. *Food Nutrition Bulletin* 24 (4 suppl): S78–90 (available at https://www.ncbi.nlm.nih.gov/pubmed/17016949/)

Woolfe, J. A. 1992. Sweetpotato: an untapped food resource. Cambridge Univ. Press and the International Potato Center (CIP). Cambridge, UK.

Yamakawa, O. and Yoshimoto, M. 2002. Sweetpotato as food material with physiological functions. *Acta Horticulturae (ISHS)* 583: 179–85.

Yen, D. E. 1974. *The Sweetpotato and Oceania: An Essay in Ethnobotany*. Bishop Museum Press, Honolulu, Hawaii.

Translating research into practice: improving cultivation in the developing world

Chapter 11

Potato production and breeding in China

Liping Jin, Chinese Academy of Agricultural Sciences, China

1 Introduction

2 Current production and consumption

3 Key trends and challenges

4 Germplasm material

5 Breeding objectives and development

6 Types of new variety

7 Virus-free seed potato production

8 Future trends

9 Where to look for further information

10 References

1 Introduction

Potatoes were first grown in China sometime between 1573 and 1619 (Zhai, 1980). According to historical records and documentary research, tetraploid cultivated potatoes were introduced from Europe and America to China by various routes at different times. At the end of the sixteenth century and the beginning of the seventeenth century, they were imported by sea to Beijing, Tianjin and other northern Chinese regions. They also travelled from the Philippines to Taiwan and Fujian, and thus spread to the provinces of Jiangsu and Zhejiang. It appears that they were brought to Shanxi province from Russia or the Kazakh Khanate (now Kazakhstan) by local businessmen; in addition, they came from Indonesia to Guangdong and Guangxi, and then spread to the regions of Yunnan, Guizhou and Sichuan (Tong and Zhao, 1991; Zhai, 1980).

The earliest written records of potatoes in China are found in a volume on products of the *Jifu tongzhi* (a local gazetteer or chronicle), which was compiled in 1682; by the middle of the eighteenth century, potatoes were widely cultivated in the regions of Beijing and Tianjin (Zhai, 1980). In his book *Spread of the American Plant*, Berthold Laufer (1938) quoted a description of potato planting from the diary of a captain who had been to Taiwan in 1650, and mentions that the potato was first introduced by the earliest pioneers from the Philippines to the coastal provinces of Taiwan, Guangdong and Fujian. Official records of Songxi County, Fujian province, published in 1700 mention that the public had been urged to plant potatoes in 1679, so between 1679 and 1700, potatoes had been

http://dx.doi.org/10.19103/AS.2017.0016.13

widely cultivated in that area. At the beginning of the nineteenth century, there were large areas devoted to potato cultivation in the provinces of Yunnan, Guizhou, Shanxi, Shaanxi and Gansu (Yang, 1978). From the mid-nineteenth century to the 1940s, the introduction of the potato continued apace through ports that were open for trade, and the number of available varieties also increased. In the three north-eastern provinces of Liaoning, Jilin and Heilongjiang, however, potato planting developed only gradually, possibly as late as the early twentieth century at the end of the Qing Dynasty.

As potatoes spread through China, they were given different names in different areas due to local dialects, the shapes of their tubers, their sources, and the method and time of introduction. These included Tu Dou (underground bean), Shanyao Dan (yam egg), Java Shu (Java tuber), Yang Yu (foreign taro), Yang Shanyu (foreign sweet potato), Di Dan (underground egg), Hongmao Shu (red hair tuber) and Holland Shu (Holland tuber). More than 20 different terms have been counted in total.

Since 1993, China has been the largest potato producer in the world. In China, given its wide distribution, multiple cultivation methods, year-round availability and good economic benefits, the potato has played a very important role in improving food security, increasing farmers' incomes, promoting agricultural modernization and developing regional economies.

2 Current production and consumption

From north to south, and low elevation to high elevation, potato production occurs all year round in China. Although potatoes are produced throughout the country, they are mainly cultivated in the cold northern regions and mountainous south-western China, where a single crop is planted, and autumn is the main harvest season. Sixty per cent of potato cultivation area is in arid and semi-arid regions, where potatoes are grown in rain-fed fields. Potatoes are usually intercropped with other crops except in the north of the country.

According to FAO statistics (http://faostat3.fao.org, accessed 20 March 2016), the Chinese potato harvesting area reached 5.65 million hectares in 2014, which was 29.43% of the global total. China produced 96.09 million metric tons of potatoes, which accounted for 24.96% of total global production. The yield was slightly below the world average, remaining at an average of 17 tons per hectare. Since 2002, harvest area, total production and yield have increased by 975 800 hectares, 25 867 million tons and 1982.2 kg per hectare, respectively (Fig. 1).

Potatoes are distributed in four agro-ecological zones (Heilongjiang Provincial Academy of Agricultural Sciences, 1994):

- the northern single-cropping zone (including the provinces of Heilongjiang, Jilin, Liaoning, Inner Mongolia, Hebei, Shanxi, Shaanxi, Ningxia, Gansu, Qinghai and Xinjiang),
- the central plain double-cropping zone (including the provinces of Hubei, Hunan, Henan, Shandong, Jiangsu, Zhejiang, Anhui and Jiangxi, and the southern parts of Liaoning, Hebei, Shanxi and Shaanxi),
- the southern winter-cropping zone (including Guangxi, Guangdong, Hainan, Fujian, Taiwan provinces, Hongkong and Macao),
- the south-western mixed-cropping zone (including Yunnan, Guizhou, Sichuan, Tibet and parts of Hunan and Hubei) (Fig. 2).

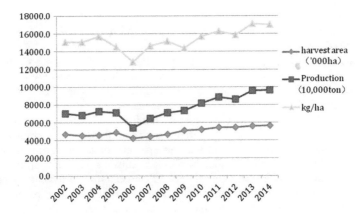

Figure 1 Progress of Chinese potato production from 2002 to 2014.

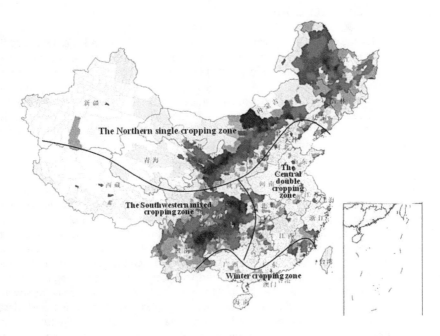

Figure 2 Agro-ecological potato production zones in China (the reddish areas are the main potato production areas).

The northern single-cropping zone and the south-western mixed-cropping zone are the two main potato-producing regions. These account for 48% and 37%, respectively, of the country's total potato planting area, and they are the main regions for both seed and commercial potato production. The central plain double-cropping zone and the southern winter-cropping zone account for 10% and 5% of the total potato planting area, respectively. They produce mainly high yields of early and mid-season potatoes which are sold mid-season and off-season as table stock making good profits for the growers

(Jin, 2013). The provinces of Sichuan, Guizhou and Gansu are the leading potato producers in China. The potato planting area for an average year in each of the above provinces is more than 600 000 hectares.

China is also the largest potato consumer in the world. Potatoes are utilized as a vegetable, staple food and raw material for processing. According to FAO statistics, the average annual per capita consumption in mainland China increased from 14 kg to 41 kg between 1994 and 2014 (http://faostat3.fao.org, accessed 20 March 2016). It is estimated that every year 61% of the total potato harvest is used for table stock as a vegetable, staple food and feed; 16% for processing into starch, starch noodles, powder, chips and French fries; 12% for seeds; and 10% is lost in storage or used for other purposes (Jin, 2013). Potatoes are used in hundreds of dishes in different cuisines as an important vegetable in the Chinese daily diet, while also being cooked with different grains as a staple food. It is expected that demand for potatoes will continue to increase due to the large and increasing population, and the need for high-quality, comprehensive nutrition.

3 Key trends and challenges

The potato is the fourth most important crop in China. It is cultivated widely across the country, both as a main food crop and as a cash crop in western areas, and thus plays a vital role in poverty alleviation, agricultural development and even regional economic progress. The livelihood of local farmers in areas stricken with extreme poverty depends on potato cultivation, which has become an important pillar industry. Potato planting accounts for over 30% of farmers' annual net income in some regions (Qu et al., 2010). China is now accelerating the development of the potato industry. Recently, the Chinese government has taken steps to support the potato industry and the potato planting area has increased year on year. It is expected that the expansion of potato planting into winter fallow paddy rice fields in the winter-cropping zone and increases in yield will contribute to the improvement of total production volumes. There are about 10 million hectares of winter fallow paddy rice fields, and it is estimated that it would be possible to increase the land available for potato production by 2 million hectares (Qu et al., 2006).

The major potato production areas commonly produce low yields due to poor production conditions, shortage of good quality seeds, poor field management and stresses such as diseases, drought and frost. In recent years, integrated cultivation methods and new technologies have been developed and applied to different production regions. Water-saving irrigation technologies are being used to retain soil moisture and collect rainwater for crop growing in arid areas. The technologies of ridging after flat sowing and plastic film mulching are promoted in the south-western zone to mitigate the effects of spring drought on potato growth, while in the high altitude mountainous region, potatoes are intercropped with corn and sweet potato. In the central plain double-cropping area, potatoes are grown in plastic greenhouses or covered by plastic mulch. In a number of areas, multiple intercropping with crops such as melons, vegetables and fruit crops is also promoted to increase farmers' profits (Li et al., 2009). In southern China, potatoes are also produced in winter fallow paddy rice fields, covered by rice straw (or rice straw plus black plastic film mulching), and are cultivated using reduced tillage techniques.

The level of mechanization in potato production in China is relatively low. Only the flat lands in the northeast, north and small parts of the northwest are suitable for mechanized production on a large scale. Where fully mechanized operation takes place, potato production companies and larger farmers have introduced large foreign machines. The majority of household farmers still use manual labour or animal power, since more than 60% of potatoes are grown in small- and medium-sized fields on sloping land in the western mountainous region with poor soil and limited facilities. With the support of research funding and the efforts of businesses, small- and medium-sized machines for plastic mulching, planting, inter-tilling, chemical spraying, harvesting and plastic recycling are being developed aimed at the western mountainous region. Research focused on the integration of cropping management and machinery has been carried out (Li, 2005; Huang et al., 2012).

The pressure of diseases is often severe in potato production fields. Late blight (*Phytophthora infestans*) and soil-borne diseases such as stem canker and black scurf (*Rhizoctonia solani*), *Fusarium* dry rot and wilt (*Fusarium* spp.), common scab (*Streptomyces scabies*) and powdery scab (*Pectobacterium atroseptica*) are the most significant problems nationwide (Tang et al., 2007). Potato viruses can also cause significant losses (Wang et al., 2011). Integrated disease control practices appropriate for different parts of China have been developed through years of research. Integrated late blight control measures consist primarily of appropriate cultivar deployment, disease forecasting, application of effective fungicides and good agricultural practices (Zhu et al., 2007). Potato production in the arid and semi-arid areas of the single-cropping region of China is also constrained by water shortages, while in the winter and double-cropping regions, low temperatures and frost damage during the growth season are often problems. These frequently result in significant losses to potato producers (Li and Jin, 2007).

4 Germplasm material

It is estimated that more than 5000 accessions of germplasm have been preserved by the Chinese national potato gene pool and breeding programmes. For example, the Institute of Vegetables and Flowers of the Chinese Academy of Agricultural Sciences has set up a low-temperature *in vitro* gene pool, conserving 2300 accessions of germplasm, including imported resources, domestic varieties and advanced clones. More than 2000 accessions have been identified and evaluated. The parent materials have been screened for biotic and abiotic stresses and good tuber qualities, as well as used for important genetic analysis, gene mapping and molecular breeding research programmes (Jin, 2006; Xu and Liu, 2011).

Pedigree analysis of potato germplasm material found that 379 cultivars were bred from 423 sources of parental material and that most parents (61.4%) were domestic varieties and clones. They were divided into eight groups including domestic varieties and breeding lines; indigenous breeds; neo-*tuberosum*; and materials introduced from North America, Europe, International Potato Center (CIP) and others. Their corresponding nuclear and cytoplasmic germplasm genetic contribution were 97.5, 107, 10.5, 16, 27.5, 71, 33 and 16.5, respectively, indicating that domestic varieties and breeding lines always play the major role in China potato breeding programmes (Duan and Jin et al., 2016). Eight varieties of Duozibai (292–20), Katahdin, Epoka, Mira, Anemone, Xiaoyezi and Zishanyao were found to have been important parents before 1980, that is, at an early stage of potato breeding in China (Xu, M. and Jin, L. 2007). Cheng (1987) reported that 80 released varieties were derived from the above-mentioned parents, accounting for 86% of the total 93 varieties bred from 1950 to 1979. Out of these, 23 varieties had been bred from Duozibai (292–20) (accounting for 24.7% of the total 93 varieties), 16 varieties from Katahdin (accounting for 17% of the total) and 14 varieties from Epoka (accounting for 15% of the total). Twenty-six varieties, such as Schwalbe, Aquila, Apta, Everest and Suskia, with high generally combining ability had been selected as good breeding parental material. Duan et al. (2009) reported that the fingerprinting of 88 cultivars approved in China from 2000 to 2007 was constructed by six pairs of SSR primers. Result of UPGMA cluster analysis of genetic similarity indicated narrow genetic basis of these cultivars.

After the seventeenth century, different potato varieties were continually introduced from Europe and America and grown in the different regions of China. Following natural and artificial selection, some varieties gradually adapted to the local conditions and became local varieties, such as 'Heba Yangyu' cultivated in the Yunnan, Guizhou and Sichuan provinces; the main cultivar 'Shenyanwo' found in Gansu and Qinghai province in the 1950s; 'Guangling Liwaihuang' widely cultivated in the single-cropping area in north China; 'Wuyangyu' planted in Sichuan, Gansu, Qinghai and Shanxi provinces; 'Xinyihongpi' grown in the winter-cropping zone of south China; and 'Fenshanyao', 'Zishaoyao', 'Daminghong' and 'Zaotudou' which were widely cultivated throughout China. These local varieties played an important role in potato production at different stages. The book *National Catalog of Potato Varieties Germplasm*, published in 1983, lists 123 native varieties (Heilongjiang Provincial Academy of Agricultural Sciences, 1983), which effectively solved the main production problems at that time. Some of these have become important parental material for potato breeding, and some still play a role in Chinese potato production today.

5 Breeding objectives and development

In general, high and stable yield, disease resistance and tuber quality are the most important breeding objectives (Jansky et al., 2009). The most important selection traits are tuber shape, shallow eyes, early maturity, high yield, disease resistance and stress tolerance. Breeding objectives vary between cultivation areas, however. In the northern single-cropping region, in which middle to late maturity varieties predominate, objectives such as resistance to late blight and black leg (particularly in the northeast) are significant. There is also emphasis on drought tolerance and resistance to soil-borne and viral diseases in the northern and north-western regions of China. Important traits for varieties grown in the central plains area are in particular early maturity and fast tuber expansion; breeding objectives also include early maturity, high yield, short dormancy and resistance to viral diseases and scab. Varieties in the high altitude area of south-western China thrive when they have high resistance to late blight, powdery scab and wart diseases, and are middle-late maturity and late maturity types. In the low altitude area of south-western China, however, middle maturity and early maturity varieties with resistance to late blight, viral disease are important. Varieties successful in south China in winter are of the early and middle maturity types, with insensitivity to day length, resistance to late blight, and ability to withstand moisture, cold and weak light.

Since the late 1930s, the methods used in China's potato breeding programmes have included introducing exotic varieties, better screening of domestic germplasm collections, inter-cultivar and inter-species crossing, and conventional breeding and molecular breeding combined into an integrated technology. According to Cheng (1987), new varieties were first developed and applied in the 1960s. Before 1983, potato variety development was focused on resistance to late blight and viral diseases, and over 90 varieties were released from potato breeding programmes, including the varieties of 'Kexin', 'Plateau' and 'Bashu' series.

In the 1980s, there was not much progress in breeding research and only 30 varieties (high-yielding table stocks) were released from national potato breeding programmes. With increased international exchange and cooperation in the 1990s, however, more potato germplasm, both for table stock and processing, was introduced to China from Europe, North America and the CIP; after 2001, potato breeding programmes received more research funding. From the 1990s onwards, the number of newly bred varieties increased significantly; at the same time, the genetic background of cultivars widened, allowing the improvement of traits such as disease resistance (Wang et al., 2005). More than 70 varieties including the 'Zhongshu', 'Jinshu', 'E Potato', 'Chunshu', 'Zhengshu', 'Longshu' and 'Qinshu' series were bred (Jin et al., 2004). Prior to 1983, 68.8% of the varieties released came from six common parents: Duozibai, Katahdin, Epoka, Mira, Anemone and Xiaoyezi. This reduced to 45% by 2005 (Xu and Jin, 2007). Exotic cultivars have also contributed greatly to the gene pool. Of the 379 varieties released before 2012, 13.7% were derived from North American stocks, 35.9% from European stocks and 17.9% from stocks of the CIP (Duan, 2013).

Chinese potato breeders have also made great efforts to resolve the problem of decreased yields due to viral infection through tuber propagation. Since the 1950s, the use of true potato seed (TPS) has been investigated as a method of producing clean seed tubers and increasing the multiplication rate (Sun, 2003; Wang et al., 2011). The variety 'Kuannae' was screened as an open-pollinated TPS producer because it produces uniform progenies with relatively high yield. In the 1970s, a nationwide network for research on TPS was established in China and uniform hybrid TPS producers were identified in 16 provinces in China, achieving TPS usage of 10 000 hectares by 1976.

There are currently more than 80 agencies engaged in potato variety breeding, but no more than 30 agencies have perennial breeding programmes. It is estimated that 299 new varieties have been released for starch processing, chipping, French fries and table stock since 2001. By the beginning of 2016, in total 521 new varieties had been released from potato breeding programmes, most of which are for table stock. In particular, the traits yield, quality, disease resistance and appearance have greatly improved in recent years (Bai et al., 2007; Ma et al., 2007; Su et al., 2007).

6 Types of new variety

Potato varieties can be categorized as native varieties or improved varieties on the basis of their sources and methods of development. They can also be classified by growth period into very early, early, early middle, middle, middle-late, late and very late varieties. Then by different usage, varieties can be roughly classified into the following types: table stock, chip processing, French fry processing, starch processing, powder and table, table and starch, fresh and processing, and special and colour types. One of those, the table stock variety is still the main breeding focus: table stock varieties account for 67.2% of total varieties bred, while 15.7% are for starch processing or of the table and starch types.

The early maturity varieties are mainly distributed in the central plain double-cropping region, the southern winter-cropping zone and the south-western low attitude double-cropping region. The middle and late maturity varieties are mainly grown in the northern single-cropping region and south-western middle and high attitude region. French fries and powder processing varieties are mainly distributed in the middle and west of Inner Mongolia, Zhangjiakou in Hebei Province, and some regions of Yunnan and Gansu provinces. High starch varieties are mainly distributed in the north-western, north-eastern and south-western middle and high attitude regions. Seed potato production is mainly carried out in northern high latitude and south-western high attitude regions.

The varieties used in Chinese potato cultivation have varied over the years and several rounds of variety substitution have been identified. Varieties such as 'Triumph', 'Katahdin' and 'Wuxia' were selected from materials introduced in the 1930s and 1940s, and played an important role in the potato production in the 1950s. Then, in the first round of variety substitution, the three late mature varieties 'Mira', 'Epoka' and 'Aquila' and one early mature variety 'Anemone' were selected from germplasm materials introduced from the former Soviet Union and Eastern Europe, becoming the main varieties used in the 1960s and 1970s. Until 1990s, 'Mira' and 'Aquila' were still planted in large scale in southwest China. In what is known as the second round of variety substitution, dozens of new varieties of 'Hutou', 'Yuejin' and 'Jinshu 2' with potato late blight resistance and high yield were bred in China and approved by the National Variety Approval Committee in the late 1960s.

Since the 1970s, national breeding programmes have bred most of the varieties used in Chinese potato production. Compared to the earlier varieties, the recently bred new varieties have better tuber appearance, higher pathogen resistance and yield, and broad adaptability. In the 1970s and 1980s, some exotic varieties, such as 'Favorita', 'Cardinal', 'Desiree' and 'CIP 24', were also introduced, and 'Favorita' is still one of the major early mature cultivars used by farmers. Meanwhile, some middle-late mature cultivars of 'Bashu 8', 'Kexin 1' and 'Gaoyuan 7', and some early mature cultivars of 'Zhengshu 2', 'Zhengshu 4' and 'Bashu 9' were released and promoted.

In the 1980s and 1990s, with the large-scale application of early maturing new varieties of 'NEA 303', 'Zhongshu 2', 'Zhengshu 5', 'Zhengshu 6' and 'Chuanyuzao', the production of early mature potatoes in central plains was rapidly developed. Since the late 1980s, some potato varieties suitable for special processing have also been selected including the introduced varieties 'Atlantic', 'Shepody', 'Agria' and 'Snowden', and fresh varieties 'Kondor' and 'Kennebec'. At present, 'Atlantic' and 'Shepody' are the main chipping, French fry and powder processing varieties. By around the year 2000, the third round of variety replacement had been completed.

After 2000, great attention was paid to the selection of special varieties for early maturity, freshness, export potential, high starch levels and processing, and nearly 300 new varieties were bred. There were particular breakthroughs in the selection of processing varieties. In 2006, two varieties of 'Zhongshu 10' and 'Zhongshu 11' for chipping were approved by the National Variety Approval Committee for the first time in China. With a number of new varieties such as 'Zhongshu 3', 'Zhongshu 5', 'Qingshu 9', 'Jizhangshu 12', 'Zhongshu 18', 'Zhuangshu 3', 'Longshu 7' and 'Emalingshu 5' being widely grown, the fourth round of variety substitution was completed around 2010.

7 Virus-free seed potato production

With the development of the potato industry and the rise in potato consumption, demand for seed potatoes has increased rapidly (Kang, 2007). Seed potato production currently mainly takes place in the northern single-cropping zone at high latitudes and the south-western mixed-cropping region at high altitudes. In China, virus-free seed potato production started in the 1970s, became more common in the mid-1980s and has only been widely used in the last ten years or so (Wang et al., 2011). The process as carried out in China includes virus removal, tissue culture, pre-basic potato seed (microtuber or minituber) production, basic potato seed production, field seed potato propagation, and quality testing and control. Pre-basic potato seeds are usually produced in greenhouses or net sheds, by transplanting *in vitro* plantlets (microtubers) into the substrate at high density, or by transplanting *in vitro* plantlets and cuttings for aeroponic culture.

Despite the promulgation of a series of standards and procedures, an effective system of quality supervision and control has not really been established in the initial seed potato production process (Bai et al., 2006). Methods of monitoring in-field seed potato production are also inadequate, so there are still great difficulties in controlling seed quality. In addition, the tissue culture and pre-basic potato seed production techniques require first-rate facilities, entail high energy consumption, and are cost and labour intensive. The result is the high cost and limited supply of commercial qualified seed potatoes. The coverage of high-quality seed potatoes in China is therefore very low, at only about 35% (with differences between regions as well), and most farmers still use seed potato they have produced themselves for the next season. The technology of virus-free seed potato production needs to be improved as soon as possible. The Chinese seed potato propagation system is shown in Fig. 3.

The seed potato production industry used to be dominated by smallholders, but now involves larger enterprises. As a result, the quality of the product is improving and the capacity for seed production is increasing: virus-free seed potato production now takes

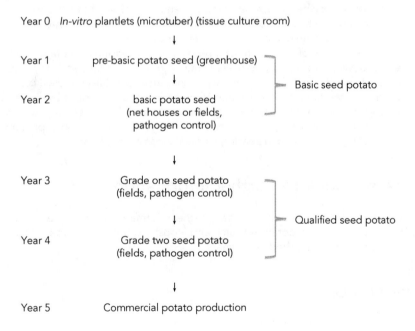

Figure 3 The seed potato propagation system in China.

place in more than 20 provinces. Enterprises usually establish their own potato seed brands. The existing large enterprises have their own complete systems for seed potato propagation, including combination management for tissue culture, net house production, field production, pathogen control and storage. The planting units of seed companies have reached up to 6000 hectares in size, on which medium- and large-size machines are utilized instead of the animals and small machines used by smallholders.

In the northerly single-cropping region, a method of large-scale machine cultivation with water-saving irrigation has been developed. Chinese government departments have also issued a series of policies on the development of the potato industry, including a series of seed quality testing and inspection standards, technical operational procedures and management methods.

8 Future trends

The Ministry of Agriculture (MOA) has paid great attention to the development of the potato industry in China and issued documents outlining its views on accelerating the development of the Chinese potato industry in 2006 and 2016. The China Potato Research System (part of the China Agriculture Research System (CARS)) and the MOA's Key Labs dedicated to potato research were built in 2008 and 2011, respectively. The MOA's Key Labs focus on basic and applied basic research related to potato biology and genetic improvement. The Potato Research System focuses on the development of applied technologies for the potato industry, such as variety breeding, seed potato propagation technology, crop management, integrated management of diseases and pests; general machinery development; development of farm machinery and post-harvest handling systems for small famers; new processed products and the delivery to the public of health, safety, nutrition and new technology messages associated with potatoes through publicity and media campaigns; training activities; and demonstrations.

In China, consumption of the potato as a vegetable is increasing as more people move from the countryside into the city, and will continue to increase further. Consumption of potato products which meet Chinese dietary habits, such as potato noodles, breads and slices, will also increase. As the population is increasing, the amount of available arable land is decreasing and water is becoming scarcer. As an all-purpose crop with a long industrial chain, the potato will play an ever increasingly important role in improving food security, alleviating poverty and increasing farmers' incomes in China.

9 Where to look for further information

Besides the listed references of this chapter, further information also can be looked for in the CARS annual report on potato research and industry progress, and the China Potato Congress proceedings published every year.

10 References

Anon (1993), 'Production' http://faostat3.fao.org (accessed 10 May 2016).
Anon (2014), 'Production' http://faostat3.fao.org (accessed 20 March 2016).

Bai, X., Du, Z., Qi, H., Fan, X., Zhang, Y. and Wang, J. (2007), 'Selection of a new potato variety Jinshu no. 5 with fine quality and disease resistance', *Chinese Potato J.*, 21(2): 128–9.

Bai, Y., Li, X., Wen, J. and Yang, M. (2006), 'Analysis and comparison of potato seed tuber production between China and Holland', *Chinese Potato J.*, 20: 357–9.

Cheng, T. (1987), 'Improvement of potato breeding in China', *Chinese Potato J.* 3: 32–5.

Duan, Y., Liu, J., Bian, C., Duan, S.C, Xu, J. and Jin, L. (2009), 'Construction of fingerprinting and analysis of genetic diversity with SSR markers for eighty-eight approved potato cultivars (*Solanum tuberosum* L.) in China', *Acta Agronomica Sinica*, 35(8): 1451–7.

Duan, S., Jin, L., Li, G., Bian, C., Xu, J., Hu, J. and Qu, D. (2016), 'Analysis of genetic relationship and Cytoplasm type on potato cultivars in China', *Acta Horticulturae Sinica*, 43(12): 2380–90.

Heilongjiang Provincial Academy of Agricultural Sciences (1983), *National Catalog of Potato Varieties Germplasm*. Heilongjiang Scientific and Technology Press, Harbin.

Heilongjiang Provincial Academy of Agricultural Sciences (1994), *Potato Culture Technology in China*. China Agricultural Press, Beijing, pp. 1–80.

Huang, J., Wang., X. and Qui, H. (2012), *Small-scale Farmers in China in the Face of Modernization and Globalization*, IIED/HIVOS, London, UK.

Jansky, S., Jin, L., Xie, K., Xie, C. and Spooner, D. (2009), 'Potato production and breeding in China', *Potato Research*, 52: 57–65.

Jin, L. (2006), *China Crops and Their Relatives*, Edited by Dong, Y. and Liu, X. China Agriculture Press, Beijing, pp. 506–59.

Jin, L. (2013), 'Status and prospect of potato industry in China'. In Chen, Y. and Qu, D. (eds), *Potato Industry and Regional Development of Countryside*. Harbin Engineering University Press, Harbin, pp. 8–18.

Jin, L., Qu, D., Xie, K., Bian, C. and Duan, S. (2004), *Potato Germplasm, Breeding Studies in China*. Proceedings of the Fifth World Potato Congress Kunming, China, pp. 175–8.

Kang, J. (2007), 'Current status of potato virus-free potato seed tuber production', *Chinese Potato J.*, 21(2): 121–2.

Li, C., He, X., Zhu, S., Zhou, H., Wang, Y. and Li, Y. (2009), 'Crop diversity for yield increase', *PLoS ONE*, 4(11), doi:10.1371/journal.pone.0008049.

Li, F. and Jin, L. (2007), 'Frost damage and prevention in potatoes'. *Guizhou Agric. Sci.*, 35(3), 121.

Li, S. (2005), 'Agricultural mechanization promotion in China', *Agri. Eng. Int.*, VII: 1–5.

Ma, H., Yin, J. and Zhang, X. (2007), 'Breeding of a new potato variety Jizhangshu no.8', *Chin. Potato J.*, 21(3): 192–3.

Mu. T., Sun, H. and Liu, X (2016), *Potato Staple Food Processing Technology*, Springer Briefs in Food, Health and Nutrition, Singapore.

Qu, D., Jin, L. and Xie, K. (2006), 'Vigorously develop the winter agriculture to lay a solid foundation for food safety'. In D. Qu and Y. Chen (eds), *Potato Industry and Winter Agriculture*, Harbin Engineering University Press, Harbin, pp. 1–6.

Qu, D., Jin, L. and Xie, K. (2010), *Ten Years Review of China Potato Industry (1998-2008)*, Chinese Agricultural Technology Press, Beijing.

Su, X. and Lai, L. (2007), 'Performance of potato variety Kexin no. 18 sown in winter and its high-yielding cultivation techniques', *Chin. Potato J.*, 21(2): 118–19.

Sun, H. (2003), *China Potato Breeding*, China Agriculture Press, Beijing.

Tong, P. and Zhao G. (1991), *The History of Potato in China*, Chinese Agricultural Scientific and Technology Press, Beijing.

Wang, B., Ma, Y., Zhang, Z., Wu, X., Wu, Y., Wang, Q. and Li, M. (2011), 'Potato viruses in China', *Crop Prot.*, 2011: 1–7, doi:10.1016/j.cropro.2011.04.001.

Wang, X., Jing, L. and Yi, H. (2005), 'Advances in breeding of potato virus-resistant cultivars', *Chinese Potato J.*, 19: 285–9.

Xie, K. and Wang, Y. (2001), 'CIP potato late blight research in China', *J. Agric. Univ. Hebei*, April 2001: 1–5.

Xu, M., Jin, L., Bian, C., Duan, S., Liu, J., Xie, K., Pan, W. and Qu, D. (2007), 'Pedigree analysis of approved potato varieties in China'. In 2007 *China Potato Congress Proceeding*, Harbin Engineering University Press, Harbin. p157–161.

Xu, J., Liu, J., Bian, C., Duan, S., Pang, W. and Jin, L. (2011), 'Identification and selection for drought resistant potato germplasm', *Chinese Potato J.*, 25(1), 1–6.

Yang, H. (1978), *History of Potato Production in China*, Xi'an Science and Technology Information Research Institute, Xi'an.

Yao, B. (2008), 'NPK fertilizer response and optimum application rate for winter potatoes', *Fuijan J. Agric. Sci.*, 23 (2): 191–5.

Zhai, Q. (1980), 'The introduction history of sweet potato and potato in North China'. In *Agricultural Science and Technology in Ancient China*, edited by Gou, C., China Agriculture Press, Beijing.

Zhu, J., Yang, Z., Zhang, F., Du, H. and Liu, D. (2007), 'Review of the genetic structure of potato late blight pathogen [*Phytophthora infestans* (de Bary)] Population', *Sci. Agric. Sinica*, 40(9): 1936–42.

Chapter 12

Improving potato cultivation to promote food self-sufficiency in Africa

Moses Nyongesa and Nancy Ng'ang'a, Kenya Agricultural and Livestock Research Organization, Kenya

1 Introduction

Statistics indicate that in sub-Saharan Africa (SSA), per capita consumption of potato increased the most in absolute terms compared to other root and tuber crops over the period 1983–1996 (Scott et al., 2000). For the period 1993–2020, the food demand for potato in SSA is projected to grow at the rate of 3.1% helped by the high status of potato as a preferred food and population growth (Scott et al., 2000). Furthermore potato fits in the local food systems as it adapts well to varied soil and climatic conditions helped by abundant labour found in these regions. However, growth in demand does not match the projected growth rate in potato production of 2.02% over the same period. The interplay of factors ensures the gains achieved are small and slow.

African agriculture, dominated by smallholder rural farmers, faces significant challenges to which successive interventions have only met limited successes. Yet, the sector represents the greatest prospects for rural development and economic growth for many countries in Africa. While highlighting potato crop improvement initiatives in the continent as an effort towards food security and poverty reduction, this chapter discusses the application of technology and the need to resolve marketing challenges as the new frontiers in the promotion of potato in the continent. Lessons learnt from interventions through PRAPACE (French acronym for Regional Potato and Sweet Potato

http://dx.doi.org/10.19103/AS.2017.0016.15

Improvement Network for East and Central Africa) imply the need for a systematic and consistent effort by governments and international donors to support growth of potato production (Crissman, et al., 2007). To realize the full benefits of technologies to target groups, other equally important factors need attention. Key issues among these are access to financial facilities and improved inputs, methods of production, knowledge and technology dissemination and well-structured markets. Growth in this sector represents enormous opportunities to spur economic growth at two levels: one, via commercial village models based on sound agribusiness principles to support rural-based processing and two, entry of intra-industry-based business models of multinational companies with large processing operations.

2 Potato production in Africa and its challenges

2.1 Potato production in Africa

Perspectives on global food supply and strategies for crop developments in Africa tend to cluster cassava, potato, sweet potatoes and yams into root and tuber crops. Yet the relative importance of individual root and tuber crops for use as food varies considerably in different African countries. Individually, potato varies greatly among African countries as a food or cash crop. The crop has been part of the food supply system in Africa since the late nineteenth century and continues to be a major food crop to date. Easily adaptable to a wide range of climatic conditions, the crop is produced in the Nile Delta regions of Egypt, tropical highlands and plateaus of East and Central Africa (Burundi, Congo, Ethiopia, Kenya, Rwanda and Uganda), the Mediterranean region of North Africa (Algeria and Egypt) and lowland tropical zones of South Africa. According to FAOSTAT (2015), potato is currently being produced in 43 African countries. Figure 1 shows a 7.6% contribution to global potato production from Africa

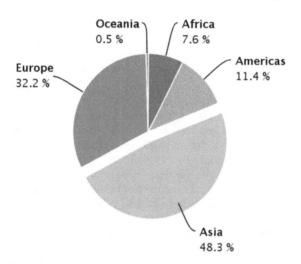

Figure 1 Share of world potato production. Source: FAOSTAT, January 2016.

annually, exceeding only Oceania. Although FAOSTAT is a widely accepted reference on crop production data, opinion is divided with regard to the reliability of potato production estimates in Africa (Low et al., 2007). The example of Nigeria and Ethiopia, where production data are grossly overestimated or underestimated, respectively, is cited to lament that any attempt at projecting production trends for this region is a futile exercise (Walker et al., 2011). Acknowledgement of this fact is urged because observations we make in this chapter are partly anchored on the FAOSTAT data having found no alternative source. Notwithstanding these observations, available data indicate that during the period between 2010 and 2014, Egypt, Algeria and Malawi were leading producers in the continent with annual production exceeding four million tons for each of these countries (Table 1). Several of these top producers including Egypt, Ethiopia, South Africa and Tanzania produced a surplus that was exported, while Algeria, Angola, Kenya and Morocco were over 96% self-sufficient but were net importers of potatoes (Table 2).

2.2 Challenges

The global population projected to hit the nine billion mark by 2050 signifies particularly unique challenges for developing countries where the most growth is anticipated (UN Secretariat, 2010). Increasing food needs will require more innovative ways of crop production to be explored as will be food crops that require less resources to cultivate. Yet, population growth is not the only challenge.

Hijmans (2003) predicted that for the period 1961–1990 and 2040–2069, there will be an average increase of 2.1–3.2°C in global temperature and 18–32% corresponding decrease in potato yields in the absence of mitigation measures. Already, farmers in Africa have become familiar with the effects of climate change. Excessively wet weather conditions that result in flooding or outbreak of crop diseases make farming challenging.

Table 1 Top potato-producing countries in Africa during the period 2010–2014

Country	Average annual production (tons)	Average acreage (ha)	Average productivity (Hg/ha)
Egypt	4 360 973	164 012	265 598
Algeria	4 188 407	141 979	294 018
Malawi	4 128 731	244 877	168 207
Kenya	2 365 036	131 174	180 855
South Africa	2 210 241	64 576	342 217
Rwanda	2 152 885	163 218	131 696
Morocco	1 745 174	56 354	310 795
Tanzania	1 558 331	191 836	81 061
Nigeria	1 159 612	263 136	44 049
Angola	735 659	101 189	72 943
Ethiopia	698 589	64 512	106 479

Source: FAOSTAT, January 2016.

Table 2 Average potato production supply for top producing countries in Africa during the period 2010–2013

	Average production (tons)	Average imports (tons)	Average exports (tons)	Average available supply (tons)	Self-sufficiency ratio (%)
Algeria	4 067 130	128 023.50	317.50	4194 836	96.9
Angola	751 707	25 102.00	0	776 809	96.6
Egypt	4 251 217	148 788.00	406 720.50	3 993 284	106.6
Ethiopia	642 779	48.00	59 681.25	583 146	110.0
Kenya	2 549 788	4 269.75	1 815.75	2 552 242	99.9
Malawi	3 993 747	575.00	228.75	3 994 093	100.0
Morocco	1 727 880	43 056.25	25 738.50	1 745 198	99.0
Nigeria	1 137 500	254.25	1 244.25	1 136 510	100.1
Rwanda	2 134 836	604.25	6 948.25	2 128 492	100.3
South Africa	2 197 072	145.75	70 280.25	2 126 937	103.3
Tanzania	1 507 663	3 101.25	24 175.00	1 486 590	101.8

Note: Available supply is merely net production plus net imports.
Source: FAOSTAT, February 2016.

Among the important diseases affecting potato production are late blight (*Phytophthora infestans*), bacterial wilt (*Ralstonia solanacearum*) and potato viruses. Being seed borne, these diseases are not only of phytosanitary significance in many countries but also pose serious challenges to seed quality management and increase costs of production. Similarly, frequent droughts have rendered many previously productive farming areas unsuitable for crop production.

Poor infrastructural networks in many African countries greatly hamper the transportation of production inputs to farms and farm produce to markets, meaning the cost of inputs is high and farmers receive low prices for their produce. It follows that increased investment in rural roads to link them to urban centres and regional markets has the potential to reverse underperformance of rural economies. Efficient transportation of farm produce for instance contributes to reduced losses while in transit and ensures the produce gets to consumers when still fresh. Processed potato products of better quality are easily achieved when the transportation system favours efficient delivery of raw materials within a short time after harvest.

However, immature agro-processing sectors in many African countries greatly limit the exploitation of opportunities that potato value chains can provide (Ugonna, 2013). Few African countries have well-organized and recognized stakeholder forums, the absence of which implies weak intra-sector linkages. The forums serve as platforms at national level to define roles and organize players including processors. These challenges require supportive policy guidelines and positive government involvement both at the national and regional levels. Recognizing the importance of agriculture as a key driver of economic development, African governments have addressed themselves to these issues through various declarations, the most recent being the Malabo Declaration (2014).

This chapter considers the following as being pillars of an effective strategy to improving potato production in Africa:

1. Development of new varieties adapted to common challenges of soil, moisture and temperature and that respond to biotic and abiotic risks
2. Streamlining and innovating seed potato production and supply systems that respond to unique African realities to deliver adequate planting materials of good quality to farmers affordably
3. Streamlining markets for fresh potatoes and increasing support towards agro-processing
4. Strengthening the capacities of national research programmes through improvement of infrastructure and training of personnel to have a critical mass of human resource with knowledge and skill to develop home-grown solutions to farming challenges in the region
5. Building on past gains and leveraging on the support from international development agencies to contribute to continental food security and developmental initiatives

These pillars resonate well with a study commissioned by the Bill & Melinda Gates Foundation (NRC, 2009), which argued for a systems approach, targeting each element of the production system, towards improved agricultural productivity.

3 Variety development and promotion

Besides the production inputs that affect crop performance, there is genetics. The right combination of genes ultimately determines the yield potential of a variety. Suitable varieties need not only be those that are well adapted to the local biotic and abiotic stresses but also those that meet the trait specifications of the consuming population. The work of the breeder is thus challenging because these are not static requirements, rather they are fast evolving and therefore newer and faster methods to deliver new varieties are needed.

Varieties grown in Africa fall into four main categories: (i) imported varieties adapted to local conditions, (ii) local varieties of unknown origin, (iii) local varieties selected from germplasm sourced from the international potato centre (CIP) – a member of the Consultative Group on International Agricultural Research (CGIAR) system and (iv) locally bred varieties. It is usual to find a complement of varieties from a combination of two or more of the previously mentioned categories. Broadly, most widely grown varieties in the north African countries of Tunisia, Algeria and Egypt are imported from Europe, whereas those grown in eastern African countries including Kenya, Ethiopia and Tanzania either have unknown origins and are identified by local names or originated from CIP. It is only in South Africa, which has a vibrant breeding programme, that locally bred potato varieties (e.g. BP1) are grown more widely than those varieties which are imported from Europe. In Nigeria, a combination of locally bred varieties, imported varieties adapted to local conditions as well as advanced clones from CIP are grown. Despite a large collection of varieties, few are grown on a commercial scale and only one or two occupy more than half of the total area under potato cultivation. Table 3 shows variety preference in eight African countries. This situation is fairly representative of many other countries in the continent.

Table 3 Popular varieties, their origin and percentage of total land area under potato production in eight African countries

Country	Most popular varieties	Source of variety (s)	Reference
South Africa	BP1 and Up-to-Date (77% of potatoes grown)		Republic of South Africa, 2003, 2013
Tunisia	Spunta (85%)	Europe	Fact sheet, 2013
Nigeria	Nicola	Europe	NRCRI, 2016
Algeria	Spunta (60% of potatoes grown)	Europe	
Malawi	Rosita, Betane (over 80%)	Local	Kateta et al., 2015
Kenya	Shangi, Dutch Robjyn, Asante	Local, Europe, CIP	
Ethiopia	Siquare, Agazar, Gudene	Local, local, CIP	Kolech et al., 2015
Tanzania	Kikondo and Arka (91%)	Local, Kenya	Mpogole, 2013

Changing consumer tastes or needs for better processing varieties alters the market share of older varieties and often result in the entry of new varieties into the markets.

Variety development in many African countries often comprises collaborative efforts between national potato programmes and international agencies linked to established potato-breeding centres based outside the African continent. Often, advanced breeding materials are donated for adaptation trials in the receiving countries leading to registration of the selected lines as varieties.

CIP, through the SSA regional office based in Nairobi, has made substantial contribution in this way in East and Central African countries, namely Kenya, Uganda, Ethiopia, Madagascar, Burundi, Rwanda and Democratic Republic of Congo. The PRAPACE network drew membership from these countries. Initially, CIP breeding programme in this region was focused on availing varieties with resistance to potato late blight (*Phytophthora infestans*), a major disease limiting potato production in the tropical highlands of SSA. However, the scope has expanded in the recent past to include development of varieties adaptable to mid-altitudes and those with heat-tolerant and late-blight resistance traits targeting the following SSA countries: Angola, Burundi, DR Congo, Ethiopia, Kenya, Tanzania, Malawi, Mozambique, Rwanda, Uganda, Nigeria and Mali.

It can be surmised that Africa is overwhelmingly dependent on external support for variety development. Generally, locally bred varieties are an exception. In Kenya, local breeding to complement the selection of CIP-sourced germplasm was recently reintroduced after many years of nil local cross-breeding. In North Africa, production for the European market is perhaps a key determinant of varieties grown and consequently potato breeding is relatively new with the Algerian potato-breeding programme introduced between 2007 and 2009.

Sustainability of a national potato programme whose variety development agenda is anchored on donor support is difficult to envisage. Instead, capacity building is a clear priority geared towards equipping local personnel with requisite breeding knowledge and skills to formulate relevant breeding objectives and implement programmes that respond to national demands. Without a doubt, breeding programmes require a lot of funding to sustain. This could partly explain the scarcity of functional breeding programmes in most African countries whose potato programmes are not only understaffed but also poorly resourced financially. In fact, root crop programmes are more severely affected by underfunding compared to cereal crop programmes (Walker, 2014). These conditions translate to work environments that encourage the departure of skilled manpower to alternative programmes that provide better career prospects. Aptly described as brain drain, this phenomenon has grown to levels that have become a serious constraint to food security efforts in many African countries.

Managers of research programmes need innovative policies to counter brain drain. It would appear that efforts at bonding staff for a period of time immediately after training for higher degrees are not enough to prevent these loses. In contrast, programmes with competitive employment terms coupled with an elaborate royalty policy that recognizes and compensates breeders for each variety developed are more likely to retain staff longer. Faced with perennial insufficient budgetary allocations, managers of national agricultural research bodies have been left with fewer options on how to mitigate these challenges. This outlook is generally valid for most SSA countries. New evidence suggests an increase in expenditure in R&D and considerable growth in human resource capacity in three countries, Nigeria, South Africa and Kenya (Beintema and Stads, 2014).

Another feature common among potato-breeding programmes in Africa is the absence or limited use of modern tools for conventional breeding. When available to the breeder, these tools yield tremendous benefits and increase the efficiency of delivery of crop varieties with desired traits. For example, prior knowledge and use of phenotypic characteristics with DNA markers greatly enhance the achievement of optimal crosses and selection of offsprings, thus shortening the crop improvement cycle (Ottoman et al., 2009). Additionally, there are obvious disadvantages in selecting candidates from a population based purely on visual observations particularly for traits with low-to-mid heritability. In this case, the breeder would be better off with molecular techniques based on single-nucleotide polymorphisms as markers to reveal variation between allelic genes. Rapid advancement in technology has made sequencing faster and affordable making it increasingly feasible for African scientists to mine the complete sequence of the potato genome for favourable traits in crop improvement. Enhancing the exploitation of these tools holds the key to quick delivery of varieties to meet the demands of the fast-changing market scenes characterized by a growing and increasingly sophisticated urban middle class in many SSA countries.

It is at the national agricultural research centres (NARCs) that germplasm for crop improvement and varieties is conserved. Long-term conservation methods such as cryopreservation and *in vitro* cultures are costly for most programmes to maintain. Germplasm is therefore maintained alternately *in situ* or in tuber form in storage on a short-term basis. The role of germplasm conservation is shared by farmers who hold varieties which they maintain on their farms aided by self-seed supply. Farmer contribution to variety/germplasm conservation in this case is inadvertent and realized only for varieties that remain in the national production system. In the long run, there is a possibility of loss of varieties that lack consumer-preferred traits.

Finally, there remain legitimate concerns from players in the agricultural sector about the implications on African agriculture, which is still dominated by small-scale growers, of the strict enforcement of the provisions of the International Treaty on Plant Genetic Resources for Food and Agriculture and the Union for the Protection of New Varieties of Plants (UPOV) agreements. For example, question about access to and the issues of farmers rights continue as plant genetic resources for food and agriculture continue to excite heated debates (Egziabher et al., 2011). Yet, it is possible to operate models designed to address the food security and developmental concerns while meeting the obligations to the UPOV convention. For example, a flexible application of UPOV regulations to allow replanting of seed potatoes of protected varieties by farmers with less than 2.5 ha land size under potato is compatible with realities of seed supply in SSA and consistent with food security agenda for the continent.

4 Systems and programs to support potato production in Africa

4.1 Seed potato supply and delivery systems

Seed potato is the single most important production input affecting potato yield. The attributes of the desired variety are locked in the seed, which also serves to deliver crucial technologies such as disease resistance to the grower. The importance of this significant production input is undermined by challenges of access to quality seed potatoes facing farmers in most of SSA countries. Generally, farmers plant seed potatoes produced on their own farms, sourced from neighbours or from markets. Repeated use of such low-quality planting material easily leads to the build-up of diseases, declining yields and food availability, and reduced incomes for farmers. This situation is a consequence of inefficient performance of seed potato production and supply systems in the affected countries. Seed potato supply in most of SSA is characterized by a continuing failure to meet national seed potato demands.

Even in countries with regulated seed systems comprising mandatory regimes of crop inspection during production as part of the certification process, only a small percentage of growers have access to stocks of quality planting material. Rigid policies that make no provisions for innovative and practical strategies to increase seed quality at farm level have aggravated the problem of suboptimal access to quality seed potatoes by farmers (Fajardo et al., 2010). Yet, increasing the supply of quality seed potato to farmers remains a key agenda.

As shown by Demo et al. (2015) through their work in five countries (Ethiopia, Kenya, Malawi, Rwanda and Uganda), rapid increase in quality seed can be achieved through a holistic approach combining increase in foundational seed potato tubers, increase in specialized early field multiplication, scaling up and out of the decentralized seed potato production, implementation of a quality control system, extensive awareness creation to support seed distribution linkages and farmer training to maintain seed quality at farm level.

By comparing various basic seed potato production systems, Mateus-Rodriquez et al. (2013) showed that the right balance between the complexity of a technology and its robustness directly affects the efficiency of production and ultimately the expected economic benefits. Flexibility is needed in the formulation of national policies that accommodate the participation of private investors in the production of quality seed potatoes at all stages of production. It is not hard to estimate the gains arising from aligning national efforts at food security with the propensity of private entrepreneurship to seize opportunities available in agriculture as demonstrated in the Ethiopian case. Here, renewed promotion of potato production has coincided with the entry of Solagrow PLC, a Dutch company, into the local seed potato industry. There is a similarity between the Solagrow PLC intervention model and the model described involving five countries as presented previously. Perhaps in anticipation of growth in the potato sector as a result of their intervention in Ethiopia, Solagrow PLC has expanded its roles to touch on each part of the value chains both downstream and upstream of the seed potato production (van de Haar, 2013).

Increased supply of quality seed potato leads directly to an increase in potato yields. Consequently, the outcome is higher quantities reaching fresh potato market and processing industry accordingly expanding profitable opportunities for many players along the value chain (Labarta, 2013). However, seed supply and delivery in many countries are predominantly informal. By its nature, the informal seed system is difficult to regulate and has persistent seed quality problems. Accordingly, strategic objectives to improve farmer access to quality seed potatoes continue to place emphasis on the formal seed system (MoALF, 2016) that accounts for less than 5% of national seed potato demand in the case of Kenya. A far more effective strategy requires concurrent intervention in the informal sector. It has been amply demonstrated that recognizing the informal seed supply system and supporting it to deliver seed of improved quality increases both yields and farmer incomes (Gildemacher, 2009, 2012). With appropriate institutional support to grant such seed semi-formal status such as Quality Declared Seed as was in Ethiopia, a large impact in food security terms is realized (FAO, 2010; Schulz et al., 2013).

Luckily, solutions to seed supply challenges are beginning to take on regional dimensions. The recent breakthrough in harmonization of regulations governing trading in seed among COMESA member countries is a significant step towards streamlining seed systems in member countries.

4.2 National Agricultural Research programmes and support

The Green Revolution of the 1960s and 1970s shows how previously food-insecure Asian and Latin American countries became net food exporters when new high-yielding wheat and rice varieties were widely adopted (IFPRI, 2002). This experience demonstrates the need for promotion of crop varieties with superior yield traits accompanied by appropriate agronomic packages. This need still informs the agenda of many National Agricultural Research Systems (NARS) in Africa, albeit with varying degrees of success. NARS in Africa typically consist of national agricultural research institutes, agriculture faculties at local universities, CGIAR agencies, non-governmental organizations, multinational corporations dealing in agro-inputs (seed, fertilizers and pesticides), local agriculture-based private companies and individual researchers. Whereas each of these organizations has a stated mandate of contributing to food and nutrition security in countries where they have operations, they seldom harmonize their strategies. This lack of focused strategies among players has inevitably led to duplicated roles and interventions and made the goal of food security in the continent become distant.

Several deliberate initiatives have been launched in the continent to keep the dream of food security in the continent alive. Launched at the turn of the new century in 2002, the Comprehensive Africa Agriculture Development Programme (CAADP) outlines strategies towards self-reliance in food and agriculture in the continent while acknowledging the enormous challenges that have to be surmounted towards this goal (CAADP, 2002). Popularly known as the Maputo Declaration of 2003, the strategy acknowledged the link between growth in the agricultural sector and economic development and required commitment from African government to allocate at least 10% of national budgets to the agricultural sector (and dedicate 4% of agricultural budget to research and development (R&D) and rural extension) and deliver 6% annual growth in the sector. A review of progress in the implementation of the Maputo Declaration after a decade shows that fewer than a fifth of the African countries have met either the 10% expenditure of the national budget on agriculture or realized 6% annual growth in the agricultural sector (One, 2014). Another unmet target by the African governments is the 4% agriculture budget on R&D. On the whole, however, another player in African agriculture, the Alliance for a Green Revolution in Africa (AGRA) established by the Bill & Melinda Gates Foundation and Rockefeller Foundation has made considerable contribution especially in soil and fertilizer management as well as capacity building in plant breeding and crop improvement.

5 Prospects for development and poverty alleviation: opportunities and challenges

Because of bulkiness, potatoes are mostly traded in national and regional markets where prices are determined at the local level. This contrasts greatly with cereal crops that are subject to global price volatility and singles out potato as a suitable food security crop for low-income communities. Indeed, a multiplicity of agencies are involved in the promotion of potato production in the continent to achieve food security and developmental goals targeting poverty reduction. Undeniably, we cannot meaningfully address poverty without dealing with agriculture, particularly productivity (Hartmann, 2003). In addition, addressing crop productivity to increase food production has several benefits: firstly, increased food supply; secondly, it directly leads to reduction in food prices in turn benefitting not only the farmer but also the urban population who will spend less on food; thirdly, increased crop production will generate employment opportunities for other players along the value chain. A lot remains to be done to realize the great potential in potato to lift many out of poverty. Invariably, a great majority of those involved in the potato sector are producers, with traders comprising the smaller part. Yet, the distribution of gains from the crop unevenly favours traders compared to the producers, largely due to the setup of marketing channels deliberately aimed at excluding the latter. Laiboni and Omiti (2014) identified a combination of poor information flow, inadequate storage and transport infrastructure, and poor physical and institutional organization of markets as contributing factors. Although mobile telephony can be leveraged to deliver market information and empower farmers, roads and storage structures to facilitate trade require heavy investment and call for reasoned and sustained public–private partnerships towards streamlining produce distribution and marketing.

Other inclusive interventions to address the existing marketing challenges such as the participatory market chain approach (PMCA) are required (Horton, 2008). The PMCA model was assessed in Uganda and found to be a valuable tool to stimulate pro-poor market innovations. Essentially, an effective intervention needs to comprise a negotiated engagement between producers and marketers (contract farming) to assure the predictability of prices as an incentive for uninterrupted supply of potato with specified characteristics. In reality, many players are needed to make the system work. Providers of financial facilities support producers by enabling them to access adequate production inputs in a timely manner while conditioning marketers/processors to pay for the produce promptly on delivery. Another set of independent players are suppliers of inputs, advisory and regulatory agencies. The involvement of all these players with their varied interests make the system complex requiring a champion to coordinate the roles of each party. In most SSA countries where industry self-regulation is undefined, this role remains a government function (FAO, 2014).

6 Future prospects

Compared with developed countries that have a long tradition of harnessing principles that underpin crop production, farming in SSA countries is characterized by low utilization of technology. It can be tempting to consider direct transfer of proven technologies already in use in temperate countries for adoption by African farmers. However, tropical conditions and unique farming challenges often render technologies designed for temperate agriculture ineffective when introduced in Africa. Two examples suffice to highlight this hindrance to technology transfer. First, consumer tastes can challenge the successful introduction and establishment of crop varieties as evidenced by the recent introduction of varieties into Kenya from the Netherlands. Not only did the majority of new varieties collapse under late blight (*Phytophthora infestans*) pressure but also the domination of the potato market by a single local variety Shangi that has delayed the uptake of the new varieties after close to three years. Additionally, the high-input management regime necessary to support the cultivation of these new varieties is incompatible with the economic realities of the majority of small-scale growers. Secondly, farming activities are done on small land parcels of less than 1 ha and this negates mechanization of crop management operations. All the benefits such as efficiency and cost reduction that accompany deployment of agricultural machinery are lost. To overcome this challenge, it is necessary to target farmer cooperatives rather than individuals in promoting the uptake of mechanization. Machinery can be jointly owned or hired through cooperatives for scheduled use on members' farms. In most instances, however, it requires organizing farmers into cooperatives. Development agencies need to identify and carefully examine suitable technologies with the potential to spur potato production by targeting soil and water management, crop variety improvement, seed potato production and delivery systems, and management of biotic constraints.

7 Conclusion

The twin challenges of climate change and population growth have made the quest for food security in Africa most urgent. Missed opportunities in the past decade by African governments at increasing funding to the agricultural sector need to provide experiences upon which to make current strategies more effective. The contribution of potato to the food system in the continent, and therefore to food security is on the rise (Thiele et al., 2013). Beyond availing food, there are opportunities in getting the crop to spur economic growth by supporting the potato value chain through a systems approach. This approach represents a paradigm shift from the obsession with increased productivity while paying little attention to the way in which improved varieties are developed or accessed; the efficiency of available seed systems in terms of delivering and distributing seed potatoes of the desirable quality; whether or not farmer market linkages exist to facilitate trade and stabilize produce prices. It is clear that although governments bear the most responsibilities in food security issues to the citizens, there are other players. By providing the infrastructure and institutional frameworks that permit investments, governments can propel the industry by embracing private–public partnerships. An agriculture-led rural economy will reduce poverty; encourage private saving and investment; stimulate domestic markets; and provide cheap, reliable food supply for urban households.

8 Where to look for further information

8.1 Guidelines for future research

The pertinent issues of climate and increasing population, urbanization and crop loses in the field and after harvest remain unresolved. Going forward, research interventions should focus on the following:

i   Climate-resilient crop varieties rich in nutrients will need to be developed and promoted;
ii Research to support evidence based policies on streamlining marketing of perishable crops;
iii Research on effective storage practices to curb seasonal price and produce quantity fluctuations and to facilitate product traceability;
iv Methodologies for rapid detection, surveillance/mapping and management of existing and emerging pest and disease threats;
v   Quality seed production and delivery systems.

8.2 Brief on organizations involved with this topic

Many agencies are involved in potato on the continent. Cumulatively, through their roles, programs and activities, they are expected to impact on the potato at local, regional and continental level. At country level are the relevant government departments responsible for agriculture to provide policy frameworks and enabling environments for potato agribusiness, NARs to lead the research agenda on the issues highlighted above; International organizations including the CG centres (CIP), Biosciences East and Central Africa (BECA-ILRI), Alliance for Green Revolution in Africa (AGRA) and Bill & Melinda Gates Foundation as knowledge hubs and platforms for capacity building in crop improvement and adoption; development partners (GIZ, Irish Aid, UKAid among others); and multinationals.

8.3 Further reading

J. Low, M. Nyongesa, M. Parker and S. Quinn (Eds). *Potato and Sweetpotato in Africa: Transforming the Value Chains for Food and Nutritional Security*. CABI, Oxfordshire, UK pp. 620.

9 References

Alliance for Commodity Trade in Eastern and Southern Africa (ACTESA). (2014). *COMESA Seed Trade Harmonization Regulations*. Lusaka, Zambia, p. 42.

Beintema, N., and Stads, G. J. (2014). Taking stock of national agricultural R&D capacity in Africa South of the Sahara. ASTI Synthesis Report. *Agricultural Science and Technology Indicators report*, http://www.ifpri.org/sites/default/files/publications/astiafricasynthesis.pdf.

Crissman, C., Anderson, P., Fuglie, K., Kapinga, R., Lemaga, B., Devaux, A., Thiele, G., langantileke, S. and Bussinik, C. (2007). Trends in the potato and sweet potato sectors in Sub-Saharan Africa and Africa and their contribution to the millennium development goals. In: Tropical root and tuber crops: Opportunities for poverty alleviation and sustainable livelihoods in Developing Countries. Proceedings of the 13th International Society for Tropical Root Crops Symposium, ISTRC, Arusha, Tanzania, pp. 8–19.

Demo, P., Lemaga, B., Kakuhenzire, R., Schulz, S., Borus, D., Barker, I., Woldegiorgis, G., Parker, M. L. and Schulte-Geldermann, E. (2015). Strategies to improve seed potato quality and supply in Sub-Saharan Africa: Experience from interventions in five countries. In: Low, J., Nyongesa, M., Quinn, S. and Parker, M. (Eds), *Potato and Sweetpotato in Africa: Transforming the Value Chains for Food and Nutrition Security*. Oxfordshire, UK: CABI, pp. 155–67.

Egziabher, T. B. G., Matos, E. and Mwila, G. (2011). The African Regional Group: Creating a Fair Play between North and South. In: Frison, C., López, F. and Esquinas-Alcázar, J. T. (Eds), *Plant Genetic Resources and Food Security: Stakeholder Perspectives on the International Treaty on Plant Genetic Resources for Food and Agriculture*. New York, US: FAO, Biodiversity International and Earthscan, pp. 41–56.

Fact sheet: Tunisian potato production (2013). Publication date: 3 August 2013, http://www.freshplaza.com/article/106730/Tunisian-potato-production.

FAO (2010). *Quality Declared Planting Material. Protocols and Standards for Vegetatively Propagated Crops*. FAO Plant Production and Protection Paper 195.

FAO (2014). *Public Sector Support in Agribusiness Development-An Appraisal of Institutional Models in Uganda*. Country Case Studies-Africa. Rome.

Gildemacher, P. R., Demo, P., Barker, I., Kaguongo, W., Woldegiorgis, G., Wagoire, W. W., Wakahiu, M., Leeuwis, C. and Struik, P. C. (2009). A description of seed potato systems in Kenya, Uganda and Ethiopia. *Am. J. Potato Res.* 86:373–82.

Gildemacher, P. R. (2012). Innovation in Seed Potato Systems in East Africa. PhD Thesis, Wageningen University, p. 184.

Hartmann, P. (2003). Roots and tuber crops and economic growth: The case of Sub-Saharan Africa. In: Tropical root and tuber crops: Opportunities for poverty alleviation and sustainable livelihoods in Developing Countries. Proceedings of the 13th International Society for Tropical Root Crops Symposium, ISTRC, Arusha, Tanzania, pp. 1–7.

Hijmans, R. J. (2003). The effect of climate change on global potato production. *Am. J. Potato Res.* 80:271–80.

Horton, D. (1987). *Potatoes: Production, Marketing and Programs for Developing Countries*. Boulder: West View Press, pp. 243.

Horton, D. (2008). *Facilitating Pro-poor Market Innovation: An Assessment of the Participatory Market Chain Approach in Uganda*. Lima, Peru: International Potato Centre (CIP), pp. 46.

International Food Policy Research Institute (IFPRI). 2002. Green Revolution: Curse or blessing? http://ifpri.org/pubs/ib/ib11.pdf.

Kateta, S. Y., Kabambe, V., Lowole, M. W. and Nalivata, P. C. (2015). Production practices of potatoes (*Solanaum tuberosum* L.) by farmers in Mzimba District of North Malawi. *Afr. J. Agric. Res.*, 10(8):797–802.

Kolech, S. A, Halseth, D., De Jong, W., Perry, K., Wolfe, D., Tiruneh, F. M. and Schulz, S. 2015. Potato variety diversity, determinants and implications for potato breeding strategy in Ethiopia. *Am. J. Potato Res.* 92(5):551–66.

Labarta, R. A. (2013). Possibilities and opportunities for enhancing availability for high quality seed potato in Ethiopia. In: Woldegiorgis, G., Schulz, S. and Berihun, B. (Eds), Seed potato tuber production and dissemination: Experiences, challenges and prospects. Proceedings of the National Workshop on seed potato tuber production and dissemination, Dahir Dar, Ethiopia, pp. 21–34

Laibuni, N. M. and Omiti, J. M. (2014). Market Structure and Price: An Empirical Analysis of Irish Potato Market in Kenya. Early Career Fellowship Program. UKAID, pp. 19.

Mateus-Rodriguez, J. R., de Haan, S., Andrade-Piedra, J. L., Maldonado, L., Haraeu, G., Baker, I., Chuquillanqiu, C., Otazu, V., Frisancho, R., Bastos, C., Pereira, S. A., Medeiros, S. A., Monteosdeoca, F. and Benitez, J. (2013). Technical and economic analysis of aeroponics and other systems for potato minituber production in Latin America. *Am. J. Potato Res.* DOI:10.1007/s12230-013-9312-5.

Ministry of Agriculture, Livestock and Fisheries (MoALF). (2016). The National Potato Strategy 2016–2020. Government of Kenya, p. 64.

Mpogole, H. (2013). Round potato production in southern highlands of Tanzania: market preferences, farmers' variety selection, and profitability. A thesis submitted in fulfillment of the requirements for the degree of doctor of philosophy of Sokoine University of agriculture, Morogoro, Tanzania.

National Root Crops Research Institute (NRCRI). (2016). Potato Programme, Kuru P. Mb. 04, Vom, Plateau State. Highlight of Research Achievements. Source: http://www.nrcri.gov.ng/pages/ipotato.htm (accessed 22 may 2016 at 21:29).

National Research Council (NRC). (2009). *Emerging Technologies to Benefit Farmers in Sub-Saharan Africa and South Asia.* Washington, DC: The National Academies Press.

ONE. (2015). *Ripe for change: The promise of Africa's Agricultural Transformation,* p. 57.

Ottoman, R. J., Hane, D. C., Brown, C. R., Yilma, S., James, S. R., Mosley, A. R., Crosslin, J. M. and Vales, M. I. (2009). Validation and implementation of marker assisted selection (MAS) for PVY resistance (Ry_{adg} gene) in a tetraploid potato breeding program. *Am. J. Potato Res.* 86:304–14.

Republic of South Africa (2003). Potato profile Fact sheet, Directorate economic analysis. January 2003. Published by Department of Agriculture. Obtainable from Resource Centre, Directorate Agricultural Information Services, Private Bag X144, Pretoria 0001, South Africa. Source: http://www.nda.agric.za/docs/factsheet/potato.htm (accessed 22 May 2016).

Republic of South Africa (2013). Potatoes production guidelines. Department of Agriculture, Forestry and Fisheries. Published by Department of Agriculture, Forestry and Fisheries, Directorate Plant Production Division: Vegetables, Private Bag X250 Pretoria 0001.

Schulz, S., Woldegiorgis, G., Hailemariam, G., Aliyi, A., van de Haar, J. and Shiferaw, W. (2013). Sustainable seed potato production in Ethiopia: From farm-saved to quality declared seed. In Woldegiorgis, G., Schulz, S. and Berihun, B. (Eds), Seed potato tuber production and dissemination: Experiences, challenges and prospects. Proceedings of the National Workshop on seed potato tuber production and dissemination, Dahir Dar, Ethiopia, pp. 60–71.

Scott, G. S., Rosegrant, M. W. and Ringler, C. (2000). *Roots and Tubers for the 21st Century: Trends, Projections, and Policy for Developing Countries.* 2020 BRIEF Policy 66. Food, Agriculture, and Environment Discussion Paper. Washington, DC, USA: International Food Policy Research Institute.

Scott, G. J., Rosegrant, M. and Bokanga, M. (2000). *Roots and tubers in the global food system: A vision statement to the year 2020* (including annex). A co-publication of the International Potato Centre (CIP), Centro Internacional de Agricultura Tropical (CIAT), International Food Policy Research Institute (IFPRI), International Institute of Tropical Agriculture (IITA), and International Plant Genetic Resources Institute (IPGRI), Printed in Lima, Peru: International Potato Centre.

Thiele, G., Theisen, K., Bonierbale, M. and Walker, T. (2010). Targeting the poor and hungry with potato science. *Potato J.* 37(3–4):31.

Ugonna, C. U., Jolaoso M. O. and Onwualu, A. P. (2013). A technical appraisal of potato value chain in Nigeria. *Int. Res. J. Agric. Sci. Soil Sci.* 3(8):291–301.

van de Haar, J. J. (2013). Opportunities and challenges for commercial seed potato production in Ethiopia. In: Woldegiorgis, G., Schulz, S. and Berihun, B. (Eds), Seed potato tuber production and dissemination: Experiences, challenges and prospects. Proceedings of the National Workshop on seed potato tuber production and dissemination, Dahir Dar, Ethiopia, pp. 1–6.

Walker, T., Thiele, G., Suarez, V. and Crissman, C. (2011). *Hindsight and Foresight about Potato Production and Consumption.* Social Sciences. Working Paper 2011–5, Lima, Peru: International Potato Center (CIP), p. 43.

Walker, T., Alene, A., Ndjeunga, J., Labarta, R., Yigezu, Y., Diagne, A., Andrade, R., Muthoni Andriatsitohaina, R., De Groote, H., Mausch, K., Yirga, C., Simtowe, F., Katungi, E., Jogo, W., Jaleta, M. and Pandey, S. (2014). Measuring the Effectiveness of Crop Improvement Research in Sub-Saharan Africa from the Perspectives of Varietal Output, Adoption, and Change: 20 Crops, 30 Countries, and 1150 Cultivars in Farmers' Fields. Report of the Standing Panel on Impact Assessment (SPIA), CGIAR Independent Science and Partnership Council (ISPC) Secretariat: Rome, Italy.

Chapter 13

Supporting smallholder women farmers in potato cultivation

Linley Chiwona-Karltun, Swedish University of Agricultural Sciences, Sweden; Maryanne Wamahiu, Stockholm University, Sweden; Chikondi Chabvuta, Actionaid International Malawi, Malawi; Dianah Ngonyama, Association of African Agricultural Professionals in the Diaspora, USA; and Paul Demo, International Potato Center (CIP), Malawi

1 Introduction

Although it can be shown that women play a critical role in smallholder agriculture in sub-Saharan Africa, this role has not been fully recognised until recently. This role applies to many crops including potatoes. This chapter explores this role and the ways it can be better supported. It begins by looking at the importance of potato cultivation to African smallholders and the key role of women in potato cultivation. The chapter then reviews the unique challenges facing women smallholders before discussing the range of strategies to support women smallholders and reflecting on their effectiveness.

2 The importance of potato cultivation to African smallholders

The importance of potato cultivation among smallholder farmers is increasing, particularly in Africa. Production in Africa has increased from 2 million tonnes in 1960 to 30 million tonnes in 2013 (Scott et al. 2013). An example is Malawi which has seen a

http://dx.doi.org/10.19103/AS.2017.0016.16

significant increase in the production of potato, both in volume and area of production (see Table 1) (Demo et al. 2007).

In countries such as Malawi, potatoes play a key role in ensuring a constant food supply to households. The crop is grown during the rainy agricultural season as well as the dry season in wetlands called *dambos*. This is usually from November to March, with potatoes ready to harvest when maize stock is depleted and the next maize crop is not yet ready to harvest. This makes potatoes an important food security crop. Figure 1 shows food availability and potato consumption in Bembeke, an extension planning area within Dedza district that is a major potato-growing area in Central Malawi. The evidence from Bembeke shows that relatively high potato consumption rates were maintained throughout the lean seasons stretching between the months of April and August when other food sources such as maize are more limited.

However, despite the importance of potatoes both as a food security and as a cash crop, it has been suggested that the research in improving potato cultivation in sub-Saharan Africa has not always kept pace with other root and tuber crops such as cassava (Scott et al. 2000, 2013). As the example of countries such as Malawi suggests, there have been attempts to improve research and development programmes in potato cultivation. The Malawian Ministry of Agriculture has been working with the International Potato Center (CIP), other non-governmental organisations, private sector partners and local farmers since 1990 to promote potato production in Malawi (Demo et al. 2007, 2015; Mviha et al. 2011; CIP-Malawi 2012; Malawi Government 2008; Demo et al. 2015). The current Malawi potato programme has had three main objectives:

1 Evaluation and selection of high-yielding disease-resistant varieties suitable for different end uses,
2 Improvement of seed potato production systems, and
3 Provision of smallholder farmers with the knowledge needed to improve production techniques.

Table 1 Annual production (MT) and yield (MT/ha) of potato in Malawi (2006–15)

Year	Potato		
	Production (MT)	Area (ha)	Yield (MT/ha)
2006	527 831	40 601	13.0
2007	594 003	40 202	14.8
2008	673 344	45 830	14.7
2009	775 629	48 332	16.0
2010	775 650	48 805	15.9
2011	928 941	52 689	17.6
2012	883 069	54 536	16.2
2013	924 969	54 597	16.9
2014	1 023 981	58 604	17.5
2015	1 065 833	61 655	17.3

Source: Ministry of Agriculture, Irrigation and Water Development, Malawi.

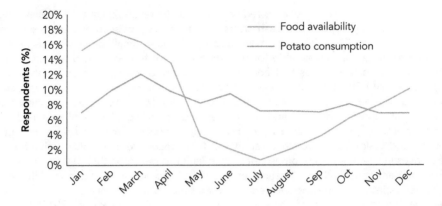

Figure 1 Potato consumption in Bembeke, Malawi.

The programme has also included elements of soil fertility and plant nutrition as well as post-harvest issues related to transportation and infrastructure designed to improve market access.

However, such attempts to improve production have not always been effective. While there are now many examples of countries in Africa with national potato breeding programmes, such as Malawi, Ethiopia and Kenya, many farmers have continued to grow local varieties using informal seed production systems, limiting the introduction and dissemination of new varieties (Demo et al. 2009, 2015; Gildemacher et al. 2011; Kolech Semagn et al. 2017). Farmers have very good reasons for maintaining their local varieties especially to maintain diversity (Brush et al. 1981). The re-propagation of relatively poor-quality seed, coupled with factors such as poor extension services, has made it harder to improve yields or combat pests and diseases which remain a major problem in potato cultivation in Africa (Kolech Semagn et al. 2017). This has highlighted the importance of engaging more effectively with smallholders, including women.

3 The role of women in potato cultivation

Research suggests that many of the smallholder farmers in sub-Saharan Africa producing food for the household are women (FAO 2011). However, access to resources such as land, capital and labour remains scarce for most women, as highlighted by the FAO study. Some studies also suggest that women constitute up to 53% of agricultural workers in sub-Saharan Africa. While figures of 60–80% female participation in farming in Africa have been questioned, participation rates are significant, ranging from 24% in countries such as Niger to 56% in Uganda and up to 80% in countries such as Cameroon (Amenyah and Puplampu 2013). These figures support the fact that reducing the gender gap could be an important pathway to contribute to development strategies, such as the sustainable development goals (SDGs) in sub-Saharan Africa and more specifically SDG-5 on gender equality and women's rights.

Experience confirms that there is a common division of labour between many male and female farmers (Amenyah and Puplampu 2013). In many cases, men predominate in food production to sell in local or regional markets (Doss 2002; Carr 2008; Wakhungu and Bunyasi 2010). In particular cash crops such as maize, tobacco and tea are largely controlled by men (Doss 2002; Dorward and Kydd 2004). Smallholder women farmers continue to be largely responsible for providing food for the household, with a greater role in producing food crops that contribute to dietary diversity, such as potatoes and vegetables (FAO 2011). Earlier research suggested women did not have much influence on crop or varietal selection (Louette and Smale 1998). More recent studies show that, through participatory research and the better understanding of varietal selection and commercialisation, women could be empowered especially in root and tuber cropping (Fischer and Qaim 2012; Oumer et al. 2014). More recent research also highlights the differences in how men and women select crop varieties based on their division of labour (Chiwona-Karltun et al. 2015; Christinck et al. 2017). A recent review of gender differences in farmer varietal preferences showed that men tend to focus more on market-related traits, while women focus more on end-use-related traits (Christinck et al. 2017). Women focus on characteristics such as storage, processing, culinary characteristics as well as cultural and ritual values (Chiwona-Karltun et al. 1998, 2015; Lope-Alzina 2007). However, despite the central role that women play in household food security, this role is not always recognised in commodity or value chains research (Dunaway 2014; Andersson et al. 2016).

The importance of women smallholders and the distinctive role they play makes it critical to engage with them to promote the broader goals of development and food security (AGRA 2012; Ngwira 2014). Targeting women smallholders is a necessary part of any attempt to address issues of social and gender inequality (Mudege et al. 2015). Moreover, evidence suggests that women farmers benefit more than men if they have access to information, knowledge and skills, as they are sometimes quicker to adopt new technologies (Wamala 2010), and that extension services are much more effective if they reach both groups (Jibat et al. 2007; Katungi et al. 2008; Quisumbing and Pandolfelli 2010). Studies on women and potato farming in Ethiopia showed that women were receptive to extension information and training from outsiders as well as sharing information that they had learned with other women (Kuma and Limenih 2015).

Another key reason to support women smallholders is their central role in food security and nutrition (Wakhungu and Bunyasi 2010; FAO 2011). In many parts of Malawi, malnutrition levels remain high with 20% of children classed as underweight and as many as 47% suffering from stunted growth (NSO 2016). In tackling this problem, CIP has worked particularly with women's groups to understand how potatoes are used in local diets, to suggest ways of improving the nutritional value of potatoes (e.g. by developing a potato recipe book targeting pregnant and lactating women and children under 5 years), and to spread the message about ways of improving nutrition (Joabe et al. 2013). A follow-up study by CIP in 2016 on the contribution of potatoes to household food security focused only on women since they were primarily responsible for food at the household level (Wamahiu et al. 2016).

4 Challenges facing women smallholders

Women smallholders face a wide range of challenges. Research suggests that women farmers often do not have title deeds to the land that they farm and have less access to credit, and that social and cultural barriers greatly impede their ability to develop their role in agriculture (Quisumbing and Pandolfelli 2010; FAO 2011). In Malawi, for example, the national land policy of 2002 requires that the registered owner of family land can only be the family/household head who is typically a man. This makes it harder for women to make their voices count in how best to manage family land and more vulnerable to being dispossessed (Ngwira et al. 2001; Ngwira 2005, 2014). Decisions on what to grow on a piece of land are still typically made by men (Kaitano and Martin 2009; Ngwira 2014; Mudege et al. 2015).

Malawi is a country that has both patrilineal and matrilineal heritage systems and inheritance of assets and resources such as land are passed on through both these systems, based on gender, age and family relations (Ngwira 2005). The patrilineal system prevails in the northern region and some parts of the central and southern regions, while the matrilineal system dominates in the southern region. In the southern region where matrilineal systems is practised, women can own, inherit and decide over land issues and even 'kick out' men from their homes, though this stronger position for women is also associated with higher rates of men abandoning their families because of their reduced influence (Kaitano and Martin 2009; Ngwira 2014).

The weaker position of women is compounded by both conscious and unconscious bias in development programmes. Many developmental programmes only engage with farmers that own land which mean they deal predominantly with men (Malawi Government 2007; Kaitano and Martin 2009; Mudege et al. 2016). One study looking at gender norms among potato farmers found that extension officers assigned demonstration plots and crops only to men that owned land and had the means to irrigate it, as women did not possess any of these assets, and did not include women in training programmes (Mudege et al. 2016). The same study highlighted other issues such as the social taboo on women attending training meetings independently without male members of the household, as well as attitudes reinforcing the view that women lacked the education and experience to benefit from training. Recent research has highlighted the negative stereotypes of women held by their husbands and (predominantly male) extension workers (Mudege et al. 2017).

A particular issue was the predominant role of men in making decisions about selling potato and other produce in local markets. Whilst women were significantly involved in cultivation, they were often excluded from involvement in marketing or sales activities (Jefremovas 1991; Kaitano and Martin 2009; Mudege et al. 2015). This has effectively excluded women from development programmes focusing on building markets for potatoes (Kuma and Limenih 2015) and other crops to improve farmer livelihoods (Fischer and Qaim 2012; Andersson et al. 2016). The situation is exacerbated by government subsidies in favour of cash crops like maize (Gilbert et al. 2002) and tobacco (Chinsinga 2012; Dimova and Gang 2017). The difficulties faced by some women in getting their voices heard are illustrated by one example from the authors' research in Malawi. One husband did not just dispute his wife's estimate of quantities and revenues generated from household potato sales, but also questioned her assessment of boiling over roasting as predominant technique in cooking potatoes, despite the fact that 74% of the other women interviewed agreed with the wife's assessment.

5 Strategies to support women smallholders

The impact of the various constraints faced by women smallholders was demonstrated in initial assessments of farmer engagement with training sessions organised by the

Malawi-CIP potato improvement programme. These identified that, although women represented 47% of those involved in growing potatoes, attendance by women at training sessions was very low (Demo et al. 2007; Mudege et al. 2016). In response, the programme was changed to ensure that all the potato programme activities, that is training, improved seed technologies, in the field targeted at least 50% of women smallholders (Demo et al. 2007, 2015). The Malawi government has also sought to move more generally to mainstream gender in all its agricultural policies and programmes (Malawi Government 2008). Donors such as the World Bank and the African Development Bank are also seeking to ensure that projects they support benefit women as well as men (Lauterbach and Matenje 2013). Research so far suggests these changes still have some way to go before they have a full impact (Quisumbing and Pandolfelli 2010; Mudege et al. 2017). Surveys suggest, for example, that, whilst women have more access to agricultural advice than before, this access does not lead to food security if there is no joint access with men that ultimately are the decision-makers (Aberman and Ragasa 2017). Women still have significantly less access and face predominantly patriarchal barriers to information than that available to men (Ager 2015).

A key strategy in supporting women smallholders is to encourage them to form their own farmer groups or clubs (Jafry et al. 2014). Collective action in the form of groups has been shown to positively affect change in groups to empower women (Barham and Chitemi 2009; Fischer and Qaim 2012). Collective actions such as women's groups provide a safe and supportive space where women can discuss common issues and solutions. The Bembeke Women Potato Group was one result of this initiative (Fig. 2). On a broader scale, the Coalition of Women Farmers has been established as a network of women farmers across Malawi (http://www.wocan.org/content/coalition-women-farmers). By allowing women to create support networks, and the opportunity to pool resources, these groups allow women farmers to start to enter the potentially more lucrative seed potato market traditionally dominated by men (Barham and Chitemi 2009; Kuma and Limenih 2015). These groups can also serve as a valuable means to disseminate nutritional information in tackling the ongoing challenges of malnutrition.

Figure 2 Photo of Bembeke potato women farmers. Credit: Paul Demo.

Belonging to farmer unions, such as the Malawi Farmers Union, creates an added advantage for women as they can more easily be linked to markets and can access extension services as a member. Research suggests farmer organisations in Malawi are becoming more supportive of women though they have not addressed wider issues such as land reform (Ager 2015). This emphasises the importance of other broader measures to promote the role of women in crop cultivation such as lobbying for financial support for women-owned business and for increased ownership of land and property by women (Malawi Government 2008). Research in other countries such as Ethiopia and Kenya has shown the impact of building women's farming groups which are then able to earn significant revenue from selling seed potatoes in local markets, increasing and diversifying household incomes and thus improving overall food security (Fischer and Qaim 2012; Oumer at al. 2014; Kuma and Limineh 2015).

6 Conclusion and future trends

Clearly, commodity value chains including those of potato must be gender responsive, and need to be aware of the many cross-cutting socio-cultural norms that impact the full participation of smallholder women and men. Access to the resource base such as land, information in the form of extension services as well as technologies that affect production are factors that, if inequitably distributed, affect the potato production base. Potato is growing in importance in Africa as both a food security crop and a way of developing markets, but several challenges need to be overcome. Selection of varieties tolerant to late blight disease remains important for smallholder farmers during the rainy season as this has detrimental effects on yields with the potential to cause complete crop failure. While some research is ongoing on improved potato seed technologies, ensuring access to these technologies still requires deeper understanding. It remains crucial to engage both women and men in access to information and utilisation. However, these activities need to be in concert with the importance of gendered varietal preferences, especially where utilisation and markets are concerned.

7 Where to look for further information

There is no one book, article or website that can completely cover all the issues on smallholder women farmers and their challenges as well as opportunities. Here, we provide some documents and websites that could guide you into further reading and inquiry into the field.

Insightful reading of current narratives and debates

Doss, C., Meinzen-Dick, R., Quisumbing, A. and Theis, S. (2018). Women in agriculture: Four myths. *Global Food Security* 16: 69–74.
Sachs, C. E. (2018). *Gendered Fields: Rural Women, Agriculture, and Environment*. Routledge, 2018.
The 'Agriculture in Africa – Telling Facts from Myths' project. http://www.worldbank.org/en/programs/africa-myths-and-facts

Resource books and resources

Gender in agriculture source book: http://documents.worldbank.org/curated/en/799571
468340869508/Gender-in-agriculture-sourcebook

Food and Agricultural Organization (FAO): http://www.fao.org/docrep/013/am307e/
am307e00.pdf

Why Women Matter in Agriculture: https://www.siani.se/wp-content/uploads/2013/09/
siani-2013-transforming-gender-relations-agriculture-africa.pdf

Mudege, N. N. and S. Walsh (2016). Gender and roots tubers and bananas seed systems:
A literature review. RTB Working Paper No. 2016-2. CGIAR Research Program on
Roots, Tubers and Bananas (RTB), Lima. ISSN:2309-6586, 26p.

Food and Agricultural Organization (2010). *Strengthening Potato Value Chains. Technical
and Policy Options for Developing Countries*. FAO, Rome.

Current programmes on gender and women in agriculture

The CGIAR Gender Strategy: http://gender.cgiar.org/genderplatform/gender-strategies/

World Farmers' Organization: http://www.wfo-oma.org/women-in-agriculture.html

Women Organizing for Change in Agriculture and Natural Resource Management
(WOCAN): http://www.wocan.org/

8 References

Aberman, N.-L. and C. Ragasa (2017). Does providing agricultural and nutrition information to both
men and women improve household food security? Discussion paper 01653. Evidence from
International Food Policy Research Institute (IFPRI), Malawi.

Ager, C. (2015). Addressing gender disparities through farming organisations in Malawi. *Journal of
Enterprising Communities: People and Places in the Global Economy* 9(4): 361–75.

AGRA (2012). Investing in sustainable agricultural growth: A five year status report. 2017. Available
from http://agra-alliance.org/download/533977c150df3/

Amenyah, I. D. and K. P. Puplampu (2013). Women in Agriculture: An assessment of the current
state of Affairs in Africa. The African Capacity Building Foundation. http://saipar.org:8080/
eprc/bitstream/handle/123456789/282/ACBF_Women%20in%20Agriculture_Feb.%202013.
pdf?sequence=1&isAllowed=y

Andersson, K., J. B. Lodin and L. Chiwona-Karltun (2016). Gender dynamics in cassava leaves value
chains: The case of Tanzania. *Journal of Gender, Agriculture and Food Security* 1(2): 84–109.

Barham, J. and C. Chitemi (2009). Collective action initiatives to improve marketing performance:
Lessons from farmer groups in Tanzania. *Food Policy* 34(1): 53–9.

Brush, S., H. Carney and Z. Huaman (1981). Genetic diversity and conservation in traditional farming
systems. *Economic Botany* 35(1): 70–88.

Carr, E. R. (2008). Men's crops and women's crops: The importance of gender to the understanding of
agricultural and development outcomes in Ghana's central region. *World Development* 36(5):
900–15.

Chinsinga, B. (2012). The political economy of agricultural policy processes in Malawi: A case study of
the fertilizer subsidy programme. Working paper 39. Future Agricultures Consortium, Brighton.
https://opendocs.ids.ac.uk/opendocs/ds2/stream/?#/documents/7332/page/1

Chiwona-Karltun, L., J. Mkumbira, J. Saka, M. Bovin, N. Mahungu and R. Rosling (1998). The
importance of being bitter – a qualitative study on cassava cultivar preference in Malawi.
Ecology of Food and Nutrition 37: 219–45.

Chiwona-Karltun, L., D. Nyirenda, C. N. Mwansa, J. E. Kongor, L. Brimer, S. Haggblade and E. O. Afoakwa (2015). Farmer preference, utilization, and biochemical composition of improved cassava (Manihot esculenta Crantz) varieties in southeastern Africa. *Economic Botany* 69(1): 42–56.

Christinck, A., E. Weltzien, F. Rattunde and J. Ashby (2017). Gender differentiation of farmer preferences for varietal traits in crop improvement: Evidence and issues. https://www.researchgate.net/publication/322790907_Gender_Differentiation_of_Farmer_Preferences_for_Varietal_Traits_in_Crop_Improvement_Evidence_and_Issues

CIP-Malawi (2012). *Improving Food Security through Enhanced Potato Productivity, Technology Development and Supply Chain in Malawi.* International Potato Center (CIP), Lilongwe.

Demo, P., J. Low and J. Mwenye (2007). Potato production in Malawi: strengths, weaknesses, opportunities and threats. In *Proceedings of the African Potato Association Conference*, Vol 7. African Potato Association Alexandria, Egypt.

Demo, P., P. Pankomera, T. Connell and N. Khumar (2009). Potential of potato farming in improving the livelihoods of small scale farmers in Malawi. In *African Crop Science conference proceedings* 9. African Crop Science Society, Kampala, Uganda, pp. 761–5.

Demo, P., B. Lemaga, R. Kakuhenzire, S. Schulz, D. Borus, I. Barker, G. Woldegiorgis, M. Parker and E. Schulte-Geldermann (2015). Strategies to improve seed potato quality and supply in sub-Saharan Africa: Experience from interventions in five countries. In: Low, J., Nyongesa, M., Quinn, S. and Parker, M. (Eds), *Potato and Sweetpotato in Africa: Transforming the Value Chains for Food and Nutrition Security.* CABI International, Oxfordshire, UK, pp. 155–67.

Dimova, R. and I. N. Gang (2017). Female engagement in commercial agriculture, interventions, and welfare in Malawi. What Works for Africa's Poorest: Programmes and Policies for the Extreme Poor, p. 107. https://www.econstor.eu/bitstream/10419/130741/1/84256991X.pdf

Dorward, A. and J. Kydd (2004). The Malawi 2002 food crisis: The rural development challenge. *The Journal of Modern African Studies* 42(3): 343–61.

Doss, C. R. (2002). Men's crops? Women's crops? The gender patterns of cropping in Ghana. *World Development* 30(11): 1987–2000.

Dunaway, W. A. (2014). Bringing commodity chain analysis back to its world-systems roots: Rediscovering women's work and households. *Journal of World-Systems Research* 20(1): 64.

FAO (2011). *The State of Food and Agriculture Year Book.* Food and Agriculture Organization, Rome.

Fischer, E. and M. Qaim (2012). Gender, agricultural commercialization, and collective action in Kenya. *Food Security* 4(3): 441–53.

Gilbert, R. A., W. D. Sakala and T. D. Benson (2002). Gender analysis of a nationwide cropping system trial survey in Malawi. *African Studies Quarterly* 6(1–2): 223–43.

Gildemacher, P. R., E. Schulte-Geldermann, D. Borus, P. Demo, P. Kinyae, P. Mundia and P. C. Struik (2011). Seed potato quality improvement through positive selection by smallholder farmers in Kenya. *Potato Research* 54(3): 253.

Jafry, T., B. Moyo and L. Mandaloma (2014). Assessment of extension and advisory methods and approaches to reach rural women – examples from Malawi. MEAS Evaluation Series March.

Jefremovas, V. (1991). Loose women, virtuous wives, and timid virgins: Gender relations and the control of resources in Rwanda. *Canadian Journal of African Studies* 25(3): 378–95.

Jibat, G., M. Belisa and H. Gudeta (2007). Promotion of participatory technology in potato farming – Ethiopia. In: Flintan, F. and Tedia, S. (Eds), *Natural Resource Management: The Impact of Gender and Social Issues.* OSSEREA/IDRC, Addis Ababa, Ethiopia, pp. 19–54.

Joabe, A., W. Kasapira, P. Demo, E. Kapalasa, T. Nyekanyeka, F. Chipungu and M. Chipanthenga (2013). *Report of a Potato Consumption Survey.* International Potato Center (CIP), Lilongwe.

Kaitano, V. and A. Martin (2009). Gender and diversity issues relating to cassava production and processing in Malawi. C:AVA Cassava Adding Value for Africa.

Katungi, E., S. Edmeades and M. Smale (2008). Gender, social capital and information exchange in rural Uganda. *Journal of International Development* 20(1): 35–52.

Kolech Semagn, A., W. De Jong, K. Perry, D. Halseth and F. Mengistu (2017). Participatory variety selection: A tool to understand farmers' potato variety selection criteria. *Open Agriculture* 2: 453.

Kuma, B. and B. Limenih (2015). Women farmers in practices: Opportunities and challenges in accessing potato production technologies in Welmera Ethiopia. *Asian Journal of Agricultural Extension, Economics and Sociology* 6(3): 149–57.

Lauterbach, C. and I. Matenje (2013). Gender, IFIs and food insecurity case study: Malawi. http://www.genderaction.org/publications/malawifoodsecurity.pdf

Lope-Alzina, D. G. (2007). Gendered production spaces and crop varietal selection: Case study in Yucatán, Mexico. *Singapore Journal of Tropical Geography* 28(1): 21–38.

Louette, D. and M. Smale (1998). Farmers' seed selection practices and maize variety characteristics in a traditionally-based Mexican community. CIMMYT economics working paper no. 98-04. CIMMYT, Mexico, D.F.

Malawi Government (2007). *National Nutrition Guidelines for Malawi*. Ministry of Health, Malawi Government, Lilongwe.

Malawi Government (2008). *Revised Gender Policy*. Ministry of Gender, Community and Child Development, Malawi.

Mudege, N. N., T. Nyekanyeka, E. Kapalasa, T. Chevo and P. Demo (2015). Understanding collective action and women's empowerment in potato farmer groups in Ntcheu and Dedza in Malawi. *Journal of Rural Studies* 42(Supplement C): 91–101.

Mudege, N. N., T. Chevo, T. Nyekanyeka, E. Kapalasa and P. Demo (2016). Gender norms and access to extension services and training among potato farmers in Dedza and Ntcheu in Malawi. *The Journal of Agricultural Education and Extension* 22(3): 291–305.

Mudege, N. N., Mdege, N., Abidin, P. and Bhatasara, S. (2017). The role of gender norms in access to agricultural training in Chikwawa and Phalombe, Malawi. *Gender, Place and Culture* 24(12): 1689–710.

Mviha, P., A. Mtukuso, M. Banda and B. Chisama (2011). A catalogue of agricultural technologies released by the Ministry of Agriculture and Food Security. Ministry of Agriculture, Malawi Government.

Ngwira, N. (2005). Women's property and inheritance rights and the land reform process in Malawi. http://www.ndr.mw:8080/xmlui/bitstream/handle/123456789/300/Women%E2%80%99s%20Property%20and%20Inheritance%20Rights%20and%20the%20Land%20Reform%20Process%20in%20Malawi1.pdf?sequence=1

Ngwira, N. (2014). Gender and poverty reduction in Malawi. http://www.un.org/esa/socdev/social/meetings/egm10/documents/Ngwira%20paper.pdf

Ngwira, N., S. Bota and M. Loevinsohn (2001). HIV/AIDS, agriculture and food security in Malawi. https://pdfs.semanticscholar.org/6e7d/7148c09eaeb218653aa99ec61436bb6daae0.pdf

NSO (2016). The 2015–16 Malawi Demographic and Health Survey (2015–16 MDHS). National Statistics Office, Zomba.

Oumer, A. M., W. G. Tiruneh and C. Y. Tizale (2014). Empowering smallholder women farmers through participatory seed potato management: Lessons from Welmera District, Ethiopia. *Journal of Sustainable Development* 7(5): 93.

Palacios-Lopez, A., L. Christiaensen and T. Kilic (2017). How much of the labor in African agriculture is provided by women? *Food Policy* 67(Suppl. C): 52–63.

Quisumbing, A. R. and L. Pandolfelli (2010). Promising approaches to address the needs of poor female farmers: Resources, constraints, and interventions. *World Development* 38(4): 581–92.

Scott, G. J., M. Rosegrant and C. Ringler (2000). Roots and tubers for the 21st century: Trends, projections, and policy options. Discussion paper 31. IFPRI, International Food Policy Research Institute, Washington DC. https://ageconsearch.umn.edu/bitstream/16243/1/dp000031.pdf

Scott, G. J., R. Labarta and V. Suarez (2013). Benchmarking food crop markets in Southern Africa: The case of potatoes and potato products 1961–2010. *American Journal of Potato Research* 90(6): 497–515.

Wakhungu, J. W. and P. Bunyasi (2010). Gender dimensions of science and technology African women in agriculture. United Nations for the Advancement of Women. http://www.un.org/womenwatch/daw/egm/gst_2010/Wakhungu-EP.2-EGM-ST.pdf

Wamahiu, M., P. Demo, F. Zotor and L. Chiwona-Karltun (2016). The contribution of Irish potato towards household dietary consumption and food security in Dedza district. International Potato Centre (CIP)-Malawi, Malawi.

Wamala, C. (2010). Does IT count? Complexities between access to and use of information technologies among Uganda's farmers. PhD Dissertation. Luleå tekniska universitet, Luleå, Sweden.

Index

CPSIA information can be obtained
at www.ICGtesting.com
Printed in the USA
BVHW02*0832050818
523474BV00010B/4/P

9 781786 761002